Siegfried Heitz

Coordinates in Geodesy

With 20 Figures

Springer-Verlag
Berlin Heidelberg New York
London Paris Tokyo

Professor Dr. Siegfried Heitz
Institute of Theoretical Geodesy
Rhenish Friedrich-Wilhelms-University
Nussallee 17
5300 Bonn 1, FRG

German edition: Siegfried Heitz,
Koordinaten auf geodätischen Bezugsflächen
© 1985 Ferd. Dümmlers Verlag, Bonn

ISBN-13: 978-3-540-50088-9 e-ISBN-13: 978-3-642-73939-2
DOI: 10.1007/978-3-642-73939-2

2132/3130-543210

PREFACE

Earth-fixed, three-dimensional coordinate systems, whose geometry is a good regional or global approximation of the geometry of the earth's gravity field, are preferred for representing the point and vertical fields of the earth's surface. Defining coordinate systems of this type is generally done based on a "geodetic reference surface", which roughly represents equipotential surfaces near the earth's surface. Points on the reference surface are assigned to the points to be depicted using a law of mapping. Projection by means of surface normals is chosen most frequently for this purpose. The accompanying "surface-normal coordinates" of a point to be described then consist of the "surface coordinates" of the foot points of the reference surface and the "height" of the point above the reference surface. This book deals with the geometric principals of these coordinates from a geodetic standpoint. I was motivated to write this book through my lectures on the same subject at the Rhenish Friedrich-Wilhelms University in Bonn for students of geodesy.

The geometry of surface-normal coordinates is determined by the intrinsic and extrinsic geometry of the reference surface, the latter being the inevitable object of these considerations. Dealing with the geometry of reference surfaces in terms of geodesy essentially consists of defining, constructing, and transforming surface coordinates. This also involves the roots of the theory of surfaces. Thus, in my opinion it is very important to first take a good look at the differential geometry of surfaces in a general way. Doing this with particular emphasis on the significance of surface coordinates, e.g., in the manner of Chapter 2.1, we have all geodetically relevant fundamentals. The theory of surfaces can be presented clearly and simply with the help of tensor calculus, this being the sole method applied here. The basics of tensor calculus can be found in the numerous and diverse textbooks on this subject.

This book differs essentially from comparable representations in other textbooks on geodesy and geodetic surveying by the fact that the considerations here are not only limited to ellipsoids of revolution as reference surfaces. Rather, all theories for arbitrary reference surfaces are treated extensively so that the results are also of more general use, e.g., by exchanging the ellipsoid of revolution for reference surfaces with higher resolution or also for determining or projecting the surfaces of buildings. On the other hand, the introduction of arbitrary reference surfaces requires the especially intensive use of the theory of surfaces, which usually plays only a subordinate role in the "geodetic geometry of ellipsoids". Interestingly enough, the close connection with differential geometry, which is necessary for arbitrary surfaces, is not only of greater general validity: it also very frequently leads to much clearer theories for coordinates on ellipsoids of revolution than the geodetic geometry of ellipsoids does.

Especially "geodesic surface coordinates" and "isothermal surface coordinates" are considered more closely as coordinates on the reference surface because they are by far the most important coordinates in geodetic applications. These coordinates are defined in Chapter 2.1. Among others, "Riemannian normal coordinates" will be discussed, which besides geodesic polar coordinates are of valuable help in calculating small geodesic triangles. Transformations between isothermal coordinates can be very clearly represented using complex-valued functions of complex variables. The basics of the complex analysis necessary for this are discussed in Chapter 2.2.

The fundamentals of differential geometry in Chapter 2.1 give the "local definition or transformation equations for surface coordinates" in the form of differential equations. Their integration leads to the "regional transformation equations" for finite coordinate differences necessary in practice. In Chapter 3, examples of these integrations are treated based on analytic setups for the solutions, for which only power series setups in the coordinates are chosen. Seen purely didactically, these methods are very suited for obtaining an initial, systematic insight into this subject. On the other hand, power series representations of this type lead to the very frequently used transformation equations as well. All representations in Chapter 3 are based on arbitrary reference surfaces.

The results of the transformations dealt with in Chapter 3 will first be applied to surface coordinates on ellipsoids of revolution in Chapter 4. Although this mainly involves discovering the general principles using an ellipsoid of revolution as example for a reference surface, nearly all the interesting possibilities for application in practical geodetic surveying are included.

While Chapters 1 to 4 are exclusively dedicated directly to coordinates on geodetic reference surfaces and their geometry, Chapter 5 goes into three-dimensional Euclidean space, which corresponds to the physical reality of geodetic methods to the greatest extent. The relationship between coordinates on a reference surface and the observations determining them is given by observation equations, which are primarily set up with the use of three-dimensional "surface-normal coordinates" defined on the basis of this reference surface, as was already mentioned at the beginning. For treating the necessary geometric principles the most important basics on arbitrary curvilinear coordinates in three-dimensional Euclidean spaces are summarized in Chapters 5.1 and 5.2. Then, surface-normal coordinates in three-dimensional Euclidean space are treated in Chapter 5.3. As examples of this, surface-normal coordinates are treated in Chapter 5.4 based on an ellipsoid of revolution as a reference surface, called in short "geodetic coordinates". Geographic and Gaussian cooordinates are considered on the reference ellipsoid. The transformation between different geodetic systems is a frequent necessity when dealing with geodetic coordinates. The currently relevant problems regarding this are treated in Chapter 5.5.

The bibliography contains only a small selection of publications which have given me ideas and/or which I have found to be very suitable supplements or alternative representations. It also contains further references to the very comprehensive literature on this subject.

NOTES ON THE ENGLISH EDITION

The contents of the first four chapters of this English edition have been changed very little from those of the original German edition ***Koordinaten auf geodaetischen Bezugsflaechen***. Only the representations of Chapter 3.2 on the inversion of power series and their application to Legendrean series in Chapter 3.3.1 have been fully rewritten. Chapter 5 has been completely rearranged and greatly expanded, now taking all geodetically relevant fundamentals of three-dimensional coordinates in Euclidean space into account. This is the reason for the broader title of the English edition ***Coordinates in Geodesy***.

I would like to thank the publishers for their work on this edition and the translator, Mr. Ralph B. Phillips, for his excellent cooperation. Furthermore, I am very grateful to Dr. Elke Stöcker-Meier and Mr. Martin Drude for critical reading the manuscript and testing several formulas and, last but not least, Mrs. Libuse Roschmann for her great support in the production of the manuscript.

Bonn, June 1988 Siegfried Heitz

CONTENTS

1. INTRODUCTION

For the regional and global representation of point fields of the earth's surface three-dimensional

$$\text{curvilinear coordinates } q^a \,, \quad a \in \{1,2,3\} \tag{1-1a}$$

are preferentially used whose coordinate surfaces and lines approximately describe the equipotential surfaces and plumb lines of the gravity field. Coordinates of this type can be defined based on a reference surface, which gives a good mean approximation of the equipotential surfaces near the earth's surface. The following has been agreed upon for this:

$$\begin{aligned}\text{reference surface} &= \text{model geoid} \\ &=: \text{coordinate surface } q^3 = 0 \,.\end{aligned} \tag{1-1b}$$

The model geoid is taken as a surface with a known intrinsic and extrinsic geometry as functions of the

$$\text{surface coordinates } u^\alpha := q^\alpha, \ \alpha \in \{1,2\} \,, \tag{1-1c}$$

so that in a rectangular, rectilinear, or Cartesian coordinate system S the position vectors of the surface points Q can be regarded as given analytical functions of u^α:

$$x_{i.Q} = x_{i.}(u^\alpha) \,. \tag{1-1d}$$

Corresponding to the explanation given above, the following has been agreed upon for the q^3-lines:

$$\begin{aligned}q^3\text{-lines} &= \text{model plumb lines} \\ &= \text{orthogonal trajectories of the model geoid.}\end{aligned} \tag{1-1e}$$

This does not completely determine the geometry of the model plumb lines and the q^3-coordinates. Insofar as only the vicinity to the earth's surface is to be represented, the following choice, for example, can be made:

$$\begin{aligned}q^3\text{-lines} &= \text{rectilinear orthogonal trajectories} \\ &\quad\ \text{of the model geoid,}\end{aligned} \tag{1-1f}$$

$$q^3 =: H = \text{height coordinates} = \text{distances from the model geoid.} \tag{1-1g}$$

Coordinates defined in this way, i.e.,

$$q^a = (u^1, u^2, H)^a = (u^\alpha, H)^a \tag{1-1h}$$

are named surface-normal coordinates.

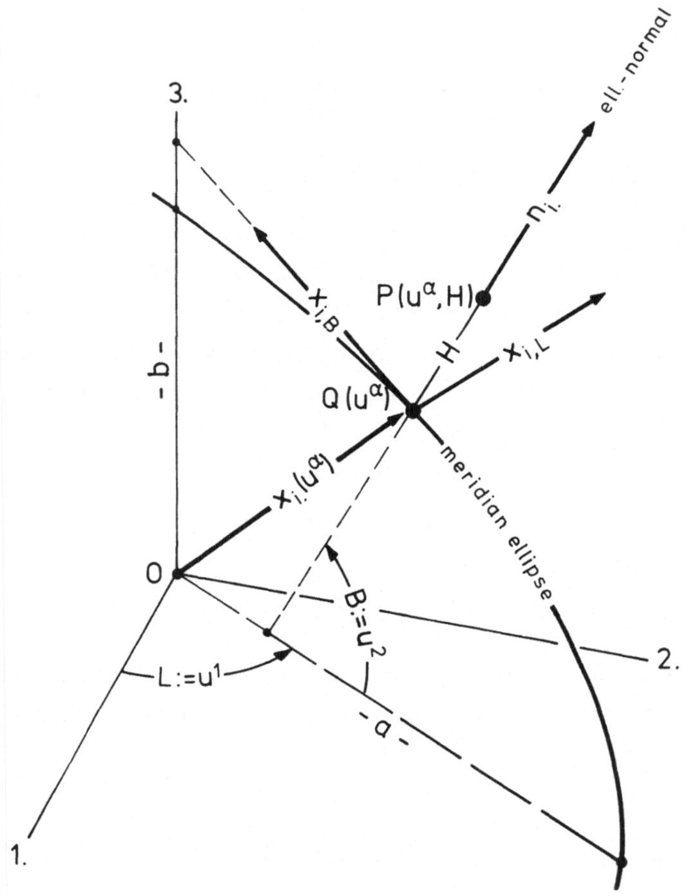

Fig. 1.1. Geodetic coordinates

Biaxial ellipsoids (ellipsoids of revolution) still play an important role as reference surfaces (1-1b) in geodetic applications. When by means of the ellipsoid normals the points P to be described are projected onto the biaxial ellipsoid, usually named reference ellipsoid, the following special surface-normal coordinates (1-1h) are obtained [see also Fig. 1.1]:

Geodetic coordinates (1-2)

$q^a =: (u^\alpha, H)$, (1-2a)

u^α = surface coordinates of perpendicular foot points Q on the reference ellipsoid, $\alpha \in \{1,2\}$, (1-2b)

H = ellipsoidal heights of the points P after (1-1g). (1-2c)

The geometric properties of the surface-normal coordinates are mainly determined by those of the surface coordinates on the reference surface after (1-1b-d) and (1-2b), the latter being the essential subject of this book. A short overview of the geodetically relevant systems of surface coordinates will first be given here. The considerations are generally based on arbitrary reference surfaces [see (1-1b)] whose position vectors after (1-1d) are described as functions of surface coordinates (1-1c). Using the surface coordinates two families of curves are determined on the surface, the so-called

Coordinate lines (1-3a)

u^1-lines: u^1 = variable, u^2 = constant,
u^2-lines: u^2 = variable, u^1 = constant

with the position vectors

$$x_{i.}(u^1) = f_{i.}(u^1;\ u^2 = \text{constant}) ,$$
$$x_{i.}(u^2) = f_{i.}(u^2;\ u^1 = \text{constant}) . \qquad (1\text{-}3b)$$

The distance between two differentially neighboring points, also called the element of arc, is given by ds and the accompanying differentials of the chosen surface coordinates (1-1c) by du^α. Then, the

First fundamental form of the theory of surfaces (1-4)

$$ds^2 = g_{\alpha\beta} du^\alpha du^\beta , \qquad (1\text{-}4a)$$

$g_{\alpha\beta} = g_{\beta\alpha}$ = covariant first fundamental tensor or metric tensor (1-4b)

is generally valid. For the Greek indices the summation convention (2-7) is valid according to which every Greek index which occurs twice in some expression, once below (subscript) and once above (superscript), is to be summed from 1 to 2. Especially for orthogonal coordinates

$$g_{12} = 0 : \quad ds^2 = g_{11}(du^1)^2 + g_{22}(du^2)^2 \qquad (1\text{-}4c)$$

is of validity. Orthogonal surface coordinate systems are preferred for most applications.

In addition to (1-3) and (1-4), the term

Geodesic curvature of a surface curve (1-5)

= curvature of the orthogonal projection of the surface curve onto the accompanying tangent plane of the surface is of primary importance in defining surface coordinates. When the surface coordinates of a point on a surface curve, their direction in this point, and their geodesic curvature as a function of a curve coordinate are given, the surface curve is thus uniquely determined. If the geodesic curvature of surface curves vanishes identically, they are then called <u>geodesic lines</u> or simply <u>geodesics</u>. Their osculating planes are normal planes of the surface.

Following this short overview of the fundamentals of the theory of surfaces the surface coordinates used in geodesy can now be discussed in more detail. There are essentially two types of surface coordinates: "geodesic surface coordinates" and "isothermal (isometric) surface coordinates". The former are defined based on the special properties of the coordinate lines [see (1-3)] as follows:

<u>Geodesic surface coordinates</u> (1-6)

u^2-lines = one-parameter family of geodesics, (1-6a)

u^1-lines = orthogonal trajectories of the u^2-lines. (1-6b)

For geodesic surface coordinates (1-6a,b)

$$g_{12} = 0 \ , \quad \partial g_{22}/\partial u^1 = 0 \ \rightarrow \ g_{22} = g_{22}(u^2) \tag{1-6c}$$

is valid. With the coordinate transformation

$$du^2 := (g_{22}(u^2))^{1/2} du^2 ,$$
$$u^2 := \int (g_{22}(u^2))^{1/2} du^2 = \text{arc length measured along the } u^2\text{-line} \tag{1-6d}$$

the first fundamental form of (1-6) is

$$ds^2 = g_{11}(u^\alpha)(du^1)^2 + (du^2)^2 \ . \tag{1-6e}$$

Geodesic surface coordinates (1-6) can be defined both as

<u>Geodesic polar coordinates</u> (1-7a)

all geodesic u^2-lines intersect at one point on the surface, i.e., the "pole" of the coordinate system,

and as

<u>Geodesic parallel coordinates</u> (1-7b)

a u^1-line with a certain value

$u^2 =: (u^2)_o$ = constant

is arbitrarily given as the "abscissa line C_o", and the geodesic u^2-lines normal to C_o are called "ordinate lines"

[see Fig. 1.2].

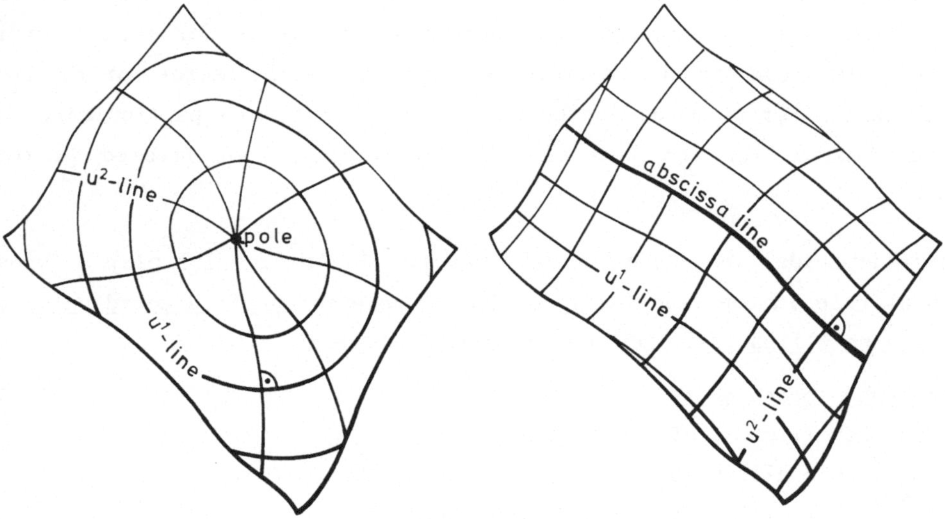

Fig. 1.2. Geodesic polar coordinates and parallel coordinates

The u^1-lines in (1-7a,b) are called <u>geodesic parallels</u>. For geodesic polar coordinates the u^1-lines are generally closed surface curves which hence are also named <u>geodesic circles around P_o</u>. Every geodesic polar coordinate system (1-7a) can obviously be interpreted as a parallel coordinate system (1-7b) when one u^1-line differing from the pole is chosen as the abscissa line. On the other hand, for geodesically parallel coordinates it cannot be generally expected that they have a pole at which all u^2-lines intersect.

Contrary to the geodesic surface coordinates [see (1-6) and (1-7)], which are defined by the particular choice of coordinate lines, the first fundamental form (1-4) is used for defining isothermal coordinates as follows:

<u>Isothermal surface coordinates</u> (1-8)

$$g_{12} = 0\ ,\quad g_{11} = g_{22} =: G\ :\qquad ds^2 = G(u^\alpha)((du^1)^2 + (du^2)^2)\ .$$

In this case, equal coordinate differentials

$$du^1 = du^2 \tag{1-8a}$$

correspond to equal arc elements

$$ds = G^{1/2}du^1 = G^{1/2}du^2 \tag{1-8b}$$

toward the orthogonal coordinate lines. Hence, in smaller areas isothermal coordinates are sufficiently accurate for use as Cartesian coordinates. $G(u^\alpha)$ can then be regarded as a constant scale factor for the area considered. Furthermore, isothermal coordinates form a particularly important basis for the laws of conformal mapping, not treated in this book.

The ellipsoidal longitude L and latitude B in Fig. 1.1 play a fundamental role in constructing geodetically relevant <u>surface coordinates on reference ellipsoids</u>. They are referred to here as

<u>Geographic coordinates</u> (1-9)

on the ellipsoid of revolution
u^1 =: L = ellipsoidal or geodetic longitude,
u^2 =: B = ellipsoidal or geodetic latitude.

For biaxial ellipsoids the position vectors $x_{i.}$ of the Q, with the exception of the poles, are reversible, unique functions of L and B, which is not generally the case for arbitrary surfaces. As a function of the geographic coordinates (1-9), the position vector of the ellipsoid of revolution after (4-8) is yielded by

$$x_{i.} = ((c/V)\cos B\cos L,\ (c/V)\cos B\sin L,\ (b/V)\sin B)\ , \tag{1-9a}$$

and the ellipsoid normals can be spanned using L and B as follows [see also (4-10)]:

$$n_{i.} = (\cos B\cos L,\ \cos B\sin L,\ \sin B)\ . \tag{1-9b}$$

The first fundamental form is

$$ds^2 = g_{11}(B)dL^2 + g_{22}(B)dB^2\ , \tag{1-9c}$$

where

$$g_{11} = R_1^2 \cos^2 B\ , \qquad g_{22} = R_2^2 \text{ and} \tag{1-9d}$$
$$R_1 = c/V\ , \quad R_2 = c/V^3\ , \qquad V^2 = 1 + e''^2 \cos^2 B\ .$$

The geographic coordinates on the ellipsoid of revolution are orthogonal and have two poles in the figure axis of the ellipsoid. They can be consequently understood as both polar and parallel surface coordinates. The B-lines or meridians are geodesics so that the L and B have the properties of geodesic surface coordinates. A comparison of (1-9d) with (1-8) shows that the geographic coordinates are not isothermal. However, they can be converted without changing the coordinate lines into

Isothermal geographic coordinates (1-10)

on the ellipsoid of revolution

L = ellipsoidal longitude, q = isothermal (isometric) latitude

using the coordinate transformation

$$dq = (g_{22}/g_{11})^{1/2} dB = (V^2 \cos B)^{-1} dB\ . \tag{1-10a}$$

Thus, the first fundamental form reads

$$ds^2 = G(B)(dL^2 + dq^2)\ , \quad G(B) = g_{11}(B) = (c/V)^2 \cos^2 B\ . \tag{1-10b}$$

The geographic coordinates of (1-9) and (1-10) are almost always used as the basis for constructing other surface coordinates by means of transformations:

$$u^\alpha = f^\alpha(L,B) = h^\alpha(L,q)\ . \tag{1-11a}$$

When u^α involves "intrinsic coordinates" defined on the surface which can be deduced from distance and/or angle measurements on the surface, the transformation functions (1-11a) are dependent only on the intrinsic geometry of the surface. This has the great advantage that the transformations of (1-11a) can be represented only as a function of the first fundamental tensors for u^α and L,B or L,q.

The geographic coordinates are related to the surface normal so that they are not of intrinsic nature. This is particularly advantageous in that the extrinsic geometry of the biaxial ellipsoid can be given relatively simply in the form of the position and normal vectors [see (1-9a,b)]. The extrinsic surface geometry is basically always necessary for establishing observation equations since both "distance and angle measurements on the surface" are available only in rare cases. Insofar as the extrinsic geometry of the reference ellipsoid is required with a high accuracy, (1-9a,b) is used in most cases. Other arbitrary surface coordinates are then first transformed into L and B using the inverse transformation of (1-11a) as follows:

$$(L,B)^{\alpha} = \bar{f}^{\alpha}(u^{\beta}) \ . \qquad (1\text{-}11b)$$

The representation of an arbitrarily dense eligible net of

$$\bar{u}^{\alpha}\text{-lines}, \quad \alpha \in \{1,2\} \qquad (1\text{-}12a)$$

on a surface corresponding to the above definition of surface coordinates should be understood as the construction of a surface coordinate system $\bar{u}^{\alpha}$. The construction of intrinsic surface coordinates explained with reference to (1-11a) is in the mathematical sense always possible when the intrinsic geometry is given in the form of the metric tensor as a function of an arbitrary surface coordinate system u^{γ}:

$$g_{\alpha\beta} = g_{\alpha\beta}(u^{\gamma}) \ . \qquad (1\text{-}12b)$$

The construction of the $\bar{u}^{\alpha}$-coordinates results in the transformation equations of the coordinates

$$\bar{u}^{\alpha} = \bar{u}^{\alpha}(u^{\beta}) \ , \quad u^{\alpha} = u^{\alpha}(\bar{u}^{\beta}) \qquad (1\text{-}12c)$$

and those of the corresponding metric tensors, which are dealt with in (2-27). The special transformation equations (1-11a,b) are examples of (1-12c).

Extrinsic surface coordinates, of which the geographic coordinates of (1-9) are an example, are related to the extrinsic geometry of the surface so that the supposition to (1-12b) is insufficient. In this case, for constructing the surface coordinates the second fundamental tensor (2-47) must be given as a function of arbitrary u^{α}-coordinates, in addition to (1-12b), or in other words, the position vector of the surface points Q [see (1-1d)] as a function of the u^{α} must be known. Again, the result has the form of (1-12c) and both fundamental tensors have to be transformed using (2-16) or (2-27).

The definition and construction of surface coordinate systems are key problems in differential geometry, particularly in the theory of surfaces. Hence, in Chapter 2.1 the basics of this theory will be presented to the extent necessary for understanding the problems arising here. In Chapter 2.2 the fundamentals of the complex analysis, which are of significance in transforming isothermal surface coordinates, are treated. Power series in the coordinates are still important for solving the differential equations of geodesic and isothermal surface coordinates discussed in Chapters 2.1.8 to 2.1.14. The application of power series is dealt with in a more general form in Chapter 3, and in Chapter 4 examples of surface coordinates on reference ellipsoids related to this are given. On the other hand, the differential equations can also be solved by numerical methods, for example by using the Runge-Kutta procedure and modifications of it. They are not dealt with in this book because they have not yet become very important in geodetic applications, and from the didactic point of view they are not as suitable as the analytical solutions by means of power series. Finally, three-dimensional coordinates in Euclidean spaces are treated in Chapter 5, first in Chapter 5.1 and 5.2 in general form and beginning with Chapter 5.3 specialized to surface-normal coordinates [see (1-1h)]. Here, much space is reserved for the geodetic coordinates [see (1-2)], which are still very important in geodesy.

2. GENERAL FUNDAMENTALS OF SURFACE CO-ORDINATES

2.1 FUNDAMENTALS OF THE THEORY OF SURFACES

2.1.1 RUDIMENTS

The differential geometry of surfaces, i.e., the theory of surfaces, deals primarily with the properties of curved surfaces on infinitesimally small regions. For this purpose, the surface to be studied is first embedded in three-dimensional Euclidean space represented by an orthogonal, rectilinear, or Cartesian coordinate system S.

The surfaces in S are then described by a

Gaussian representation of the position vectors (2-1a)

of the surface points

$x_{i.} = x_{i.}(u^{\alpha})$, $i \in \{1,2,3\}$, $\alpha \in \{1,2\}$

as functions of independent

Surface coordinates (2-1b)

u^{α} , $\alpha \in \{1,2\}$.

The following symbols are generally used for the partial derivatives of the components of the position vectors $x_{i.}(u^{\alpha})$ and arbitrary scalar position functions $f(u^{\alpha})$ with respect to the surface coordinates

$$\partial x_{i.}/\partial u^{\alpha} =: x_{i,\alpha} , \qquad \partial^2 x_{i.}/(\partial u^{\alpha}\partial u^{\beta}) =: x_{i,\alpha\beta} ,$$
$$\partial f/\partial u^{\alpha} =: f_{,\alpha} , \qquad \partial^2 f/(\partial u^{\alpha}\partial u^{\beta}) =: f_{,\alpha\beta} . \qquad (2\text{-}2)$$

Two families of coordinate lines [see (1-3)] on the surface are determined with the surface coordinates. At every point the two families of coordinate lines have two accompanying tangent vectors, i.e.,

$$x_{i,\alpha} = \partial x_{i.}/\partial u^{\alpha} , \quad \alpha \in \{1,2\} , \qquad (2\text{-}3a)$$

for which in the case of independent surface coordinates

$$\epsilon_{ijk.}x_{j,1}x_{k,2} \neq 0_{i.} \qquad (2\text{-}3b)$$

must be true. In general, tangent vectors are not unit vectors. The normal vector of the surface is defined as the unit vector

$$n_{i.} = \epsilon_{ijk.} x_{j,1} x_{k,2} / |\epsilon_{lmn.} x_{m,1} x_{n,2}| \quad , \tag{2-4}$$

whose existence is dependent on (2-3b) [see also (2-25)].

Local studies in surface theory are, as a rule, related to the

<u>Moving trihedron of the surface</u> (2-5)

$$x_{i,1} \; , \; x_{i,2} \; , \; n_{i.} \quad .$$

While $n_{i.}$ and thus the <u>tangential plane</u> spanned by $x_{i,1}$ and $x_{i,2}$ are invariably connected with the surface, $x_{i,1}$ and $x_{i,2}$ are dependent upon the particular selection of the surface coordinates.

Arbitrary <u>surface curves</u> can be represented by the surface coordinates as functions of a curve coordinate t, i.e.,

$$u^{\alpha} = u^{\alpha}(t) \quad , \tag{2-6a}$$

whereby the position vectors of the surface curves become

$$y_{i.}(t) = x_{i.}(u^{\alpha}(t)) \quad . \tag{2-6b}$$

$$dy_{i.}/dt = x_{i,\alpha}\, du^{\alpha}/dt \tag{2-6c}$$

is a tangent vector on the curve whose absolute value is not usually one. In (2-6c) the index α occurs twice whereby we use the following operation:

<u>Summation convention</u> (2-7)

Every Greek index which occurs twice in some expression, once below (subscript) and once above (superscript), is to be summed from 1 to 2; if a Greek index occurring in this fashion is not to be summed, it is to be placed in parentheses.

This has already been pointed out in (1-4). Convention (2-7) is analogously valid for <u>Cartesian tensors</u> in which only subscripts occur. As a rule, they are denoted by Latin letters and terminated by a point [see Heitz 1980-1983; Jeffreys 1974].

Two systems of surface coordinates (2-1b), i.e., u^{α} and $\bar{u}^{\alpha}$, will now be considered. Here, the following is valid:

<u>Transformations between surface coordinates</u> (2-8)

u^{α} und $\bar{u}^{\alpha}$

are obtained by the transformation functions

$$u^{\alpha} = u^{\alpha}(\bar{u}^{\beta}) \tag{2-8a}$$

and the accompanying inverse transformations

$$\bar{u}^{\alpha} = \bar{u}^{\alpha}(u^{\beta}) \ . \tag{2-8b}$$

During the transformations (2-8) <u>scalar position functions behave as invariants</u>:

$$f(u^{\alpha}) = f(u^{\alpha}(\bar{u}^{\beta})) = \bar{f}(\bar{u}^{\beta}) \ . \tag{2-9}$$

With the symbols

$$\partial u^{\alpha}/\partial \bar{u}^{\beta} =: u^{\alpha}_{,\bar{\beta}} \ , \qquad \partial \bar{u}^{\alpha}/\partial u^{\beta} =: \bar{u}^{\alpha}_{,\beta} \tag{2-10}$$

corresponding to (2-2) the following <u>transformation equations for partial derivatives of scalar position functions</u> resulting from (2-8) are obtained based on (2-9):

$$f_{,\alpha} = \bar{f}_{,\beta}\, \bar{u}^{\beta}_{,\alpha} \ , \qquad \bar{f}_{,\bar{\alpha}} = f_{,\beta}\, u^{\beta}_{,\bar{\alpha}} \ . \tag{2-11}$$

For example, if the first transformation equation is given, the second equation also exists only when the functional determinant does not vanish, i.e.,

$$|\bar{u}^{\beta}_{,\alpha}| = \begin{vmatrix} \bar{u}^{1}_{,1} & \bar{u}^{1}_{,2} \\ \bar{u}^{2}_{,1} & \bar{u}^{2}_{,2} \end{vmatrix} \neq 0 \ . \tag{2-12}$$

In addition to the transformation laws (2-9) and (2-11), the law for coordinate differentials is important as well. Proceeding from (2-8a,b) the <u>transformation equations for coordinate differentials</u> are obtained as follows:

$$du^{\alpha} = u^{\alpha}_{,\bar{\beta}}\, d\bar{u}^{\beta} \ , \qquad d\bar{u}^{\alpha} = \bar{u}^{\alpha}_{,\beta}\, du^{\beta} \ . \tag{2-13}$$

The paired values of $f_{,\alpha}$ and du^{α} which are assigned to a surface point P can be understood as <u>surface vectors relative to the u^{α}-system</u>. The same is true for the paired values $\bar{f}_{,\alpha}$ and $d\bar{u}^{\alpha}$ assigned to the same surface point P relative to the $\bar{u}^{\alpha}$-system. The transformation laws of $f_{,\alpha}$ [see (2-11)] are named <u>covariant</u>, and those of u^{α} [see (2-13)] <u>contravariant</u>. With respect to arbitrarily defined paired values these transformation laws are generalized in the following way:

<u>Covariant and contravariant vectors of a surface</u> (2-14)

The sets of two quantities a_α or a^α form covariant or contravariant vectors, respectively, relative to the u^α-system of a surface when they are transformed into one and the same point in the transition to another $\bar{u}^\alpha$-system according to (2-11) and (2-13), respectively:

$$\bar{a}_\alpha = u^\beta_{,\bar{\alpha}}\, a_\beta \,, \qquad a_\alpha = \bar{u}^\beta_{,\alpha}\, \bar{a}_\beta \,, \tag{2-14a}$$

$$\bar{a}^\alpha = \bar{u}^\alpha_{,\beta}\, a^\beta \,, \qquad a^\alpha = u^\alpha_{,\bar{\beta}}\, \bar{a}^\beta \,. \tag{2-14b}$$

The transformations of (2-14) are obtained with the <u>transformation matrices</u>

$$u^\alpha_{,\bar{\beta}} = \partial u^\alpha / \partial \bar{u}^\beta \,, \qquad \bar{u}^\alpha_{,\beta} = \partial \bar{u}^\alpha / \partial u^\beta \,, \tag{2-15a}$$

which correspond to the rotation matrices in the transformation of Cartesian vectors [see Heitz 1980-1983; Jeffreys 1974]. For the transformation matrices (2-15a) the following conditions hold:

$$u^\alpha_{,\bar{\beta}}\, \bar{u}^\beta_{,\gamma} = \delta^\alpha_\gamma = \begin{cases} 1 & \text{for } \alpha = \gamma \\ 0 & \text{for } \alpha \neq \gamma \end{cases} \,. \tag{2-15b}$$

Extensions of the notion of vectors are the following definitions:

<u>Covariant and contravariant tensors of the second rank of a surface</u> (2-16)

The sets of four quantities $a_{\alpha\beta}$ or $a^{\alpha\beta}$ or a^α_β form covariant, contravariant, or mixed tensors of the second rank, respectively, relative to the u^α-system of a surface when they are transformed into one and the same point in the transition to another $\bar{u}^\alpha$-system accoording to

$$\bar{a}_{\alpha\beta} = u^\gamma_{,\bar{\alpha}}\, u^\delta_{,\bar{\beta}}\, a_{\gamma\delta} \,, \qquad a_{\alpha\beta} = \bar{u}^\gamma_{,\alpha}\, \bar{u}^\delta_{,\beta}\, \bar{a}_{\gamma\delta} \,, \tag{2-16a}$$

$$\bar{a}^{\alpha\beta} = \bar{u}^\alpha_{,\gamma}\, \bar{u}^\beta_{,\delta}\, a^{\gamma\delta} \,, \qquad a^{\alpha\beta} = u^\alpha_{,\bar{\gamma}}\, u^\beta_{,\bar{\delta}}\, \bar{a}^{\gamma\delta} \,, \tag{2-16b}$$

$$\bar{a}^\beta_\alpha = u^\gamma_{,\bar{\alpha}}\, \bar{u}^\beta_{,\delta}\, a^\delta_\gamma \,, \qquad a^\beta_\alpha = \bar{u}^\gamma_{,\alpha}\, u^\beta_{,\bar{\delta}}\, \bar{a}^\delta_\gamma \,. \tag{2-16c}$$

<u>Tensors of rank $n > 2$</u> are defined by analogous laws of transformation with n-fold multiplication with the corresponding transformation matrices.

Theorems (2-14) and (2-16) have no relationship to the embedment of the surface in three-dimensional Euclidean space. They are related solely to the intrinsic surface geometry, which will be discussed in more detail in the following chapter. In addition to the intrinsic surface geometry, the extrinsic surface geometry substantiated by (2-1a) is also important in many applications. Among other things, transformations of "surface vectors" [see (2-11)] and "surface tensors" [see (2-13)] into the three-dimensional Euclidean space are necessary for this. Transformations of this type are treated in Chapter 2.1.3.

2.1.2 FIRST FUNDAMENTAL FORM

For the arc elements ds of a surface curve

$$ds^2 = dy_{i.} dy_{i.}$$

is valid. Together with (2-6c) the

First fundamental form (2-17)

$$ds^2 = g_{\alpha\beta} du^\alpha du^\beta \qquad (2\text{-}17a)$$

with

$$g_{\alpha\beta} = g_{\beta\alpha} = x_{i,\alpha} x_{i,\beta} \qquad (2\text{-}17b)$$

= first covariant fundamental tensor or metric tensor,

$$g = |g_{\alpha\beta}| = g_{11} g_{22} - g_{12}^2 \qquad (2\text{-}17c)$$

is obtained as given in (1-4). In addition to $g_{\alpha\beta}$, the first contravariant fundamental tensor $g^{\alpha\beta}$ is also used, which is defined as the inverse of $g_{\alpha\beta}$, i.e.,

$$g_{\alpha\gamma} g^{\gamma\beta} = \delta^\beta_\alpha = \begin{cases} 1 & \text{for } \alpha = \beta \\ 0 & \text{for } \alpha \neq \beta \end{cases} . \qquad (2\text{-}18a)$$

Hereby, we obtain

$$g^{11} = g_{22}/g , \qquad g^{12} = g^{21} = -g_{12}/g , \qquad g^{22} = g_{11}/g . \qquad (2\text{-}18b)$$

In the case of orthogonal coordinate lines the tangent vectors [see (2-3a)] are perpendicular to one another so that

$$g_{12} = g^{12} = x_{i,1} x_{i,2} = 0 , \qquad (2\text{-}19a)$$

and (2-17c) and (2-18b) become

$$g^{11} = 1/g_{11} \quad , \quad g^{22} = 1/g_{22} \quad , \quad g = g_{11}g_{22} \quad . \tag{2-19b}$$

The first fundamental form (2-17a) is invariant with respect to transformations after (2-2). Thus, for a $\bar{u}^{\alpha}$-system of surface coordinates

$$ds^2 = \bar{g}_{\alpha\beta}d\bar{u}^{\alpha}d\bar{u}^{\beta} \tag{2-20a}$$

with

$$\bar{g}_{\alpha\beta} = \bar{x}_{i,\bar{\alpha}}\bar{x}_{i,\bar{\beta}} \tag{2-20b}$$

is valid. If the first fundamental tensor (2-17b) is given for a system of surface coordinates u^{α}, the distance measurement or metric in the surface and thereby the entire <u>intrinsic geometry of the surface</u> can be obtained using the first fundamental form. Hence, the first fundamental tensor is also referred to as the <u>metric tensor</u>. The intrinsic surface geometry is equivalent to Riemannian geometry in two-dimensional, generally curved, or non-Euclidean spaces.

<u>Quantities of the intrinsic surface geometry</u>, are for example,

- distances between surface points measured along prescribed surface curves, (2-21a)
- angles between the tangents of two surface curves at their points of intersection, (2-21b)
- surface areas of closed surface curves and (2-21c)
- geodesic curvatures of surface curves [see (1-5)] . (2-21d)

If P_a and P_b are two points on a surface curve as represented in (2-6),

$$s_{ab} = \int_a^b ds = \int_a^b (g_{\alpha\beta}(du^{\alpha}/dt)(du^{\beta}/dt))^{1/2}dt \tag{2-22}$$

is obtained for the distance between these two points according to example (2-21a). In example (2-21b) two surface curves C_n, $n = 1,2$, with the curve coordinates t_n which intersect at point P are considered. Their equations are

$$u_n^{\alpha} = u^{\alpha}(t_n) \quad , \tag{2-23a}$$

whereby the tangent vectors in P, i.e.,

$$z_{i.n} = dy_{i.n}/dt_n = x_{i,\alpha}du^{\alpha}/dt_n \ , \tag{2-23b}$$

are yielded with the following absolute values:

$$|z_{i.n}| = (g_{\alpha\beta}(du^{\alpha}_n/dt_n)(du^{\beta}_n/dt_n)) \quad . \tag{2-23c}$$

For the cosinus of the angle φ between the two tangent vectors (2-23b) we obtain

$$\cos\varphi = g_{\alpha\beta}(du^{\alpha}_1/dt_1)(du^{\beta}_2/dt_2)/(|z_{i.1}|\cdot|z_{j.2}|) \ . \tag{2-23d}$$

The absolute values of (2-23c) are dependent only on $g_{\alpha\beta}$ so that this is also true for φ according to (2-23d).

It can also be shown that the absolute value of the vector product of the tangent vectors $x_{i,\alpha}$ is a quantity of the intrinsic geometry. Regarding the expansion theorem of vector algebra [see Heitz 1980-1983, (2-55); Jeffreys 1974, Chap.I,(55)] we obtain

$$\begin{aligned} |\epsilon_{ijk.}x_{j,1}x_{k,2}|^2 &= \epsilon_{ijk.}\epsilon_{ilm.}x_{j,1}x_{k,2}x_{l,1}x_{m,2} \\ &= g_{11}g_{22} - g_{12}^2 = g \ . \end{aligned} \tag{2-24}$$

Thereby, (2-4) becomes

$$n_{i.} = \epsilon_{ijk.}x_{j,1}x_{k,2}g^{-1/2} \ . \tag{2-25}$$

The element of area can be represented by means of the parallelepipedal product of the trihedron of (2-5) as follows:

$$do = \epsilon_{ijk.}n_{i.}x_{j,1}x_{k,2}du^1du^2 \ , \tag{2-26a}$$

and together with (2-25) we obtain

$$do = g^{1/2}du^1du^2 \ . \tag{2-26b}$$

Equation (2-26b) is the proof for example (2-21c). Example (2-21d) will be discussed in Chapter 2.1.5.

The following <u>transformations of $g_{\alpha\beta}$ and $g^{\alpha\beta}$</u> are obtained from the coordinate transformations (2-8):

$$\begin{aligned} \bar{g}_{\alpha\beta} &= u^{\gamma}_{,\bar{\alpha}}\, u^{\delta}_{,\bar{\beta}}\, g_{\gamma\delta} \ , & g_{\alpha\beta} &= \bar{u}^{\gamma}_{,\alpha}\, \bar{u}^{\delta}_{,\beta}\, \bar{g}_{\gamma\delta} \ , \\ \bar{g}^{\alpha\beta} &= \bar{u}^{\alpha}_{,\gamma}\, \bar{u}^{\delta}_{,\delta}\, g^{\gamma\delta} \ , & g^{\alpha\beta} &= u^{\alpha}_{,\bar{\gamma}}\, u^{\beta}_{,\bar{\delta}}\, \bar{g}^{\gamma\delta} \ . \end{aligned} \tag{2-27}$$

A comparison with (2-16a,b) shows that $g_{\alpha\beta}$ is a covariant and $g^{\alpha\beta}$ is a contravariant tensor of the second rank, as postulated in (2-17) and (2-18).

In considering (2-27) it can be seen that the quantities of the intrinsic geometry are invariant forms with respect to coordinate transformations (2-8). This means that the results obtained in the u^α-system, e.g., (2-22) to (2-26), are still formally valid in the transition to the $\bar{u}^\alpha$-system when u^α, $g_{\alpha\beta}$, and $g^{\alpha\beta}$ are provided with crossbars.

For the vector product of the tangent vectors of the $\bar{u}^\alpha$-lines

$$\epsilon_{ijk}.\bar{x}_{j,\bar{1}}\bar{x}_{k,\bar{2}} = \epsilon_{ijk}.x_{j,\alpha}u^\alpha_{,\bar{1}}\, x_{k,\beta}u^\beta_{,\bar{2}} \tag{2-28a}$$
$$= \epsilon_{ijk}.x_{j,1}x_{k,2}(u^1_{,\bar{1}}\, u^2_{,\bar{2}} - u^1_{,\bar{2}}\, u^2_{,\bar{1}})$$

results. Based on (2-24), from the square of this vector equation we obtain

$$\bar{g} = g\,(u^1_{,\bar{1}}\, u^2_{,\bar{2}} - u^1_{,\bar{2}}\, u^2_{,\bar{1}})^2 = g\,|u^\alpha_{,\bar{\beta}}|^2 \tag{2-28b}$$

and conversely

$$g = \bar{g}\,(\bar{u}^1_{,1}\, \bar{u}^2_{,2} - \bar{u}^1_{,2}\, \bar{u}^2_{,1})^2 = \bar{g}\,|\bar{u}^\alpha_{,\beta}|^2 \quad . \tag{2-28c}$$

Herein, the functional determinants corresponding to (2-12) are

$$|u^\alpha_{,\beta}| \;,\quad |\bar{u}^\alpha_{,\beta}| \quad . \tag{2-28d}$$

We designate

$$a_\alpha = g_{\alpha\beta}a^\beta \tag{2-29a}$$

as the covariant vector a_α belonging to the contravariant vector a^β. The inversion of (2-29a) results from inner multiplication with the first fundamental tensor or metric tensor considering (2-18a) giving

$$a^\alpha = g^{\alpha\beta}a_\beta \;. \tag{2-29b}$$

The "lowering and raising" of tensor indices by means of the metric tensors applies also to the tensors of higher ranks, e.g.,

$$a^\beta_\alpha = g_{\alpha\gamma}a^{\gamma\beta} \;,\qquad a_{\alpha\beta} = g_{\alpha\gamma}g_{\beta\delta}a^{\gamma\delta} \;, \tag{2-29c}$$
$$a^\alpha_\beta = g^{\alpha\gamma}a_{\gamma\beta} \;,\qquad a^{\alpha\beta} = g^{\alpha\gamma}g^{\beta\delta}a_{\gamma\delta} \;.$$

The scalar product of the two contravariant vectors a^α and b^β is defined by

$$c = g_{\alpha\beta} a^\alpha b^\beta \ . \qquad (2\text{-}30a)$$

Using the accompanying covariant vectors a_α and b_β we have

$$c = g^{\alpha\beta} a_\alpha b_\beta = a_\alpha b^\alpha = a^\alpha b_\alpha \ . \qquad (2\text{-}30b)$$

The square of the absolute value of a vector is

$$|a^\alpha|^2 = g_{\alpha\beta} a^\alpha a^\beta = g^{\alpha\beta} a_\alpha a_\beta = a_\alpha a^\alpha \ . \qquad (2\text{-}30c)$$

If γ is the angle between the vectors a^α and b^β, then

$$c = g_{\alpha\beta} a^\alpha b^\beta = g^{\alpha\beta} a_\alpha b_\beta = a_\alpha b^\alpha = a\ b\ \cos\gamma \qquad (2\text{-}30d)$$

[cf. (2-23)].

For <u>isothermal surface coordinates</u> [see (1-8)] we have

$$g_{\alpha\beta} = g^{1/2}\ \delta^\beta_\alpha \ , \qquad g = \det|g_{\alpha\beta}| \ , \qquad (2\text{-}31a)$$

and together with (2-18)

$$g^{\alpha\beta} = g^{-1/2} \delta^\beta_\alpha \ . \qquad (2\text{-}31b)$$

(2-29) thus becomes

$$a_\alpha = g^{1/2}\ a^\alpha \ , \qquad a^\alpha = g^{-1/2} a_\alpha \ . \qquad (2\text{-}31c)$$

If $g = 1$, in particular, the covariant and contravariant components of a vector are then equal to one another, e.g.,

$$a_\alpha = a^\alpha \ . \qquad (2\text{-}31d)$$

2.1.3 COVARIANT AND CONTRAVARIANT BASES

The definitions for vectors and tensors [see (2-14) and (2-16)] are based solely on intrinsic surface geometry. In this section it will be shown that these so-called surface tensors can also be represented in the tangential plane of the surface points with reference to two-dimensional Cartesian coordinate systems. Hence, we can conclude that surface tensors also correspond to Cartesian tensors in three-dimensional Euclidean space in which the surface is embedded.

At every surface point a <u>covariant basis</u>

$$b_{i.\alpha} := x_{i,\alpha} \quad = g_{\alpha\beta} b^{\beta}_{i.} \tag{2-32a}$$

and a <u>contravariant basis</u>

$$b^{\alpha}_{i.} \quad := g^{\alpha\beta} x_{i,\beta} = g^{\alpha\beta} b_{i.\beta} \tag{2-32b}$$

are defined. All four Cartesian basis vectors (2-32a,b) lie in the tangential plane of the surface point. The absolute values of the basis vectors yield

$$|b_{i.\alpha}| = (g_{(\alpha\alpha)})^{1/2} \ , \qquad |b^{\alpha}_{i.}| = (g^{(\alpha\alpha)})^{1/2} \ , \tag{2-32c}$$

and the following relations are valid:

$$b_{i.\alpha} b_{i.\beta} = g_{\alpha\beta} \ , \qquad b^{\alpha}_{i.} \ b^{\beta}_{i.} = g^{\alpha\beta} \ , \tag{2-32d}$$

$$b_{i.\alpha} b^{\beta}_{i.} \quad = \delta^{\beta}_{\alpha} = \left\{ \begin{matrix} 1 \\ 0 \end{matrix} \right. \text{for} \begin{matrix} \alpha = \beta \\ \alpha \neq \beta \end{matrix} \ . \tag{2-32e}$$

In transforming the surface coordinates [see (2-8)] the basis vectors are transformed as follows:

$$\bar{b}_{i.\bar{\alpha}} = u^{\beta}_{,\bar{\alpha}} \ b_{i.\beta} \ , \qquad b_{i.\alpha} = \bar{u}^{\bar{\beta}}_{,\alpha} \ \bar{b}_{i.\beta} \ , \tag{2-33a}$$

$$\bar{b}^{\bar{\alpha}}_{i.} \quad = \bar{u}^{\bar{\alpha}}_{,\beta} \ b^{\beta}_{i.} \ , \qquad b^{\alpha}_{i.} \quad = u^{\alpha}_{,\bar{\beta}} \ \bar{b}^{\bar{\beta}}_{i.} \ . \tag{2-33b}$$

A comparison with (2-14a,b) shows that the covariant and contravariant basis-vector components for $i = 1, 2, \text{ or } 3$ are covariant and contravariant "surface vectors" in the sense of theorem (2-14), respectively.

In the case of <u>orthogonal surface coordinates</u> [see (2-19)], (2-32a,b) become

$$b_{i.\alpha} = g_{\alpha(\alpha)} \ b^{(\alpha)}_{i.} \ , \qquad b^{\alpha}_{i.} \quad = g^{\alpha(\alpha)} \ b_{i.(\alpha)} \ . \tag{2-34a}$$

Thus, covariant and contravariant basis vectors are collinear and also oriented in the same direction because $g_{(\alpha\alpha)}, \ g^{(\alpha\alpha)} > 0$. Yet, their absolute values (2-32c) are only equal when

$$g_{11} = g^{22} = 1 \ . \tag{2-34b}$$

In the tangential planes of the surface points two, generally <u>oblique, rectilinear parallel coordinate systems</u>, i.e.,

$$S_{+} \quad \text{with the basis} \quad b_{i.\alpha} \ , \tag{2-35a}$$

$$S^{+} \quad \text{with the basis} \quad b^{\alpha}_{i.} \quad , \tag{2-35b}$$

are spanned by the covariant basis [see (2-32a)] and the contravariant basis [see (2-32b)]. In these systems a vector a_i of the tangential plane can be represented by

$$a_{i.} = b_{i.\alpha} a^{\alpha} = b^{\alpha}_{i.} a_{\alpha} \; . \tag{2-36a}$$

a^{α} are the vector components in S_+ measured in units of $|b_{i.\alpha}|$, and a_{α} are the vector components in S^+ measured in units of $|b^{\alpha}_{i.}|$. The inverse transformations of (2-36a) become

$$a^{\alpha} = b^{\alpha}_{i.} a_{i.} \; , \qquad\qquad a_{\alpha} = b_{i.\alpha} a_{i.} \; . \tag{2-36b}$$

If in (2-36b) $a_{i.}$ is an arbitrary vector which does not lie in the tangential plane, it will then be projected vertically onto the tangential plane with (2-36b). This can be easily recognized when S is oriented so that the third axis points in the direction of the surface normal of the test point. Then,

$$b_{3.\alpha} = b^{\alpha}_{3.} = 0 \; , \tag{2-36c}$$

whereby the component $a_{3.}$ perpendicular to the tangential plane remains undetermined. A subsequent transformation (2-36a) then no longer yields the original vector $a_{i.}$.

In the transition to other surface coordinates $\bar{u}^{\alpha}$, $a_{i.}$ is invariant whereby the first equations of (2-36a,b) assume the forms of

$$a_{i.} = \bar{b}_{i.\alpha} \bar{a}^{\alpha} \; , \qquad\qquad \bar{a}^{\alpha} = \bar{b}^{\alpha}_{i.} a_{i.} \; . \tag{2-37a}$$

Observing (2-33b), the second equation of (2-37a) can be written as

$$\bar{a}^{\alpha} = \bar{u}^{\alpha}_{,\beta} b^{\beta}_{i.} a_{i.} \; ,$$

whereby together with (2-36b)

$$\bar{a}^{\alpha} = \bar{u}^{\alpha}_{,\beta} a^{\beta} \tag{2-37b}$$

results. In the same way we obtain

$$\bar{a}_{\alpha} = u^{\beta}_{,\bar{\alpha}} a_{\beta} \; . \tag{2-37c}$$

Accordingly, the vector components

a^{α} in S_+ and a_{α} in S^+

are transformed after (2-14b,a) so that the following theorem is true:

Covariant and contravariant vectors of a surface (2-38)

a_α and a^α as defined in theorem (2-14) can be interpreted as vector components in the S^+ and S_+ systems of the tangential plane [see (2-35)] so that with reference to their embedment in three-dimensional Euclidean space S the transformation equations (2-36a,b) are also applicable to them, in addition to (2-14a,b).

Corresponding to (2-36a) a Cartesian tensor of the second rank in S can be defined by

$$a_{ij.} = b_{i.\alpha} b_{j.\beta}\, a^{\alpha\beta} = b^{\alpha}_{i.}\, b^{\beta}_{j.}\, a_{\alpha\beta} = b^{\alpha}_{i.}\, b_{j.\beta}\, a^{\beta}_{\alpha} \qquad (2\text{-}39a)$$

based on the tensor components $a^{\alpha\beta}$, $a_{\alpha\beta}$, and a^{β}_{α} in S^+ and/or in S_+. The inverse transformations of (2-39a) become

$$a^{\alpha\beta} = b^{\alpha}_{i.}\, b^{\beta}_{j.} a_{ij.}\,, \quad a_{\alpha\beta} = b_{i.\alpha} b_{j.\beta} a_{ij.}\,, \quad a^{\beta}_{\alpha} = b_{i.\alpha} b^{\beta}_{j.}\, a_{ij.} \;. \qquad (2\text{-}39b)$$

A general presupposition in (2-39b) is that $a_{ij.}$ is a tensor spanned on the respective tangential plane, as is the case, for example, according to (2-39a). If this is not the case, $a_{ij.}$ is projected onto the tangential plane using (2-39b) so that a subsequent transformation (2-39a) does not yield the original tensor $a_{ij.}$ [see remark to (2-36c) as well].

Corresponding to (2-37) and based on (2-39) the validity of the following theorem can be shown, which is a generalization of theorem (2-38):

Covariant and contravariant tensors of the second rank of a surface (2-40)

$a_{\alpha\beta}$, $a^{\alpha\beta}$, and a^{β}_{α} as defined in theorem (2-16) can be interpreted as tensor components in the S^+ and S_+ systems of the tangential plane [see (2-35)] so that with reference to their embedment in three-dimensional Euclidean space S the transformation equations of (2-39a,b) are also applicable to them, in addition to (2-16a,b).

(2-39) and theorem (2-40) can be extended by the respective n-fold inner multiplication with the corresponding base vectors to tensors of rank n > 2.

$a_{i.}$ and $d_{i.}$ are two vectors on the tangential plane of a surface point. Considering (2-36a) we obtain the following for the (general) *product of the vectors* corresponding to (2-39a):

$$a_{i.}d_{j.} = b_{i.\alpha}b_{j.\beta}\,a^{\alpha}d^{\beta} = b^{\alpha}_{i.}\,b^{\beta}_{j.}\,a_{\alpha}d_{\beta} = b^{\alpha}_{i.}\,b_{j.\beta}\,a_{\alpha}d^{\beta} \ . \tag{2-41a}$$

With the contraction $i = j$, (2-41a) becomes the *scalar product*

$$c = a_{i.}d_{i.} = g_{\alpha\beta}a^{\alpha}d^{\beta} = g^{\alpha\beta}a_{\alpha}d_{\beta} = a^{\alpha}d_{\alpha} = a_{\alpha}d^{\alpha} \ , \tag{2-41b}$$

which is in agreement with (2-30a,b,d).

For *transforming $\delta_{ij.}$ in the tangential plane* of a surface point according to (2-39b) due to (2-32d,e) we obtain

$$\delta_{\alpha\beta} = g_{\alpha\beta} \ , \quad \delta^{\alpha\beta} = g^{\alpha\beta} \ , \quad \delta^{\alpha}_{\beta} \ \text{after (2-32e)} \ . \tag{2-42}$$

Let $e_{i.}$ be a unit vector lying in the tangential plane of a surface point. We want to determine a unit vector $f_{i.}$ which also lies in the tangential plane and is perpendicular to $e_{i.}$ and thus

$$e_{i.}\,f_{i.} = 0 \ . \tag{2-43a}$$

For solving this problem we need the normal vector [see (2-25)] which after (2-32a) takes the form of

$$n_{i.} = \epsilon_{ijk.}b_{j.1}b_{k.2}g^{-1/2} \ . \tag{2-43b}$$

If $n_{i.} - e_{i.} - f_{i.}$ forms a positively oriented system, then

$$f_{i.} = \epsilon_{ijk.}n_{j.}e_{k.} = g^{-1/2}\epsilon_{ijk.}\epsilon_{jlm.}b_{l.1}b_{m.2}e_{k.} \ . \tag{2-43c}$$

After applying the development theorem of vector algebra [see Heitz 1980-1983, (2-55); Jeffreys 1974, (55)] and subsequently exchanging the indices this equation becomes

$$f_{i.} = -\epsilon_{ij.}e_{j.} \ , \tag{2-43d}$$

whereby

$$\epsilon_{ij.} = (b_{i.1}b_{j.2} - b_{i.2}b_{j.1})\,g^{-1/2} = -\epsilon_{ji.} \tag{2-43e}$$

is the *ϵ-tensor of the second rank*. By transforming (2-43e) according to the second of (2-39b) we obtain the *covariant ϵ-tensor* of the second rank, i.e.,

$$\epsilon_{\alpha\beta} = b_{i.\alpha}b_{j.\beta}\epsilon_{ij.} = (g_{\alpha1}g_{\beta2} - g_{\alpha2}g_{\beta1})\,g^{-1/2} = \begin{bmatrix} 0 & 1 \\ -1 & 0 \end{bmatrix} g^{1/2} \ . \tag{2-44a}$$

The contravariant ϵ-tensor of the second rank becomes

$$\epsilon^{\alpha\beta} = b^{\alpha}_{i.}\, b^{\beta}_{j.}\, \epsilon_{ij.} = g^{\alpha\gamma} g^{\beta\delta} \epsilon_{\gamma\delta} \tag{2-44b}$$
$$= (\delta^{\alpha}_{1}\, \delta^{\beta}_{2} - \delta^{\alpha}_{2}\, \delta^{\beta}_{1})\, g^{-1/2} = \begin{bmatrix} 0 & 1 \\ -1 & 0 \end{bmatrix} g^{1/2} ,$$

and

$$\epsilon^{\beta}_{\alpha} = b_{i.\alpha} b^{\beta}_{j.}\, \epsilon_{ij.} = g^{\beta\gamma} \epsilon_{\alpha\gamma} = g_{\alpha\gamma} \epsilon^{\gamma\beta} \tag{2-44c}$$
$$= (g_{\alpha 1} \delta^{\beta}_{2} - g_{\alpha 2} \delta^{\beta}_{1})\, g^{-1/2} = \begin{bmatrix} -g_{12} & g_{11} \\ -g_{22} & g_{12} \end{bmatrix} g^{1/2}$$

is the mixed ϵ-tensor of the second rank. From (2-44a-c)

$$\epsilon^{\alpha\gamma} \epsilon_{\beta\gamma} = \delta^{\alpha}_{\beta} \qquad \text{and} \qquad \epsilon^{\alpha}_{\gamma}\, \epsilon^{\gamma}_{\beta} = -\delta^{\alpha}_{\beta} \tag{2-44d}$$

result. With (2-44a-c), (2-43d) can be written in the covariant and contravariant forms, i.e.,

$$f_{\alpha} = \epsilon_{\alpha\beta} e^{\beta} = \epsilon^{\beta}_{\alpha}\, e_{\beta} , \qquad f^{\alpha} = \epsilon^{\alpha\beta} e_{\beta} = \epsilon^{\alpha}_{\beta}\, e^{\beta} . \tag{2-44e}$$

2.1.4 EQUATIONS OF GAUSS AND WEINGARTEN

For studying surfaces in the vicinities of surface points partial derivatives of the moving trihedron [see (2-5)]

$$x_{i,\alpha} , \quad n_{i.} , \quad \alpha \in \{1,2\} \tag{2-45a}$$

are necessary with respect to the surface coordinates u^{α} as linear combinations of this trihedron. Due to (2-32)

$$x_{i,\alpha} = b_{i.\alpha} = g_{\alpha\beta} b^{\beta}_{i.} . \tag{2-45b}$$

The results are made up of the fundamental equations of Gauss

$$x_{i,\alpha\beta} = b_{i.\alpha,\beta} = \Gamma^{\gamma}_{\alpha\beta} x_{i,\gamma} + L_{\alpha\beta} n_{i.} = \Gamma^{\gamma}_{\alpha\beta} b_{i.\gamma} + L_{\alpha\beta} n_{i.} \tag{2-46a}$$

and Weingarten

$$n_{i,\beta} = -L^{\gamma}_{\beta}\, x_{i,\gamma} = -L^{\gamma}_{\beta}\, b_{i.\gamma} \tag{2-46b}$$

[see also (2-49)]. These equations are at first purely formal representations when the use of the letter L in both equations is disregarded. $L_{\alpha\beta}$ is the so-called second fundamental tensor of the surface theory for which we obtain

$$L_{\alpha\beta} = n_{i.} x_{i,\alpha\beta} \tag{2-47a}$$

by inner multiplication of (2-46a) with $n_{i.}$. From $n_{i.}x_{i,\alpha} = 0$ we have

$$n_{i,\beta}x_{i,\alpha} + n_{i.}x_{i,\alpha\beta} = 0$$

so that

$$L_{\alpha\beta} = n_{i.}x_{i,\alpha\beta} = -n_{i,\beta}x_{i,\alpha} = -n_{i,\alpha}x_{i,\beta} \tag{2-47b}$$

as well. In (2-46b) the mixed second fundamental tensor appears, i.e.,

$$L^{\gamma}_{\beta} = g^{\gamma\alpha}L_{\alpha\beta} \quad . \tag{2-47c}$$

This result is obtained by inner multiplication of (2-46b) with $x_{i,\delta}$ and observing (2-17b), (2-18a), and (2-47b). The inner product of (2-46a) with $x_{i,\delta}$ yields

$$\Gamma_{\alpha\beta|\delta} := x_{i,\alpha\beta}x_{i,\delta} = \Gamma^{\gamma}_{\alpha\beta}g_{\gamma\delta} \tag{2-48a}$$

with the <u>Christoffel's symbols</u>

of the first kind: $\Gamma_{\alpha\beta|\gamma} = x_{i,\alpha\beta}x_{i,\gamma}$ and

of the second kind: $\Gamma^{\gamma}_{\alpha\beta} = g^{\gamma\delta}\Gamma_{\alpha\beta|\delta}$. (2-48b)

Christoffel's symbols are functions solely of the first fundamental tensor. From

$$g_{\alpha\beta,\gamma} := \partial g_{\alpha\beta}/\partial u^{\gamma} = x_{i,\alpha\gamma}x_{i,\beta} + x_{i,\alpha}x_{i,\beta\gamma} = \Gamma_{\alpha\gamma|\beta} + \Gamma_{\beta\gamma|\alpha} \tag{2-48c}$$

the result is

$$\Gamma_{\alpha\beta|\gamma} = (g_{\alpha\gamma,\beta} + g_{\beta\gamma,\alpha} - g_{\alpha\beta,\gamma})/2 \quad . \tag{2-48d}$$

The equations of Gauss and Weingarten (2-46) are founded on the covariant basis vectors (2-32a). If the contravariant basis vectors (2-32b) are used instead, we then obtain

$$b_{i.\alpha,\beta} = g_{\alpha\gamma}b^{\gamma}_{i,\beta} + g_{\alpha\gamma,\beta}b^{\gamma}_{i.} = \Gamma_{\alpha\beta|\gamma}b^{\gamma}_{i.} + L_{\alpha\beta}n_{i.}$$

for (2-46a) due to (2-48a). Observing (2-48d) the result is

$$g_{\alpha\gamma}b^{\gamma}_{i,\beta} = -\Gamma_{\beta\gamma|\alpha}b^{\gamma}_{i.} + L_{\alpha\beta}n_{i.} \quad .$$

After inner multiplication with $g^{\alpha\delta}$ and exchanging the indices we finally get the <u>equations of Gauss for the contravariant basis vectors</u>, i.e.,

$$b^{\alpha}_{i,\beta} = -\Gamma^{\alpha}_{\beta\gamma}b^{\gamma}_{i.} + L^{\alpha}_{\beta}\, n_{i.} \quad . \tag{2-49a}$$

Inserting (2-32a) in (2-46b) yields the <u>equations of Weingarten with contravariant basis vectors</u>, i.e.,

$$n_{i,\beta} = -L_{\beta\gamma} b^{\gamma}_{i.} \quad . \tag{2-49b}$$

If the first and second fundamental tensors of a surface are given as functions of the surface coordinates u^{γ}, i.e.,

$$g_{\alpha\beta} = g_{\alpha\beta}(u^{\gamma}) \quad , \qquad L_{\alpha\beta} = L_{\alpha\beta}(u^{\gamma}) \quad , \tag{2-50a}$$

the equations of Gauss and Weingarten [see (2-46)] then represent a system of differential equations of the second order for the position vector $x_{i.}(u^{\gamma})$. It can be shown that the position vector and thus the surface, excepting movements of the spatial coordinate system (six integration constants), can be determined using these differential equations. Thus, the equations of Gauss and Weingarten [see (2-46)] are important to the surface theory, just as the equations of Frenet are important to the curve theory [see Chapter 5.2.3], in which curvature and torsion of the space curve occur in place of the fundamental tensors of the surface [see (2-50a)]. However, arbitrary functions cannot be selected for the six possible components of the symmetrical tensors of (2-50a). These functions must fulfill integrability conditions in the form of the equations of Gauss, i.e.,

$$R^{\mu}_{\alpha\beta\gamma} := \Gamma^{\mu}_{\alpha\beta,\gamma} - \Gamma^{\mu}_{\alpha\gamma,\beta} + \Gamma^{\upsilon}_{\alpha\beta}\Gamma^{\mu}_{\upsilon\gamma} - \Gamma^{\upsilon}_{\alpha\gamma}\Gamma^{\mu}_{\upsilon\beta} = L_{\alpha\beta}L^{\mu}_{\gamma} - L_{\alpha\gamma}L^{\mu}_{\beta} \quad , \tag{2-50b}$$

and the equations of Mainardi and Codazzi, i.e.,

$$L_{\alpha\beta,\gamma} - L_{\alpha\gamma,\beta} + \Gamma^{\upsilon}_{\alpha\beta}L_{\upsilon\gamma} - \Gamma^{\upsilon}_{\alpha\gamma}L_{\upsilon\beta} = 0 \quad . \tag{2-50c}$$

We arrive at these equations through the symmetry conditions for the third partial derivatives of the position vectors, i.e.,

$$x_{i,\alpha\beta\gamma} = x_{i,\alpha\gamma\beta} \quad ,$$

which have to be derived based on (2-46). The statements of (2-50a-c) contain the so-called fundamental theorem of the surface theory.

The first expression of (2-50b) is a mixed form of the covariant Riemannian curvature tensor, i.e.,

$$R_{\upsilon\alpha\beta\gamma} = g_{\mu\upsilon}R^{\mu}_{\alpha\beta\gamma} \quad . \tag{2-50d}$$

Thereby, (2-50b) can also be written in the following way:

$$L_{\alpha\beta}L_{\gamma\upsilon} - L_{\alpha\gamma}L_{\beta\upsilon} = R_{\upsilon\alpha\beta\gamma} \quad . \tag{2-50e}$$

The Riemannian curvature tensor is solely a function of the first fundamental tensor and its partial derivatives, which is obvious from (2-50b,d) and considering (2-48). Due to the symmetry of the second fundamental tensor it ensues from (2-50e) that the Riemannian curvature tensor in both the first and both the second indices is skew-symmetrical, i.e.,

$$R_{\alpha\upsilon\beta\gamma} = -R_{\upsilon\alpha\beta\gamma} = R_{\upsilon\alpha\gamma\beta} = -R_{\alpha\upsilon\gamma\beta} \; . \tag{2-50f}$$

In particular, based on (2-50e) we obtain

$$R_{2112} = L_{11}L_{22} - (L_{12})^2 = |L_{\alpha\beta}| =: L \; . \tag{2-50g}$$

2.1.5 COVARIANT DERIVATIVES OF SURFACE VECTORS

We arrive at the term covariant derivative via the partial differentiation of vectors $a_{i.}$ of the tangential planes, in short tangential vectors, with respect to the surface coordinates u^α. From (2-36a), i.e.,

$$a_{i.} = b_{i.\alpha}a^\alpha = b_{i.}^{\alpha}\,a_\alpha \; ,$$

we have

$$a_{i,\beta} = b_{i.\alpha}a^\alpha{}_{,\beta} + b_{i.\alpha,\beta}a^\alpha = b_{i.}^{\alpha}\,a_{\alpha,\beta} + b_{i,\beta}^{\alpha}a_\alpha \; .$$

If (2-46a) or (2-49a) is used in the partial differentiation of the basis vectors, we obtain in the moving trihedrons of the surface

$$b_{i.\alpha} \, , \; n_{i.} \qquad \text{or} \qquad b_{i.}^{\alpha} \, , \; n_{i.} \tag{2-51}$$

the following representations for $a_{i,\beta}$:

$$a_{i,\beta} = a^\gamma{}_{;\beta}\, b_{i.\gamma} + L_{\alpha\beta}a^\alpha\, n_{i.} = a_{\gamma;\beta}\, b_{i.}^{\gamma} + L_{\beta}^{\alpha}a_\alpha\, n_{i.} \; . \tag{2-52}$$

Herein,

$$a^\gamma{}_{;\beta} := a^\gamma{}_{,\beta} + \Gamma^\gamma_{\alpha\beta}a^\alpha \tag{2-53a}$$

is the covariant derivative of the contravariant surface vektor a^α, and

$$a_{\gamma;\beta} := a_{\gamma,\beta} - \Gamma^\alpha_{\beta\gamma}a_\alpha \tag{2-53b}$$

is the covariant derivative of the covariant surface vector a_α. It can be shown that the quantities of (2-53a,b) are mixed and pure covariant tensors of rank two, respectively. Herein lies the essential difference between the covariant derivatives of a surface vector and its partial derivatives

$$a^{\gamma}_{,\beta} , \quad a_{\gamma,\beta} ,$$

which are not tensors. The partial derivatives of scalar position functions $f(u^{\alpha})$, on the other hand, are also covariant derivatives, i.e.,

$$f_{,\alpha} = f_{;\alpha} . \tag{2-53c}$$

The accompanying attribute of being a tensor is proven with (2-11) and (2-14).

The <u>total differential of a tangential vector $a_{i.}$</u> together with (2-52) becomes

$$\begin{aligned} da_{i.} = a_{i,\beta}du^{\beta} &= (a^{\gamma}_{;\beta}b_{i.\gamma} + L_{\beta\gamma}a^{\gamma}n_{i.})\, du^{\beta} \\ &= (a_{\gamma;\beta}b^{\gamma}_{i.} + L^{\gamma}_{\beta}\, a_{\gamma}n_{i.})\, du^{\beta} . \end{aligned} \tag{2-54}$$

The accompanying contravariant and covariant components are obtained according to (2-36) as follows:

$$Da^{\alpha} := da_{i.}b^{\alpha}_{i.} = a^{\alpha}_{;\beta}du^{\beta} , \quad Da_{\alpha} := da_{i.}b_{i.\alpha} = a_{\alpha;\gamma}du^{\beta} . \tag{2-55}$$

They are designated <u>absolute differentials of a^{α} and a_{α}</u>, respectively.

2.1.6 MEASURES OF THE CURVATURE OF SURFACE CURVES AND SURFACES

Here we will regard a surface curve C as a function of the arc length s. Based on the curve equations

$$u^{\alpha} = u^{\alpha}(s) \tag{2-56a}$$

the position vector of C in the spatial coordinate system S becomes

$$x_{i.}(s) = x_{i.}(u^{\alpha}(s)) . \tag{2-56b}$$

In the following the first and second derivatives with respect to the arc length are labelled with ' und '' :

$$\begin{aligned} dx_{i.}/ds &=: x'_{i.} , \qquad & d^2x_{i.}/ds &=: x''_{i.} , \\ du^{\alpha}/ds &=: u^{\alpha'} , & d^2u^{\alpha}/ds^2 &=: u^{\alpha''} . \end{aligned} \tag{2-57}$$

According to (5-28a) $e_{ij.}$ is the moving trihedron of the space curve C.

The first of Frenet's equations (5-30), j = 1 , yields

$$x''_{i.} = e'_{i1.} = \kappa\, e_{i2.} \,, \tag{2-58a}$$

whereby

$$\kappa = (x''_{i.}\, x''_{i.})^{1/2} \tag{2-58b}$$

is the <u>curvature of the space curve C</u>. In studying C as a surface curve we span the <u>curvature vector $x''_{i.}$</u> of C by the moving trihedron of the surface as well [see (2-5) and (2-45)]. We then introduce the orthogonal unit vector to $n_{i.}$ and $e_{i1.} = x'_{i.}$, i.e.,

$$m_{i.} = \epsilon_{ijk.}\, n_{j.}\, x'_{k.} = -\epsilon_{ij.}\, x'_{j.} \,, \tag{2-59}$$

the second form of which ensues according to (2-4). The following representation is made for (2-58a):

$$x''_{i.} = \kappa\, e_{i2.} = \kappa_g\, m_{i.} + \kappa_n\, n_{i.} \,. \tag{2-60a}$$

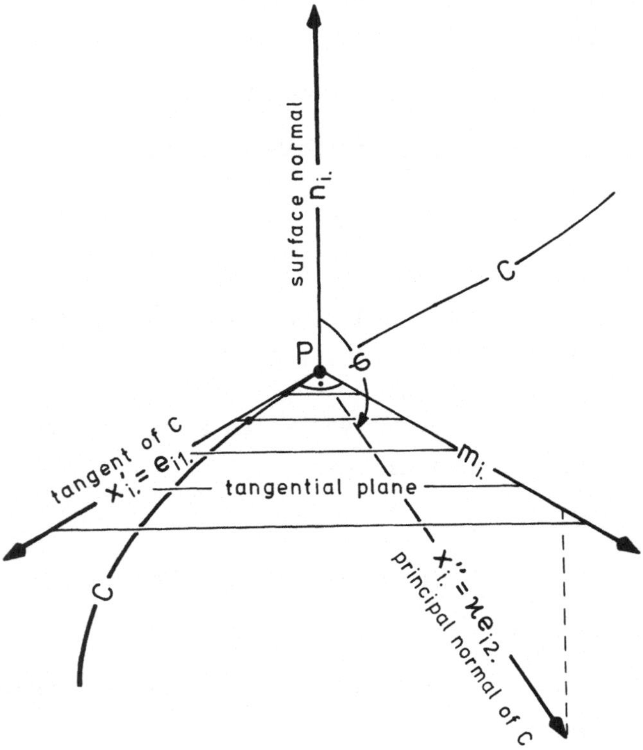

Fig. 2.1. Tangent and normals of a surface curve C

The coefficients of $m_{i.}$ and $n_{i.}$ occurring herein are designated as follows:

$$\kappa_g = \text{geodesic curvature of the surface curve } C\ ,$$
$$\kappa_n = \text{normal curvature of the surface in the direction } x'_{i.}\ . \tag{2-60b}$$

Regarding (2-59) the inner product of (2-60a) and $m_{i.}$ or $n_{i.}$ result in

$$\kappa_g = \epsilon_{ijk.} x'_{i.} x''_{j.} n_{k.}\ , \qquad \kappa_n = x''_{i.} n_{i.}\ . \tag{2-61}$$

Due to (2-54a,b) we obtain with $a_{i.} =: x'_{i.}$

$$x''_{i.} = (Du^{\gamma'}/ds)\, b_{i.\gamma} + L_{\alpha\beta} u^{\alpha'} u^{\beta'} n_{i.}\ . \tag{2-62}$$

A comparison with (2-60a) first yields the expression for normal curvature equivalent to (2-61), i.e.,

$$\kappa_n = x''_{i.} n_{i.} = L_{\alpha\beta} u^{\alpha'} u^{\beta'}\ . \tag{2-63}$$

The right side multiplied by ds^2 is the so-called <u>second fundamental form</u>. Another result of the comparison of (2-60a) with (2-62) is

$$\kappa_g m_{i.} = (Du^{\gamma'}/ds)\, b_{i.\gamma}\ .$$

After inner multiplication with $b_{i.}^{\delta}$ and an exchange of the indices this equation becomes

$$\kappa_g m^{\gamma} = Du^{\gamma'}/ds = u^{\gamma''} + \Gamma^{\gamma}_{\alpha\beta} u^{\alpha'} u^{\beta'}\ . \tag{2-64}$$

m^{γ} and $u^{\delta'}$ are two unit vectors perpendicular to one another so that from (2-44e) and (2-30) we obtain

$$\epsilon_{\delta\gamma} u^{\delta'} m^{\gamma} = m_{\gamma} m^{\gamma} = 1\ .$$

Therefore, the inner product of (2-64) with $\epsilon_{\delta\gamma} u^{\delta'}$ produces the following equation for <u>geodesic curvature</u>:

$$\kappa_g = \epsilon_{\delta\gamma} u^{\delta'} Du^{\gamma'}/ds = \epsilon_{\delta\gamma} u^{\delta'} (u^{\gamma''} + \Gamma^{\gamma}_{\alpha\beta} u^{\alpha'} u^{\beta'})\ . \tag{2-65}$$

Consequently, the geodesic curvature is solely a function of the first fundamental tensor, i.e., a quantity of the intrinsic surface geometry, which has already been pointed out [see (2-21d)].

The previous considerations have been based on (2-56) and, therefore, on the arc length s as curve coordinate. Using the abbreviations

$$du^{\alpha}/dt =: u^{\alpha\cdot}\ , \qquad d^2u^{\alpha}/dt^2 =: u^{\alpha\cdot\cdot} \tag{2-66}$$

and via the first fundamental form (2-17a) we obtain the following for an arbitrary curve coordinate t:

$$ds = (g_{\alpha\beta} u^{\alpha\cdot} u^{\beta\cdot})^{1/2} dt\ . \tag{2-67}$$

If this expression is substituted for ds in (2-65), the more general equation for geodesic curvature then becomes

$$\kappa_g = \epsilon_{\delta\gamma} u^{\delta\cdot}(u^{\gamma\cdot\cdot} + \Gamma^{\gamma}_{\alpha\beta} u^{\alpha\cdot} u^{\beta\cdot})(g_{\mu\upsilon} u^{\mu\cdot} u^{\upsilon\cdot})^{-3/2} \ . \qquad (2\text{-}68)$$

Let

$$\varphi = \text{angle}(e_{i2.}, n_{i.}) \qquad (2\text{-}69a)$$

be the angle between the principal normal of C and the surface normal [see Fig. 2.1]. Thereby,

$$e_{i2.} m_{i.} = \sin\varphi \ , \qquad e_{i2.} n_{i.} = \cos\varphi \ , \qquad (2\text{-}69b)$$

and by inner multiplication of (2-60a) with $m_{i.}$ or $n_{i.}$ in place of (2-61) we obtain

$$\kappa_g = \kappa \sin\varphi \ , \qquad \kappa_n = \kappa \cos\varphi \ . \qquad (2\text{-}70)$$

The second equation of (2-70) is the so-called <u>theorem of Meusnier</u>. From (2-70) we have

$$\kappa_g^2 + \kappa_n^2 = \kappa^2 \ . \qquad (2\text{-}71)$$

2.1.7 NORMAL AND PRINCIPAL CURVATURES OF SURFACES

The <u>normal curvature of a surface</u> is obtained with (2-63) and by observing (2-66) and (2-67):

$$\kappa_n = L_{\alpha\beta} u^{\alpha'} u^{\beta'} = (L_{\alpha\beta} u^{\alpha\cdot} u^{\beta\cdot})/(g_{\gamma\delta} u^{\gamma\cdot} u^{\delta\cdot}) \ . \qquad (2\text{-}72)$$

κ_n is at one surface point solely a <u>function of the direction</u> determined by $u^{\alpha'}$ or $u^{\alpha\cdot}$ or the ratio

$$\upsilon = du^2/du^1 \ . \qquad (2\text{-}73)$$

If κ_n is not constant at one surface point, there are then two <u>principal directions</u>, i.e.,

$$\upsilon_1 = (du^2/du^1)_1 \ , \qquad \upsilon_2 = (du^2/du^1)_2 \ , \qquad (2\text{-}74a)$$

for which κ_n assumes extreme values, i.e.,

$$d\kappa_n/d\upsilon = 0 \ . \qquad (2\text{-}74b)$$

These extreme values are the so-called <u>principal curvatures</u>

$$\kappa_{n1} = 1/R_1 , \qquad \kappa_{n2} = 1/R_2 \tag{2-74c}$$

at a surface point; R_1 and R_2 are called the principal curvature radii.

(2-72) together with (2-73) can be written as

$$\kappa_n = (L_{11} + 2L_{12}\upsilon + L_{22}\upsilon^2)/(g_{11} + 2g_{12}\upsilon + g_{22}\upsilon^2) . \tag{2-75}$$

Due to (2-74b) we then obtain a quadratic equation for υ which after multiplication with $(du^1)^2$ reads

$$(L_{11}g_{12} - L_{12}g_{11})(du^1)^2 + (L_{11}g_{22} - L_{22}g_{11})\, du^1 du^2 + (L_{12}g_{22} - L_{22}g_{12})(du^2)^2 = 0 . \tag{2-76}$$

The two solutions for (2-76), i.e., υ_1 and υ_2, are the principal directions of (2-74a). If υ_1 and υ_2 are inserted in (2-56) or (2-72), the accompanying principal curvatures [see (2-74c)] are then obtained. We can also arrive at these curvatures in the following manner. (2-75) can be rewritten as

$$(L_{11} - \kappa_n g_{11}) + 2\,(L_{12} - \kappa_n g_{12})\,\upsilon + (L_{22} - \kappa_n g_{22})\,\upsilon^2 = 0 . \tag{2-77a}$$

For every value of κ_n there are generally two solutions for υ due to this quadratic equation in υ. However, only one solution is permitted for the principal curvatures [see (2-74c)] in each case so that the discriminant of (2-77a) must vanish. This leads to the quadratic equation in κ_n:

$$\kappa_n^2 - g^{\alpha\beta}L_{\alpha\beta}\kappa_n + L/g = 0 \tag{2-77b}$$

[g and L after (2-17c) and (2-50g)]. The solutions to this equation are the principal curvatures [see (2-74c)] for which the following relationships are valid:

$$K = \kappa_{n1}\kappa_{n2} = (1/R_1)(1/R_2) = L/g$$
$$= \text{Gaussian curvature} , \tag{2-78a}$$

$$H = (\kappa_{n1} + \kappa_{n2})/2 = (1/R_1 + 1/R_2)/2 = g^{\alpha\beta}L_{\alpha\beta}/2$$
$$= \text{mean curvature} . \tag{2-78b}$$

Together with (2-50g) for L, (2-78a) becomes the theorema egregium of Gauss, i.e.,

$$K = R_{2112}/g = -g_{1\mu}R^{\mu}_{212}/g = g_{2\mu}R^{\mu}_{112}/g , \tag{2-79a}$$

which states that Gaussian curvature is only a function of the first fundamental tensor. (2-79a) can also be written as

$$K = -(1/4)\epsilon^{\alpha\beta}\epsilon^{\gamma\delta}R_{\alpha\beta\gamma\delta} \tag{2-79b}$$

which is evident considering (2-44b) and (2-50f). The invariance of K in coordinate transformations ensues directly from the tensor product of (2-79b). For rectangular surface coordinates (2-79a) and (2-79b) can be converted into

$$K = -(1/2)g^{-1/2}((g^{-1/2}g_{22,1})_{,1} + (g^{-1/2}g_{11,2})_{,2}) \; . \tag{2-79c}$$

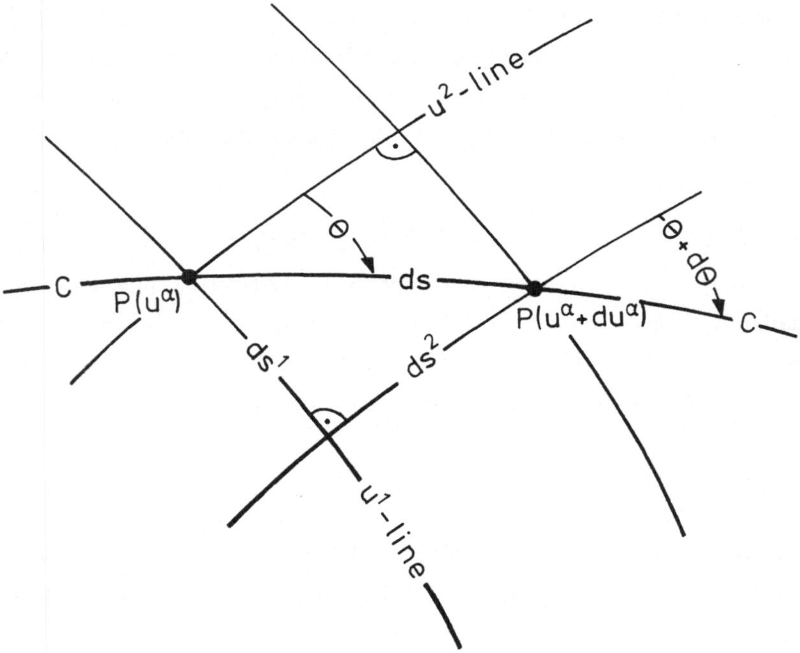

***Fig.* 2.2.** Orthogonal coordinate lines

<u>Curvature lines</u> (2-80)

are surface curves whose tangents at every point are identical to one of the directions of principal curvature.

Consequently, they are generally two independent families of curvature lines on a surface. If the surface coordinates u^α are selected so that the coordinate lines coincide with the curvature lines, i.e.,

$$u^1\text{-lines} \equiv \text{curvature lines} \quad \kappa_{n1} = 1/R_1 \, ,$$
$$u^2\text{-lines} \equiv \text{curvature lines} \quad \kappa_{n2} = 1/R_2 \, . \tag{2-81a}$$

(2-76) must then be true for both $du^1 = 0$ and $du^2 = 0$. This is only possible when

$$L_{12}g_{22} - L_{22}g_{12} = 0 \, , \qquad L_{11}g_{12} - L_{12}g_{11} = 0 \, ,$$

stipulating

$$g_{12} = L_{12} = 0 \tag{2-81b}$$

for (2-81a). Accordingly, the two families of curvature lines and as well as thus the <u>principal directions (2-74a) are orthogonal</u> to one another. The measures of curvature of (2-78) together with (2-81b) become

$$K = (L_{11}L_{22})/(g_{11}g_{22}) \, , \qquad H = (L_{11}/g_{11} + L_{22}/g_{22})/2 \, , \tag{2-81c}$$

leading to

$$\kappa_{n1} = 1/R_1 = L_{11}/g_{11} \, , \qquad \kappa_{n2} = 1/R_2 = L_{22}/g_{22} \, . \tag{2-81d}$$

Together with (2-81) the first form of (2-72) can be written in the following manner:

$$\kappa_n = \kappa_{n1}g_{11}(u^{1'})^2 + \kappa_{n2}g_{22}(u^{2'})^2 \, . \tag{2-82a}$$

κ_n is the normal curvature for the dirctions given by

$$u^{1'} = du^1/ds \qquad \text{or} \qquad u^{2'} = du^2/ds \, . \tag{2-82b}$$

This can also be described using the <u>direction angle θ</u> between the arc element and the curvature line u^2 = variable [see Fig. 2.2]. Thereby, we obtain

$$(g_{11})^{1/2}u^{1'} = \sin\theta \, , \qquad (g_{22})^{1/2}u^{2'} = \cos\theta \, , \tag{2-82c}$$

with which (2-82a) becomes the <u>theorem of Euler</u>, i.e.,

$$\kappa_n = \kappa_{n1}\sin^2\theta + \kappa_{n2}\cos^2\theta \quad \text{bzw.} \quad 1/R = (\sin^2\theta)/R_1 + (\cos^2\theta)/R_2 \, . \tag{2-83}$$

2.1.8 SURFACE CURVES WITH GIVEN GEODESIC CURVATURE

Analogous to the moving trihedron for space curves [see (5-28)] every point P on a surface curve C is assigned to a

<u>Moving two-dimensional basis</u> (2-84)

which is spanned by the two orthonormal vectors:

$$x'_{i.} = dx_{i.}/ds = b_{i.\alpha}u^{\alpha'} \quad , \tag{2-84a}$$

$$m_{i.} = \epsilon_{ijk.}n_{j.}x'_{k.} = -\epsilon_{ij.}x'_{j.} = b_{i.\alpha}m^{\alpha} \tag{2-84b}$$

[cf. (2-56) ff]. <u>Frenet's equations for surface curves</u> are valid for this basis:

$$Du^{\alpha'}/ds = -\kappa_g\epsilon^{\alpha}_{\beta}\,u^{\beta'} \quad , \qquad Dm^{\alpha}/ds = -\kappa_g\epsilon^{\alpha}_{\beta}\,m^{\beta} \;. \tag{2-85}$$

According to (2-44e) the following is true:

$$m^{\alpha} = -\epsilon^{\alpha}_{\beta}\,u^{\beta'} \quad , \qquad u^{\alpha'} = \epsilon^{\alpha}_{\beta}\,m^{\beta} \;.$$

Inserting the first expression into the left side of (2-64) directly yields the first equation of (2-85). When the second expression is substituted for $u^{\alpha'}$ in the first equation of (2-85), we obtain

$$\epsilon^{\alpha}_{\beta}\,Dm^{\beta}/ds = -\kappa_g m^{\alpha}$$

due to Ricci's theorem [see Sokolnikoff 1964, 35] which results in the vanishing of the covariant derivatives of the ϵ-tensors. Inner multiplication of the equation above with $\epsilon^{\gamma}_{\alpha}$ gives the second equation of (2-85).

Using the first differential equation (2-85) the coordinates of a surface curve C as a function of the arc length s, i.e.,

$$u^{\alpha} = u^{\alpha}(s) \quad , \tag{2-86a}$$

are uniquely determined when the geodesic curvature of C as a function of s, i.e,

$$\kappa_g = \kappa_g(s) \quad , \tag{2-86b}$$

and the initial values of

$$(u^{\alpha})_a = \text{coordinates of a point } P_a \in C \;, \tag{2-86c}$$

$$(u^{\alpha'})_a = (du^{\alpha}/ds)_a \quad \text{for } \alpha = 1 \text{ or } 2 \tag{2-86d}$$

are given. The two possible derivatives in (2-86d), i.e., $\alpha = 1$ or 2, are dependent on one another since according to (2-17a)

$$g_{\alpha\beta}u^{\alpha'}u^{\beta'} = 1 \, . \tag{2-86e}$$

In addition to this initial value problem, the boundary value problem of the first differential equation of (2-85) is also of practical interest in which the boundary values

$$(u^{\alpha})_a \, , \quad (u^{\alpha})_b = \text{coordinates of two points } P_a, \, P_b \in C \tag{2-87}$$

are given in place of (2-86c,d).

2.1.9 GEODESIC LINES

Geodesic lines (2-88)

are defined here as surface curves with vanishing geodesic curvature: $\kappa_g = 0$.

Thus, proceeding from (2-61a) it follows that with regard to geodesic lines the principal normals are always collinear to the surface normals, i.e.,

$$x''_{i.} \parallel n_{i.} \, . \tag{2-89}$$

The osculating planes of geodesic lines are thus perpendicular to the surface at all points. The first of Frenet's equations (2-85) merges into the differential equation of geodesic lines, i.e.,

$$Du^{\gamma'}/ds = u^{\gamma''} + \Gamma^{\gamma}_{\alpha\beta}u^{\alpha'}u^{\beta'} = 0 \, , \quad \gamma \in \{1,2\} \, . \tag{2-90a}$$

If the metric tensor and thus Christoffel's symbols are given as functions of the surface coordinates u^{α}, then (2-90a) is a system of two differential equations of the second order for the surface coordinates of the geodesic as functions of the arc length:

$$u^{\alpha} = u^{\alpha}(s) \, . \tag{2-90b}$$

The functions of (2-90b) are generally determined with (2-90a) up to three integration constants, related to given values of s.

The following initial and boundary value problems for (2-90a), also named geodetic problems, are of particular geodetic importance. The so-called direct or first geodetic problem deals with the

Initial value problem of geodesic lines (2-91)

Given are the surface coordinates of a point P_a

$$(u^\alpha)_a \,, \tag{2-91a}$$

the direction of a geodesic line C_g extending from P_a to a second point P_b, for example, determined using

$$(u^{\alpha'})_a = (du^\alpha/ds)_a \quad \text{and} \tag{2-91b}$$

$$s = \text{distance } P_a - P_b \text{ along } C_g \,. \tag{2-91c}$$

Required are the surface coordinates of the point P_b

$$(u^\alpha)_b \tag{2-91d}$$

and the direction of the geodesic line C_g in P_b, for example, determined using

$$(u^{\alpha'})_b = (du^\alpha/ds)_b \,. \tag{2-91e}$$

This is an initial value problem to (2-90) in which the three initial values (2-91a,b) for $s = 0$ are given as integration constants. The two possible derivatives in (2-91b) are dependent on one another due to (2-86e).

The inverse or second geodetic problem involves the

First boundary value problem of geodesic lines (2-92)

Given are the surface coordinates of two points P_a, P_b

$$(u^\alpha)_a \,, \qquad (u^\alpha)_b \,. \tag{2-92a}$$

Required is a

$$\text{geodesic line } C_g, \quad P_a, P_b \in C_g \,. \tag{2-92b}$$

We particularly want to calculate

$$s = \text{distance } P_a - P_b \text{ along } C_g \quad \text{and} \tag{2-92c}$$

$$(u^{\alpha'})_a = (du^\alpha/ds)_a \,, \qquad (u^{\alpha'})_b = (du^\alpha/ds)_b \,. \tag{2-92d}$$

In this first boundary value problem to (2-90) two integration constants $(u^\alpha)_a$ are given for $s = 0$. We could give one of the coordinates $(u^\alpha)_a$

as a third integration constant for another value $s \neq 0$. This is however not the case in problem (2-92): here, both coordinates of P_b are known, and s is thus the required quantity.

The <u>third geodetic problem</u> is the

<u>Second boundary value problem of geodesic lines</u> (2-93)

<u>Given</u> are the surface coordinates of a point P_a

$$(u^\alpha)_a \tag{2-93a}$$

and a

surface curve C , $\quad P_a \notin C$. (2-93b)

<u>Required</u> is a

geodesic line C_g , $\quad P_a \in C_g$, (2-93c)

intersecting C vertically. We particularly want to calculate the surface coordinates of the intersection point P_b of C and C_g, i.e.,

$(u^\alpha)_b$, $\quad P_b = C \cap C_g$ and (2-93d)

s = distance $P_a - P_b$ along C_g . (2-93e)

Several considerations on <u>geodesic lines in orthogonal surface coordinates u^α</u> will now follow [see Fig. 2.2; here, however, coordinate lines will not be assumed to be lines of curvature]. For the metric tensor (2-19) is valid, with which Christoffel's symbols of the second kind [see (2-48b,d)] become

$$\begin{aligned}
&\Gamma^1_{11} = g_{11,1}/(2g_{11}) , \qquad \Gamma^1_{12} = g_{11,2}/(2g_{11}) , \\
&\Gamma^1_{22} = -g_{22,1}/(2g_{11}) , \\
&\Gamma^2_{11} = -g_{11,2}/(2g_{22}) , \qquad \Gamma^2_{12} = g_{22,1}/(2g_{22}) , \\
&\Gamma^2_{22} = g_{22,2}/(2g_{22}) .
\end{aligned} \tag{2-94}$$

In accordance with (2-82c)

$$u^{\alpha'} = ((g_{11})^{-1/2}\sin\theta , \quad (g_{22})^{-1/2}\cos\theta) \tag{2-95a}$$

with

θ = direction angle of the surface curve C relative to the u^2-line (2-95b) is true for arbitrary surface curves C in orthogonal surface coordinates [see Fig. 2.2].

The differential equations of a geodesic line C_g [see (2-90a)] can be expressed in orthogonal surface coordinates as follows:

$$2g_{11}u^{1''} + g_{11,1}(u^{1'})^2 + 2g_{11,2}u^{1'}u^{2'} - g_{22,1}(u^{2'})^2 = 0 \ ,$$
$$2g_{22}u^{2''} - g_{11,2}(u^{1'})^2 + 2g_{22,1}u^{1'}u^{2'} + g_{22,2}(u^{2'})^2 = 0 \ . \qquad (2\text{-}96)$$

If (2-95a) is differentiated with respect to s and is subsequently multiplied by $2g_{11}$, we obtain

$$2g_{11}u^{1''} = -g_{11,1}(u^{1'})^2 - g_{11,2}u^{1'}u^{2'} + 2g^{1/2}u^{2'}\theta' \ .$$

This is true for arbitrary surfaces curves. The left side can be eliminated for geodesic lines using the first equation of (2-96), leading to the following equations:

$$\theta' = d\theta/ds = -g^{-1/2}(g_{11,2}u^{1'} - g_{22,1}u^{2'})/2 \ , \qquad (2\text{-}97a)$$

$$d\theta = \theta' ds = -g^{-1/2}(g_{11,2}du^1 - g_{22,1}du^2)/2 \ . \qquad (2\text{-}97b)$$

The objects of the following studies are geodesic lines on surfaces of revolution. For this purpose cylindrical coordinates corresponding to Fig. 2.3 are introduced with reference to a Cartesian coordinate system S for the points P to be described:

L = length of P relative to plane 1.-3. ,
r = distance between P and the axis 3. , (2-98a)
z = height of P above plane 1.-2. .

Together with definition (2-98a) the position vector of a point P in S becomes

$$x_{i.} = (r\cos L,\ r\sin L,\ z) \ , \qquad (2\text{-}98b)$$

and conversely

$$L = \arctan(x_{2.}/x_{1.}) \ , \quad r = (x_{1.}^2 + x_{2.}^2)^{1/2} \ , \quad z = x_{3.} \ . \qquad (2\text{-}98c)$$

The axis 3. of S should be the symmetry axis of the surface of revolution and used primarily as surface coordinates, i.e.,

$$u^1 =: L \ , \qquad u^2 =: r \ . \qquad (2\text{-}99a)$$

To the extent that it appears practical the surface coordinate

$$B = \text{latitude of } P = \text{angle between } n_{i.} \text{ and the plane } 1.-2. \qquad (2\text{-}99b)$$

will be used. The coordinate lines of a surface of revolution are designated as

$$\begin{aligned} &L\text{-lines} = \text{latitude or parallel circles} , \\ &r\text{-lines} = B\text{-lines} = \text{meridians} . \end{aligned} \qquad (2\text{-}100)$$

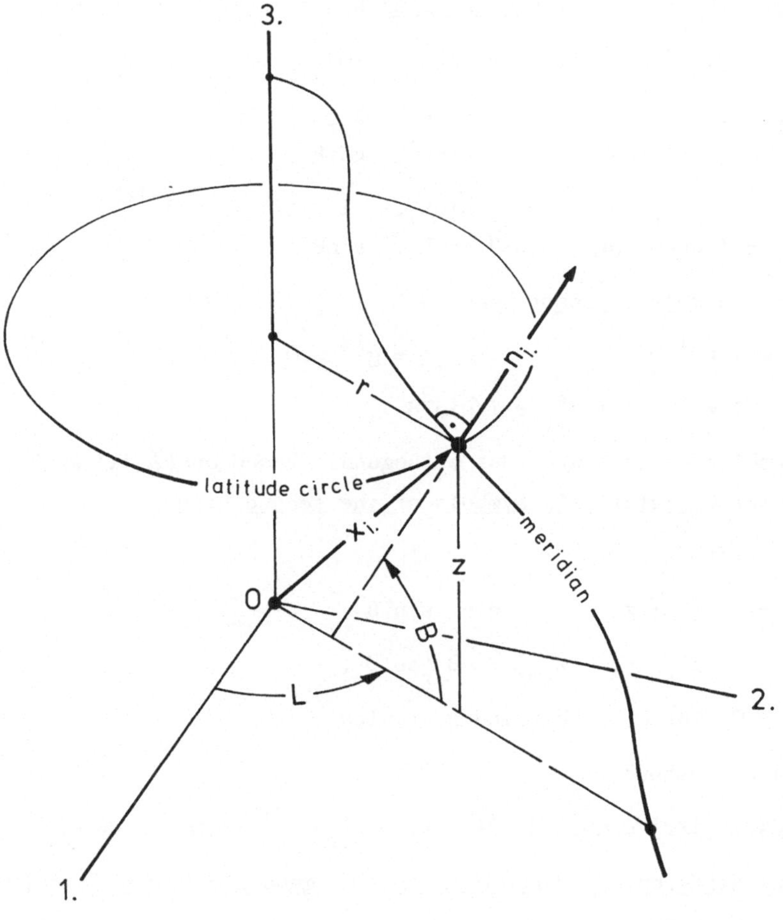

Fig. 2.3. Surfaces of revolution

For the position vectors of surface points as functions of the coordinates (2-99a) we obtain

$$x_{i.}(L,r) = (r\cos L,\ r\sin L,\ z(r))\ , \tag{2-101}$$

whereby for L = constant

$$z = z(r) \tag{2-102a}$$

is the <u>meridian equation</u>.

$$dz/dr =: z_{,r} = -\cot B\ , \qquad d^2z/dr^2 =: z_{,rr} = B_{,r}/\sin^2 B \tag{2-102b}$$

is true. If the outer normal is selected for $n_{i.}$ corresponding to Fig. 2.2, then the <u>moving trihedron (2-5)</u> has the following form:

$$x_{i,1} = (\ -r\sin L\ ,\quad r\cos L\ ,\quad 0\ )\ , \tag{2-103a}$$

$$\begin{aligned} x_{i,2} &= (\ \cos L\ ,\quad \sin L\ ,\quad z_{,r}\ ) \\ &= (\ \cos L\ ,\quad \sin L\ ,\ -\cot B\ )\ , \end{aligned} \tag{2-103b}$$

$$\begin{aligned} n_{i.} &= (\ -z_{,r}\cos L\ ,\ -z_{,r}\sin L\ ,\quad 1\ )(1 + z^2_{,r})^{-1/2} \\ &= (\ \cos B\cos L\ ,\ \cos B\sin L\ ,\ \sin B\ )\ , \end{aligned} \tag{2-103c}$$

so that the <u>metric tensor</u> becomes

$$\begin{aligned} g_{11} &= 1/g^{11} = r^2\ , \qquad g_{12} = g^{12} = 0\ , \\ g_{22} &= 1/g^{22} = 1 + z^2_{,r} = 1/\sin^2 B\ . \end{aligned} \tag{2-104}$$

The coordinate lines are thus orthogonal. Based on (2-94) with (2-104) we obtain for <u>Christoffel's symbols of the second kind</u>

$$\begin{aligned} \Gamma^1_{12} &= 1/r\ , \\ \Gamma^2_{11} &= -r/(1 + z^2_{,r}) = -r\sin^2 B\ , \\ \Gamma^2_{22} &= z_{,r}z_{,rr}/(1 + z^2_{,r}) = -B_{,r}\cot B\ , \\ \Gamma^\gamma_{\alpha\beta} &= 0 \quad \text{for all other index triples}\ . \end{aligned} \tag{2-105}$$

(2-95a) corresponds to

$$dL/ds = (1/r)\sin\theta,\quad dr/ds = (1 + z^2_{,r})^{-1/2}\cos\theta = \sin B\cos\theta\ , \tag{2-106}$$

and the <u>differential equations of the geodesic lines</u> [see (2-96)] for surfaces of revolution are

$$r\, d^2L/ds^2 + 2(dL/ds)(dr/ds) = 0\ , \tag{2-107a}$$

$$d^2r/ds^2 - r\sin^2 B\,(dL/ds)^2 - B_{,r}\cot B\,(dr/ds)^2 = 0\ . \tag{2-107b}$$

(2-107a) is equivalent to the differential equation

$$d(r^2 dL/ds)/ds = 0 \ ,$$

whereby

$$r^2 dL/ds = \text{constant} \tag{2-108a}$$

follows. (2-108a) together with (2-106) passes into the <u>equation of Clairaut</u>, i.e.,

$$r \sin\theta = \text{constant} \ . \tag{2-108b}$$

Observing (2-106) differentiation with respect to s yields

$$d\theta/ds = -(1/r) \tan\theta \ dr/ds = -\sin B \ dL/ds \tag{2-109a}$$

or the so-called <u>equation of Bessel</u>, i.e.,

$$d\theta = -\sin B \ dL \ . \tag{2-109b}$$

Based on (2-97) we directly obtain the second form as well.

2.1.10 GEODESIC SURFACE COORDINATES

In the following we will examine

<u>Geodesic surface coordinates</u> (2-110)

$$\bar{u}^2\text{-lines} = \text{one-parameter family of geodesic lines} \ , \tag{2-110a}$$

$$\bar{u}^1\text{-lines} = \text{orthogonal trajectories of } \bar{u}^2\text{-lines} = \text{geodesic parallels} \ , \tag{2-110b}$$

which were defined in (1-6) and here are designated by $\bar{u}^\alpha$ instead of u^α. Due to the position vector $\bar{x}_{i.}(\bar{u}^\alpha)$ the metric tensor generally becomes

$$\bar{g}_{\alpha\beta} = \bar{x}_{i,\bar{\alpha}} \bar{x}_{i,\bar{\beta}} \ , \qquad \bar{x}_{i,\bar{\mu}} := \partial \bar{x}_{i.} / \partial \bar{u}^\mu \ . \tag{2-110c}$$

The geodesic surface coordinates are by definition orthogonal so that (2-19) is true:

$$\bar{g}_{12} = \bar{g}^{12} = 0 \ , \quad \bar{g}^{11} = 1/\bar{g}_{11} \ , \quad \bar{g}^{22} = 1/\bar{g}_{22} \ , \quad \bar{g} = \bar{g}_{11}\bar{g}_{22} \ . \tag{2-110d}$$

For $\bar{u}^2$-lines $d\bar{u}^1 = 0$ so that based on (2-90a) we obtain for $\gamma = 1$

$$\bar{\Gamma}^1_{22} \bar{u}^{2\prime} \bar{u}^{2\prime} = 0 \qquad \rightarrow \qquad \bar{\Gamma}^1_{22} = -\bar{g}^{11} \bar{g}_{22,1}/2 = 0 \ ,$$

whereby

$$\bar{g}_{22,1} = 0 \qquad \rightarrow \qquad \bar{g}_{22} = \bar{g}_{22}(\bar{u}^2) \tag{2-110e}$$

follows. Thus, $\bar{g}_{22}$ is solely a function of $\bar{u}^2$. For geodetic applications

$$\bar{u}^2 := \text{arc lengths measured along the geodesic } \bar{u}^2\text{-lines} \qquad (2\text{-}110f)$$

corresponding to (1-6d,e) is generally chosen so that

$$\bar{g}_{22} = 1 \; : \quad ds^2 = \bar{g}_{11}(\bar{u}^\alpha)(d\bar{u}^1)^2 + (d\bar{u}^2)^2 \; . \qquad (2\text{-}110g)$$

We will now go into <u>transformations (2-8) between geodesic surface coordinates (2-110) and arbitrary surface coordinates</u>, the latter being designated by

$$u^\alpha \; , \quad \alpha \in \{1,2\} \; . \qquad (2\text{-}111a)$$

Due to the position vector $x_{i.}(u^\alpha)$ the accompanying metric tensor becomes

$$g_{\alpha\beta} = x_{i,\alpha} x_{i,\beta} \; , \qquad x_{i,\mu} := \partial x_{i.} / \partial u^\mu \; . \qquad (2\text{-}111b)$$

It will be taken as a known function of u^α.

First, <u>$u^\alpha - \bar{u}^\alpha$ transformations for geodesic polar coordinates $\bar{u}^\alpha$</u> [see (1-7a)] will be treated for which

$$(u^\alpha)_o = \text{coordinates of the pole } P_o \; , \qquad (2\text{-}112a)$$

$$\nu_o^\alpha = \text{reference direction in } P_o \; , \quad |\nu_o^\alpha| = 1 \qquad (2\text{-}112b)$$

are given, and which are defined as follows for a test point P:

$$\bar{u}^1 =: \beta = \text{direction angle of the geodesic line } P_o - P \text{ relative to } \nu_o^\alpha \; , \qquad (2\text{-}112c)$$

$$\bar{u}^2 =: r = \text{length of the geodesic line } P_o - P \qquad (2\text{-}112d)$$

[see Fig. 2.4]. For the <u>coordinate lines</u> the following is valid:

$$\begin{aligned} \beta\text{-lines} &= \text{geodesic circles around } P_o \; , \\ r\text{-lines} &= \text{geodesic lines through } P_o \; ; \end{aligned} \qquad (2\text{-}112e)$$

the geodesic circles are "geodesically parallel".

ν_o^α is a contravariant surface vector in P_o with the length of one, i.e.,

$$g_{\alpha\beta} \nu_o^\alpha \nu_o^\beta = 1 \; , \qquad (2\text{-}113a)$$

which can be arbitrarily chosen. We will assume that ν_o^α is the tangent vector of the u^2-line in P_o so that

$$\nu_o^\alpha =: \nu_{o2}^\alpha = (g_{22})^{-1/2} \delta_2^\alpha \; . \qquad (2\text{-}113b)$$

In addition to υ_{o2}^{α}, the contravariant tangent vector of the u^1-line in P_o is introduced, i.e.,

$$\upsilon_{o1}^{\alpha} = (g_{11})^{-1/2}\delta_1^{\alpha} \; . \tag{2-113c}$$

For the angle φ between υ_{o1}^{α} and υ_{o2}^{α} we obtain

$$\begin{aligned} \cos\varphi &= g_{\alpha\beta}\upsilon_{o1}^{\alpha}\,\upsilon_{o2}^{\beta} = g_{12}/(g_{11}g_{22})^{1/2} \; , \\ \sin\varphi &= (1-\cos^2\varphi)^{1/2} = g^{1/2}/(g_{11}g_{22})^{1/2} \; . \end{aligned} \tag{2-113d}$$

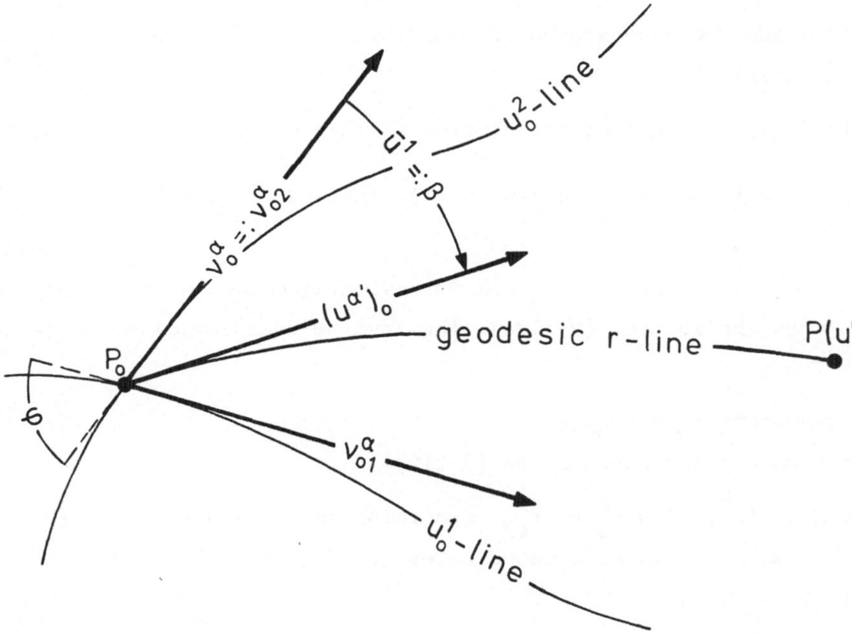

Fig. 2.4. Geodesic polar coordinates $\bar{u}^{\alpha} = (\beta, r)$

Since the tangent vector $(u^{\alpha'})_o$ on the r-line is also a unit vector, the equations

$$g_{\alpha\gamma}(u^{\alpha'})_o\upsilon_{o2}^{\gamma} = \cos\beta \; , \qquad g_{\alpha\gamma}(u^{\alpha'})_o\upsilon_{o1}^{\gamma} = \cos(\varphi - \beta) \tag{2-113e}$$

are true, from which together with (2-113b-d)

$$(u^{\alpha'})_o = ((g_{22}/g)^{1/2}\sin\beta \; , \; (g_{11}/g)^{1/2}\sin(\varphi - \beta)) \tag{2-113f}$$

follow. In the case of orthogonal surface coordinates u^{α}

$$g_{12} = 0 \; , \qquad\qquad \varphi = \pi/2 \; ,$$

these equations are identical to (2-95a) with $\theta =: \beta$, i.e.,

$$(u^{\alpha'})_o = ((g_{11})^{-1/2}\sin\beta \ , \ (g_{22})^{-1/2}\cos\beta \ ) \ . \qquad (2\text{-}113g)$$

In constructing geodesic polar coordinates two inverse transformation problems are decisive. First, there is the

<u>$\bar{u}^{\alpha} \rightarrow u^{\alpha}$ transformation problem</u> (2-114)

for geodesic polar coordinates (2-112):

<u>Given</u> are $(u^{\alpha})_o$ and $\nu^{\alpha}_o =: \nu^{\alpha}_{o2}$ according to (2-112a,b) and (2-113b) and the geodesic polar coordinates (2-112c,d) of a surface point P. (2-114a)

<u>Required</u> are the surface coordinates (2-111) of P . (2-114b)

This is a direct geodetic problem (2-91). Instead of (2-91b)

$$\bar{u}^1 =: \beta \qquad (2\text{-}114c)$$

is given here, whereby $(u^{\alpha'})_o$ can be calculated using (2-113f,g), and (2-91c) takes the form of (2-112d). The inverse transformation of (2-114) is the

<u>$u^{\alpha} \rightarrow \bar{u}^{\alpha}$ transformation problem</u> (2-115)

for geodesic polar coordinates (2-112):

<u>Given</u> are $(u^{\alpha})_o$ and $\nu^{\alpha}_o =: \nu^{\alpha}_{o2}$ according to (2-112a,b) and (2-113b) and the surface coordinates (2-111) of a surface point P. (2-115a)

<u>Required</u> are the geodesic polar coordinates (2-112c,d) of P. (2-115b)

This problem is equivalent to an inverse geodetic problem (2-92). Here, (2-92c) corresponds to (2-110f). $\bar{u}^1$ [see (2-114c)] can be calculated using $(u^{\alpha'})_o$ [see (2-92d)] according to (2-113e).

We will now treat <u>u^{α}- $\bar{u}^{\alpha}$ transformations for geodesic parallel coordinates $\bar{u}^{\alpha}$</u> [see (1-7b)] for which

$$(\bar{u}^{\alpha})_o = 0 = \text{coordinates of the origin } P_o \ , \qquad (2\text{-}116a)$$

$$C_o = \text{abscissa line} = \text{geodesic line through } P_o \ , \qquad (2\text{-}116b)$$

$$\nu^{\alpha}_{Co} = \text{tangent vector of } C_o \text{ in } P_o \ , \quad |\nu^{\alpha}_{Co}| = 1 \ , \qquad (2\text{-}116c)$$

and the following definitions for a test point P are given:

$\bar{u}^1$ = abscissa of P (2-116d)
= distance between the foot point Q of the geodesic vertical of P on C_o and P_o measured along C_o,

$\bar{u}^2$ = ordinate of P (2-116e)
= length of the geodesic line Q - P

[see Fig. 2.5, v^α_{o1} and v^α_{o2} after (2-113b,c)]. The coordinate lines are

$\bar{u}^1$-lines = geodesic parallels to C_o ,
$\bar{u}^2$-lines = geodesic verticals on C_o . (2-116f)

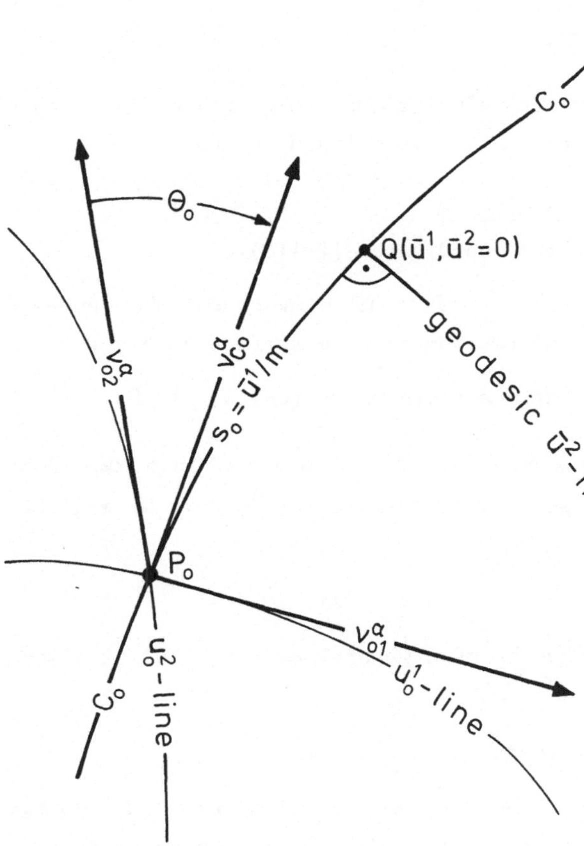

Fig. 2.5. Geodesic parallel coordinates $\bar{u}^\alpha$

For

θ_o = direction angle of C_o at P_o relative to v^α_{o2} (2-117a)

[see (2-113b)] we can also particularly assume that

$\theta_o = 0 \quad \rightarrow \quad \nu^{\alpha}_{Co} = (g_{22})^{-1/2}\delta^{\alpha}_{2}$. (2-117b)

The path of the abscissa C_o can as such be arbitrarily determined, for example, using the given

$\kappa_g = \kappa_g(s)$ = geodesic curvature of C_o (2-118)

corresponding to (2-86). In (2-116b) only the practically most important case of $\kappa_g = 0$ is given. In place of (2-116d)

$\bar{u}^1 = m\, s_o$, (2-119a)

whereby

m = constant factor , s_o = distances $P_o - Q$ along C_o (2-119b)

is also frequently used.

The former of the two transformations, which are decisive in constructing geodesic parallel corrdinates, is defined by the

$\bar{u}^{\alpha} \rightarrow u^{\alpha}$ transformation problem (2-120)

for geodesic parallel coordinates (2-116):

Given are $(u^{\alpha})_o$ and ν^{α}_{Co} after (2-116a-c) and the geodesic parallel coordinates (2-116d,e) of a surface point P. (2-120a)

Required are the surface coordinates (2-111) of P . (2-120b)

The solution to this problem can be divided into two direct geodetic problems (2-91). First, the coordinates $(u^{\alpha})_Q$ of Q will be determined taking

$(u^{\alpha'})_a =: \nu^{\alpha}_{Co}$, $s =: \bar{u}^1/m$ (2-120c)

for (2-91b,c), whereby the more general case (2-119) is assumed. By doing so, we also obtain

$(u^{\alpha'})_{CQ}$ = tangent vector on C_o in Q (2-120d)

according to (2-91e). In this way nothing basically changes also regarding the postulate of (2-118). Now the more general initial value problem (2-86) is to be solved for C_o, instead of (2-91). Proceeding from (2-120d)

$(u^{\alpha'})_{QP}$ = tangent vector on a $\bar{u}^2$-line in Q (2-120e) becomes

$$(u^{\alpha'})_{QP} = \pm\epsilon^{\alpha}_{\beta}\,(u^{\beta'})_{CQ} \qquad \text{(2-120f)}$$

according to (2-43) and (2-44). The direct geodetic problem (2-91) can then be solved again for the geodesic line Q - P assuming

$$(u^{\alpha'})_a =: (u^{\alpha'})_{QP}\,, \qquad s =: \bar{u}^2\,, \qquad \text{(2-120g)}$$

with which we obtain the coordinates (2-111) of P, among other results. In addition to (2-120), the inverse transformation for constructing geodesic parallel coordinates is also of importance and is defined by the

<u>$u^\alpha \to \bar{u}^\alpha$ transformation problem</u> (2-121)

for geodesic parallel coordinates (2-116):

<u>Given</u> are $(u^\alpha)_o$ and ν^α_{Co} after (2-116a-c) and the surface coordinates (2-111) of a surface point P . (2-121a)

<u>Required</u> are the geodesic parallel coordinates (2-116d,e) of P . (2-121b)

Solving the second boundary value problem (2-93) for the geodesic line Q - P, also called the <u>third geodetic problem</u>, postulating

$$(u^\alpha)_a =: (u^\alpha)_P\,, \qquad C =: \bar{u}^2\text{-line } Q - P \qquad \text{(2-121c)}$$

in (2-93a,b) leads to the answers of (2-93d,e), i.e.,

$$(u^\alpha)_b =: (u^\alpha)_Q\,, \qquad s =: \bar{u}^2\,. \qquad \text{(2-121d)}$$

Finally, the initial boundary value problem (2-92) with

$$\left.\begin{aligned}(u^\alpha)_a &=: (u^\alpha)_o \text{ after (2-116a)}\,,\\ (u^\alpha)_b &=: (u^\alpha)_Q \text{ after (2-121d)}\end{aligned}\right. \qquad \text{(2-121e)}$$

in (2-92a) is solved for the geodesic line P_o - Q. Combined with (2-119) the answer desired here [see (2-92c)] is

$$s =: s_o = \bar{u}^1/m\,. \qquad \text{(2-121f)}$$

In the more general case of (2-118) this is done in a corresponding manner based on (2-86) and (2-87).

2.1.11 SPECIAL STUDIES OF GEODESIC POLAR COORDINATES

For the <u>geodesic polar coordinates</u> (2-112c,d), which are designated here by

$$u^{\alpha} = (\beta, r) \, , \tag{2-122a}$$

together with

$$g_{11} =: h^2 \tag{2-122b}$$

and according to (2-110) the first fundamental form becomes

$$ds^2 = h^2 \, d\beta^2 + dr^2 \, . \tag{2-122c}$$

Consequently, the metric tensor has the the form of

$$g_{\alpha\beta} = h^2 \, \delta^1_\alpha \, \delta^1_\beta + \delta^2_\alpha \, \delta^2_\beta \, , \tag{2-122d}$$

whereby for the Christoffel's symbols of the second kind (2-48b) we obtain

$$\begin{aligned} &\Gamma^1_{11} = h^{-1} h_{,1} \, , \quad \Gamma^2_{11} = -h \, h_{,2} \, , \quad \Gamma^1_{12} = h^{-1} h_{,2} \, , \\ &\Gamma^2_{12} = \Gamma^1_{22} = \Gamma^2_{22} = 0 \, . \end{aligned} \tag{2-122e}$$

The <u>r-lines</u> are geodesic lines so that their geodesic curvature vanishes, i.e.,

$$\kappa_{gr} = 0 \, . \tag{2-123a}$$

For the <u>β-lines</u>, i.e., the so-called geodesic circles,

$$\beta' = h^{-1} \, , \qquad\qquad r' = 0 \tag{2-123b}$$

are true, whereby its geodesic curvature together with (2-122e) becomes

$$\kappa_{g\beta} = -h^{-1} \, h_{,2} \tag{2-123c}$$

based on (2-65).

(2-79c) with (2-122d) yields the following expression for <u>Gaussian curvature</u>:

$$K = -h^{-1} \, h_{,22} \, . \tag{2-124}$$

In the <u>vicinity of the pole P_0 (r = 0)</u> the geodesic polar coordinates (2-122a) merge into plane polar coordinates with

$$h = (g_{11})^{1/2} = r \qquad \text{for} \qquad r \to 0 \, . \tag{2-125a}$$

As a consequence,

$$h_o = 0 \,, \qquad (h_{,1})_o = 0 \,, \qquad (h_{,2})_o = 1 \qquad \text{(2-125b)}$$

follow, in which the index o should indicate $r = 0$. Based on (2-124)

$$h_{,22} = \partial^2 h/\partial r^2 = -K\,h$$

is true, and differentiation with respect to $r = u^2$ yields

$$h_{,222} = \partial^3 h/\partial r^3 = -K\,h_{,2} - K_{,2} h \,.$$

At P_o $(r = 0)$ these two expressions become

$$(h_{,22})_o = 0 \,, \qquad (h_{,222})_o = -K_o \,. \qquad \text{(2-125c)}$$

The first terms of a Taylor series for h in r with the reference point P_o are

$$h = r - (1/6)K_o r^3 + \ldots \qquad \text{(2-125d)}$$

due to (2-125a-c). Thereby, the circumference U of a geodesic circle with r = constant becomes

$$U = \int_0^{2\pi} h\,d\beta = 2\pi\,r - (1/3)K_o\pi\,r^3 + \ldots$$

so that

$$K_o = \lim_{r \to 0} 3(2\pi\,r - U)/(\pi\,r^3) \,. \qquad \text{(2-126a)}$$

For the area F of the geodesic circle we obtain combined with (2-125d)

$$F = \int_0^{2\pi}\int_0^{r} h\,d\beta\,dr = \pi\,r^2 - (1/12)K_o\pi\,r^4 + \ldots \,,$$

out of which

$$K_o = \lim_{r \to 0} 12(\pi\,r^2 - F)/(\pi\,r^4) \qquad \text{(2-126b)}$$

llows.

The <u>differential equations of geodesic lines</u> in geodesic polar coordinates will now be given, for which we will refer back to (2-96). Observing (2-122d) we obtain

$$h\,\beta'' + (h_{,1}\beta' + 2h_{,2}r')\beta' = 0 \,, \qquad r'' - h\,h_{,2}(\beta')^2 = 0 \,. \qquad \text{(2-127a)}$$

For the differentiation of

θ = direction angle of the geodesic line relative to the r-lines (2-127b)

with respect to the arc length we obtain

$$\theta' = -h_{,2}\,\beta' \tag{2-127c}$$

after (2-97). Dividing by β' yields

$$\theta_{,1} = \partial\theta/\partial\beta = -h_{,2}\ . \tag{2-127d}$$

The equations for the arbitrary surface curves of (2-95) here are

$$\beta' = h^{-1}\sin\theta\ , \qquad r' = \cos\theta\ . \tag{2-127e}$$

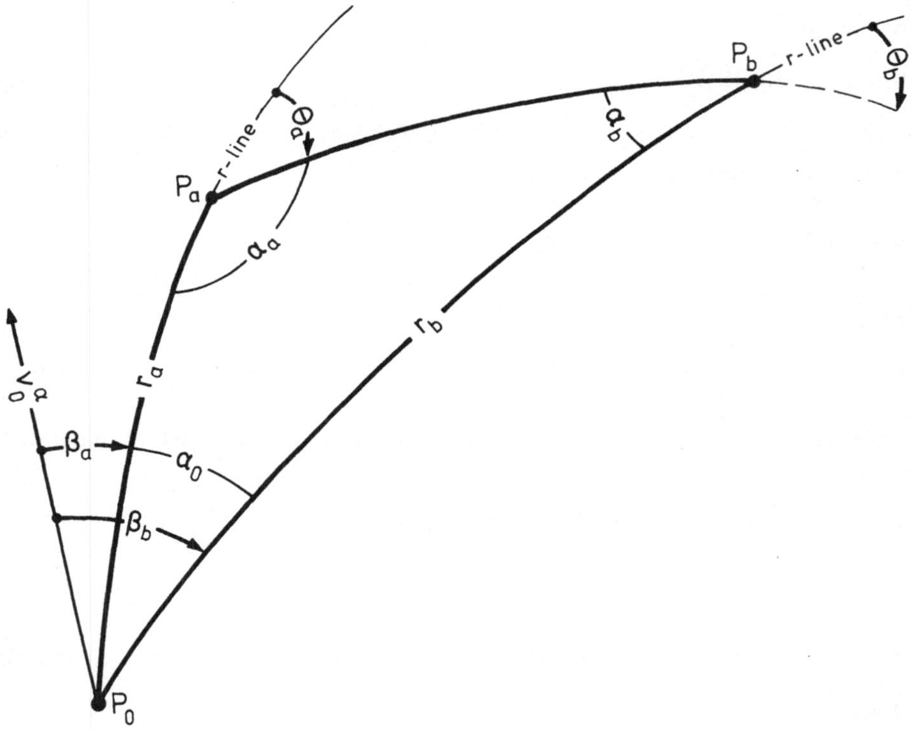

Fig. **2.6.** Geodesic triangle

In the following we will derive an expression for the <u>angular sum of geodesic triangles</u>. Geodesic triangles are regions of a surface bound by three geodesic lines. Their study proves to be especially simple using geodesic polar coordinates whose pole P_0 coincides with one of the corners of the triangle. The other two corners are represented by P_a and

P_b, corresponding to Fig. 2.6. The sides of the triangle $P_o - P_a$ and $P_o - P_b$ are each an r-line with the constant coordinates β_a and β_b. The so-called <u>total curvature of the geodesic triangle</u>, which is the same as the surface integral of Gaussian curvature, becomes with regard to (2-124)

$$K_T := \int_{\Delta F} K\, dF = -\int_{\beta_a}^{\beta_b} d\beta \int_0^{r(\beta)} (\partial^2 h/\partial r^2) dr \;, \tag{2-128a}$$

ΔF = area of the geodesic triangle,
df = h $d\beta$ dr = element of area.

For the integral over r we obtain due to (2-125b)

$$\int_0^{r(\beta)} (\partial^2 h/\partial r^2) dr = [\partial h/\partial r]_{r=0}^{r(\beta)} = (\partial h/\partial r)_{r(\beta)} - 1 \;.$$

If (2-127d) is substituted for $\partial h/\partial r = h_{,2}$ here, the following thus results from (2-128a):

$$K_T = \int_{\beta_a}^{\beta_b} (1 + \partial\theta/\partial\beta) d\beta = \beta_b - \beta_a + \theta_b - \theta_a \;. \tag{2-128b}$$

The interior angles of the geodesic triangle at P_o, P_a, and P_b are represented by α_o, α_a, and α_b, respectively. With them the following is true:

$$\int_{\Delta F} K\, dF = \alpha_o + \alpha_a + \alpha_b - \pi =: \epsilon \;. \tag{2-128c}$$

The right side of the equation is referred to as the <u>excess ϵ of the geodesic triangle</u>. Thus, (2-128c) states that the total curvature of a geodesic triangle is equal to its excess. This is a special statement of the theorem of Gauss and Bonnet. If

$$K = K_o = 1/(R_{1o} R_{2o}) = \text{constant in } \Delta F \tag{2-128d}$$

is sufficiently accurate for geodesic triangles, (2-128c) then yields

$$\epsilon = K_o \Delta F \;. \tag{2-128e}$$

This equation strictly applies to geodesic triangles on spheres, on which the geodesic lines and thus the sides of the triangles are great circles. Hence, ϵ_o is referred to as the <u>spherical excess</u> of a geodesic triangle.

2.1.12 RIEMANNIAN NORMAL COORDINATES

Riemannian normal coordinates v^α of a point P are defined based on geodesic polar coordinates u^α [see (2-122)] using transformation equations such as

$$v^\alpha = t^\alpha r \qquad \text{with} \qquad t^\alpha = (\sin\beta,\ \cos\beta)\ . \tag{2-129a}$$

For the inverse transformation we obtain

$$\beta = \arctan(v^1/v^2)\ , \qquad r = ((v^1)^2 + (v^2)^2)^{1/2}\ . \tag{2-129b}$$

The quantities related to v^α are provided with a cross bar in the following. Thereby, the first fundamental form in v^α generally reads

$$ds^2 = \bar{g}_{\alpha\beta} dv^\alpha dv^\beta\ , \qquad \bar{g}_{\alpha\beta} = x_{i,\bar{\alpha}} x_{i,\bar{\beta}}\ . \tag{2-129c}$$

Due to (2-129a,b), for the transformation matrices (2-15a) between the normal coordinates and the polar coordinates, we obtain

$$\begin{aligned} v^\alpha_{,1} &= (\cos\beta,\ -\sin\beta) r = (v^2,\ -v^1)\ , \\ v^\alpha_{,2} &= (\sin\beta,\ \ \cos\beta) \ = v^\alpha/r = t^\alpha\ , \end{aligned} \tag{2-129d}$$

$$\beta_{,\bar{\alpha}} = v^\alpha_{,1}/r^2\ , \qquad r_{,\bar{\alpha}} = v^\alpha_{,2}\ . \tag{2-129e}$$

Proceeding from (2-122d) we can now calculate the metric tensor $\bar{g}_{\alpha\beta}$ of the Riemannian normal coordinates as follows:

$$\bar{g}_{\alpha\beta} = (\beta_{,\bar{\alpha}}\ ,\ r_{,\bar{\alpha}})^\gamma\ (\beta_{,\bar{\beta}}\ ,\ r_{,\bar{\beta}})^\delta\ g_{\gamma\delta}\ . \tag{2-129f}$$

The individual components are

$$\begin{aligned} \bar{g}_{11} &= (v^1/r)^2 + (h\ v^2/r^2)^2 = \sin^2\beta + (h/r)^2\cos^2\beta\ , \\ \bar{g}_{22} &= (v^2/r)^2 + (h\ v^1/r^2)^2 = \cos^2\beta + (h/r)^2\sin^2\beta\ , \\ \bar{g}_{12} &= (v^1 v^2/r^2)\ (1 - (h/r)^2) = \sin\beta\ \cos\beta\ (1 - (h/r)^2)\ . \end{aligned} \tag{2-129g}$$

In the vicinity of the pole P_o, where

$$r = 0 \qquad \rightarrow \qquad v^\alpha = 0\ , \tag{2-130a}$$

the expressions derived with (2-125) are valid for h. Accordingly, at the pole we have

$$(\bar{g}_{11})_o = (\bar{g}_{22})_o = 1\ , \qquad (\bar{g}_{12})_o = 0 \tag{2-130b}$$

so that (2-129c) at P_o is expressed as

$$ds^2 = (dv^1)^2 + (dv^2)^2 \ . \tag{2-130c}$$

Consequently, the coordinates v^α are orthogonal at the pole, and dv^α equals the accompanying arc elements. In the vicinity of the pole v^α hence has approximately the same properties as plane Cartesian coordinates.

The r-lines of the polar coordinates are geodesic lines through the pole P_o. Due to (2-129a) the tangent vector of such an r-line in v^α becomes

$$v^{\alpha'} := \partial v^\alpha/\partial r = t^\alpha = (\sin\beta \ , \ \cos\beta) \tag{2-131a}$$

so that

$$t^{\alpha'} := \partial t^\alpha/\partial r = v^{\alpha''} = 0 \ . \tag{2-131b}$$

Thus, t^α is the contravariant tangent vector on the r-line (β = constant) independent of r with a length of one:

$$\bar{g}_{\alpha\beta} t^\alpha \ t^\beta = |t^\gamma|^2 = 1 \ . \tag{2-131c}$$

Due to (2-131a,b), (2-90a) yields the following equation for geodesic lines through P_o:

$$\bar{\Gamma}^\gamma_{\alpha\delta} t^\alpha \ t^\delta = 0 \qquad \text{for} \qquad \beta = \text{constant} \ , \tag{2-131d}$$

whereby the Christoffel's symbols refer to v^α and $\bar{g}_{\alpha\beta}$. Contrary to the assumption for (2-129) let

$$u^\alpha = \text{arbitrary surface coordinates} \tag{2-131e}$$

with the metric tensor $g_{\alpha\beta}$. Then,

$$u^{\alpha'} := \partial u^\alpha/\partial r \ , \qquad |u^{\alpha'}| = 1 \tag{2-131f}$$

is the contravariant tangent vector on a geodesic r-line through P_o. According to (2-14) the transformation equations

$$t^\alpha = v^\alpha_{,\beta} u^{\beta'} \ , \qquad u^{\alpha'} = u^\alpha_{,\bar{\beta}} t^\beta \tag{2-131g}$$

are valid. Proceeding on the assumptions for (2-113) [see Fig.2.4], at P_o β is the direction angle of the r-line with reference to the u^2-line. Furthermore, orthogonal u^α-coordinates are presumed so that (2-113g) is true. Thereby, the following equations ensue from the first equations of (2-131g) at P_o

for $\beta = 0$: $t^\alpha = \delta_2^\alpha = (v^\alpha_{,2})_o (g_{22})_o^{-1/2}$,

for $\beta = \pi/2$: $t^\alpha = \delta_1^\alpha = (v^\alpha_{,1})_o (g_{11})_o^{-1/2}$,

and thus for $g_{12} = 0$

$$(v^\alpha_{,\beta})_o = \delta_\beta^\alpha \, (g_{(\beta\beta)})_o^{1/2} \; . \tag{2-131h}$$

Analogously, the second equation of (2-131g) yields

$$(u^\alpha_{,\bar\beta})_o = \delta_\beta^\alpha \, (g_{(\beta\beta)})_o^{-1/2} \; . \tag{2-131i}$$

Together with (2-131i) transforming the normal coordinates v^α into the covariant basis of u^α at P_o yields

$$\bar v^\alpha := (u^\alpha_{,\bar\beta})_o v^\beta = (u^{\alpha'})_o r = (g_{(\alpha\alpha)})_o^{-1/2} v^\alpha \; . \tag{2-131j}$$

The first two equations are also true for oblique coordinates u^α at P_o; $g_{12} = 0$ is only presumed for the last equation.

2.1.13 ISOTHERMAL SURFACE COORDINATES

Whereas the definition of geodesic surface coordinates (2-110) is based on the properties of coordinate lines, the first fundamental tensor is primarily used for isothermal surface coordinates. Corresponding to (1-8),

<u>Isothermal surface coordinates</u> (2-132)

are surface coordinates $\bar u^\alpha$ with the first fundamental form

$$ds^2 = \bar G(\bar u^\alpha) \, ((d\bar u^1)^2 + (d\bar u^2)^2) \; , \tag{2-132a}$$

and accordingly

$$\bar g_{11} = \bar g_{22} = \bar g^{1/2} =: \bar G \; , \qquad \bar g^{11} = \bar g^{22} = 1/\bar G \; , \tag{2-132b}$$

$$\bar g_{12} = \bar g^{12} = 0 \tag{2-132c}$$

are valid for the metric tensor in keeping with (2-110c).

Due to (2-132b) the same coordinate differentials correspond to the same distance differentials in the directions of the coordinate lines (1-8a,b). Isothermal surface coordinates are orthogonal in accordance with (2-132c).

We will now go into transformations (2-8) between isothermal surface coordinates $\bar{u}^{\alpha}$ and arbitrary surface coordinates u^{α}. Hereby, the metric tensor is transformed according to (2-27), and due to (2-132b,c) we obtain from the first equation of (2-132b,c) the following:

$$g^{\gamma\delta} \bar{u}^{1}_{,\gamma} \; \bar{u}^{2}_{,\delta} = \bar{g}^{12} = 0 \,, \tag{2-133a}$$

$$g^{\gamma\delta} \bar{u}^{(\alpha)}_{,\gamma} \; \bar{u}^{(\alpha)}_{,\delta} = \bar{g}^{(\alpha\alpha)} = 1/\bar{G} \,, \qquad \alpha = 1 \;\text{ or }\; 2 \,. \tag{2-133b}$$

In (2-133a) we see that $\bar{u}^{1}_{,\gamma}$ and $\bar{u}^{2}_{,\delta}$ are two covariant surface vectors orientated perpendicular to one another. Based on (2-133b) their absolute values become

$$|\bar{u}^{(\alpha)}_{,\gamma}| = \bar{G}^{-1/2} \,, \qquad \alpha = 1 \;\text{ or }\; 2 \,. \tag{2-133c}$$

According to (2-44e)

$$\bar{u}^{1}_{,\alpha} \, \bar{G}^{1/2} = \pm \, \epsilon^{\beta}_{\alpha} \, \bar{u}^{2}_{,\beta} \, \bar{G}^{1/2}$$

is then valid, from which we obtain the differential equations of isothermal surface coordinates after multiplying by $\bar{G}^{-1/2}$, i.e.,

$$\bar{u}^{1}_{,\alpha} = \pm \, \epsilon^{\beta}_{\alpha} \, \bar{u}^{2}_{,\beta} \qquad \text{or} \qquad \bar{u}^{2}_{,\alpha} = \pm \, \epsilon^{\beta}_{\alpha} \, \bar{u}^{1}_{,\beta} \,. \tag{2-134a}$$

Proceeding from the first equation of (2-27a) we obtain the differential equations equivalent to (2-134a) for the inverse transformation $u^{\alpha}(\bar{u}^{\beta})$ in the same manner:

$$u^{\alpha}_{,\bar{1}} = \pm \, \epsilon^{\alpha}_{\beta} \, u^{\beta}_{,\bar{2}} \qquad \text{or} \qquad u^{\alpha}_{,\bar{2}} = \pm \, \epsilon^{\alpha}_{\beta} \, u^{\beta}_{,\bar{1}} \,. \tag{2-134b}$$

The differential equations (2-134a) and (2-134b) are necessary and sufficient conditions for the fact that the surface coordinates $\bar{u}^{\alpha}$ are isothermal.

The theorem of Ricci [see Sokolnikoff 1964, 35] states that the covariant derivatives of metric tensors in Riemannian space vanish, i.e.,

$$g_{\alpha\beta;\delta} = 0 \,, \qquad g^{\alpha\beta}_{\;\;\;;\delta} = 0 \,, \tag{2-135a}$$

so that

$$\epsilon_{\alpha\beta;\delta} = 0 \tag{2-135b}$$

is also true. $\bar{u}^{1}_{,\alpha}$ and $\bar{u}^{2}_{,\alpha}$ can each be regarded as partial derivatives of the scalar position functions, i.e.,

$$\bar{u}^{1} = \bar{u}^{1}(u^{\beta}) \,, \qquad \bar{u}^{2} = \bar{u}^{2}(u^{\beta}) \,, \tag{2-136a}$$

thus forming covariant surface vectors. Hence, their covariant derivatives can be calculated with (2-53b). Due to (2-135) the covariant derivative of (2-134a) after following inner multiplication by $g^{\alpha\delta}$ is

$$g^{\alpha\delta}\,\bar{u}^{1}_{,\alpha;\delta} = \pm\,\epsilon^{\beta\delta}\,\bar{u}^{2}_{,\beta;\delta}\ .$$

The covariant derivatives are symmetrical with regard to their indices, as seen in (2-53). Thus, the $\epsilon_{\alpha\beta}$ factor on the right side is also a symmetrical tensor with regard to α and β, with which the right side of the foregoing equation vanishes identically. We hence obtain

$$\Delta\bar{u}^{\beta} := g^{\alpha\delta}\,\bar{u}^{\beta}_{,\alpha;\delta} = g^{\alpha\delta}\,\bar{u}^{\beta}_{,\delta;\alpha} = 0\ , \qquad \beta\in\{1,2\}\ . \tag{2-136b}$$

Herein, Δ is the so-called second differential operator of Beltrami, which becomes the Laplacian operator in plane Cartesian coordinates [see Kreyszig 1975, Sect. 77; Heitz 1980-1983, (12-65)]. Both functions in (2-136a) thus satisfy the Beltramian differential equations of (2-136b). In contrast to (2-134a), the partial differential equations of the second order (2-136b) only involve necessary conditions. If, for example, the function $\bar{u}^{1}(u^{\beta})$ satisfies (2-136b), then $\bar{u}^{2}(u^{\beta})$ can be calculated for a point P starting at P_o with the line integral

$$\bar{u}^{2} = (\bar{u}^{2})_o + \int_{P_o}^{P}\bar{u}^{2}_{,\alpha}du^{\alpha} = (\bar{u}^{2})_o \pm \int_{P_o}^{P}\epsilon^{\beta}_{\alpha}\,\bar{u}^{1}_{,\beta}du^{\alpha} \tag{2-137}$$

observing (2-134b). Due to (2-136b) this line integral is independent of the path.

The differential equations of (2-134a,b) generally have a unique solution when the transformation equations of (2-136a) for a surface curve C_o are given. This way of defining isothermal surface coordinates can be used for nearly all geodetic applications, which is why it is discussed separately here. Regarding the properties of constructed isothermal coordinates there is a wide range of variation due to the available choices of C_o and the transformation laws along C_o. C_o is given as

$$C_o = \text{surface curve} \tag{2-138a}$$

with the equations in the u^{α}-system and the $\bar{u}^{\alpha}$-system

$$u^{\alpha}_o = u^{\alpha}(t)\ , \qquad \bar{u}^{\alpha}_o = \bar{u}^{\alpha}(t) \tag{2-138b}$$

as functions of a curve coordinate t. The first equation determines the geometry of the surface curve C_o in the u^α-system, and the second equation contains the initial values of the differential equations for isothermal coordinates (2-134a,b) along C_o. Equating the first fundamental forms of both coordinates yields

$$ds^2 = \bar{G}\,((d\bar{u}^1)^2 + (d\bar{u}^2)^2) = g_{\alpha\beta}du^\alpha du^\beta \ .$$

For points on C_o we then have

$$\bar{G}_o = (g_{\alpha\beta})_o(du_o^\alpha/dt)(du_o^\beta/dt)/((d\bar{u}_o^1/dt)^2 + (d\bar{u}_o^2/dt)^2) \ . \qquad (2\text{-}138c)$$

Using (2-138c) the metric tensor of the isothermal coordinates along C_o is determined as well. More specifically, the following assumptions are made, for example, for C_o:

C_o = coordinate line of the $\bar{u}^\alpha$-system (2-138d)

and/or

C_o = geodesic line . (2-138e)

A frequent choice for the curve coordinate t is

$t = m\,s_o$, m = constant factor , s_o = arc length of C_o . (2-138f)

Based on (2-138) we obtain the

<u>$u^\alpha \rightarrow \bar{u}^\alpha$ transformation problem</u> (2-139)

for isothermal surface coordinates (2-132):

<u>Given</u> are the equations for C_o [see (2-138b)] and the coordinates u^α of a surface point P; (2-139a)

<u>Required</u> are the isothermal surface coordinates $\bar{u}^\alpha$ of P; (2-139b)

and the inverse of this, i.e., the

<u>$\bar{u}^\alpha \rightarrow u^\alpha$ transformation problem</u> (2-140)

for isothermal surface coordinates (2-132):

<u>Given</u> are the equations for C_o [see (2-138b)] and the isothermal coordinates $\bar{u}^\alpha$ of a surface point P; (2-140a)

<u>Required</u> are the surface coordinates u^α of P. (2-140b)

The arbitrary surface coordinates u^α with a known metric tensor corresponding to the assumptions made with reference to (2-111) are given in (2-139) and (2-140).

If the surface coordinates u^α are also isothermal so that

$$g_{11} = g_{22} = g^{1/2} , \quad g^{11} = g^{22} = g^{-1/2} , \quad g_{12} = g^{12} = 0 \tag{2-141a}$$

are true analogous to (2-134), then the differential equations of (2-134) become Cauchy-Riemannian differential equations, i.e.,

$$\bar{u}^1_{,1} = \bar{u}^2_{,2} , \qquad \bar{u}^1_{,2} = -\bar{u}^2_{,1} , \tag{2-141b}$$

or vice versa, i.e.,

$$u^1_{,\bar{1}} = u^2_{,\bar{2}} , \qquad u^1_{,\bar{2}} = -u^2_{,\bar{1}} . \tag{2-141c}$$

(2-136b) takes the form of Laplacian differential equations, i.e.,

$$\Delta\bar{u}^\alpha := \bar{u}^\alpha_{,\beta\beta} = \partial^2\bar{u}^\alpha/(\partial u^1)^2 + \partial^2\bar{u}^\alpha/(\partial u^2)^2 = 0 , \tag{2-141d}$$

and the inverse is also true, i.e.,

$$\Delta u^\alpha := u^\alpha_{,\bar{\beta}\bar{\beta}} = \partial^2 u^\alpha/(\partial\bar{u}^1)^2 + \partial^2 u^\alpha/(\partial\bar{u}^2)^2 = 0 . \tag{2-141e}$$

In complex analysis it is shown that real and imaginary parts of an analytic function of a complex variable also fulfill Cauchy-Riemannian differential equations (2-141b). At this point only the most important result obtained needs to be mentioned, namely, that the transformations of (2-139) and (2-140) between the isothermal coordinates u^α and $\bar{u}^\alpha$ can always be represented by analytic functions in the form of

$$\bar{u}^1 + i\,\bar{u}^2 = f(u^1 + i\,u^2) , \qquad i = (-1)^{1/2} . \tag{2-142}$$

This will be discussed in more detail in Chapters 2.2 and 3.5.

2.1.14 SPECIAL STUDIES OF ISOTHERMAL SURFACE COORDINATES

Here we will first consider under which conditions it is possible to convert orthogonal surface coordinates u^α into isothermal surface coordinates $\bar{u}^\alpha$ while retaining the coordinate lines. The transformation equations then have the form of

$$\bar{u}^1 = \bar{u}^1(u^1) , \qquad \bar{u}^2 = \bar{u}^2(u^2) \tag{2-143a}$$

so that

$$\bar{u}^1_{,2} = 0 , \qquad \bar{u}^2_{,1} = 0 \tag{2-143b}$$

and in the u^α-system

$$g_{12} = g^{12} = 0 \qquad \text{(2-143c)}$$

by definition. Thus, the following two equations ensue from (2-133b):

$$g^{(\alpha\alpha)}(\bar{u}^{(\alpha)}_{,(\alpha)})^2 = (1/g_{(\alpha\alpha)})(\bar{u}^{(\alpha)}_{,(\alpha)})^2 = 1/\bar{G} .$$

Dividing equation $\alpha = 2$ by equation $\alpha = 1$ together with

$$(\bar{u}^1_{,1})^2 =: f_1(u^1) , \qquad (\bar{u}^2_{,2})^2 =: 1/f_2(u^2) \qquad \text{(2-144)}$$

produces the necessary and sufficient condition for a transformation (2-143a), i.e.,

$$g_{11}(u^\alpha)/g_{22}(u^\alpha) = f_1(u^1)\ f_2(u^2) . \qquad \text{(2-145)}$$

Equivalently, a partial differential equation of the second order satisfied by g_{11}/g_{22} can be given. From (2-145) we obtain

$$\partial(g_{11}/g_{22})/\partial u^1 = f_2\ \partial f_1/\partial u^1 , \quad \partial(g_{11}/g_{22})/\partial u^2 = f_1\ \partial f_2/\partial u^2 \qquad \text{(2-146a)}$$

and

$$\partial^2(g_{11}/g_{22})/(\partial u^1 \partial u^2) = (\partial f_1/\partial u^1)(\partial f_2/\partial u^2) . \qquad \text{(2-146b)}$$

Considering (2-146a) an initial form of the desired differential equation is acquired from (2-146b), i.e.,

$$(g_{11}/g_{22})\partial^2(g_{11}/g_{22})/(\partial u^1 \partial u^2) = (\partial(g_{11}/g_{22})/\partial u^1)(\partial(g_{11}/g_{22})/\partial u^2) ,$$

which is equivalent to the two following forms:

$$\partial((g_{11}/g_{22})^{-1}\partial(g_{11}/g_{22})/\partial u^1)/\partial u^2 = 0 , \qquad \text{(2-147a)}$$

$$\partial^2 \ln(g_{11}/g_{22})/(\partial u^1 \partial u^2) = 0 . \qquad \text{(2-147b)}$$

Just like (2-145), satisfying (2-147a) or (2-147b) is the necessary and sufficient condition for transformation (2-143a).

As examples of (2-145) and (2-147), the meridians and latitude circles of surfaces of revolution (2-100) with the accompanying surface coordinates (2-99a), i.e.,

$$u^\alpha = (L, r)^\alpha , \qquad \text{(2-148a)}$$

will be considered. It should be checked whether the meridians and latitude circles can also be coordinate lines of isothermal coordinates $\bar{u}^\alpha$ permitting transformation (2-143a). Based on (2-104) we directly obtain

$$g_{11}/g_{22} = r^2/(1 + z^2_{,r}) = f_1\ f_2(r) , \qquad f_1 = 1 = \text{constant} . \qquad \text{(2-148b)}$$

As we see immediately, (2-148b) fulfills the condition for (2-145). Due to

$$\partial(g_{11}/g_{22})/\partial u^1 = \partial(g_{11}/g_{22})/\partial L = 0 \tag{2-148c}$$

the differential equation (2-147) is also clearly fulfilled. Since in this case the transformation

$$\bar{u}^1 = u^1 = L \tag{2-149a}$$

satisfies the Beltramian differential equation (2-136b), the accompanying second transformation equation (2-143a) becomes according to (2-137)

$$\bar{u}^2 = (\bar{u}^2)_o - \int_{P_o}^{P} \epsilon_{21} g^{11}\, \bar{u}^1_{,1} du^2 ,$$

$$\bar{u}^2 = (\bar{u}^2)_o + \int_{P_o}^{P} ((1 + z^2_{,r})^{1/2}/r)\, dr =: \bar{r} . \tag{2-149b}$$

The first fundamental form for the isothermal surface coordinates

$$\bar{u}^\alpha = (L, \bar{r})^\alpha \tag{2-149c}$$

of the surface of revolution reads

$$ds^2 = \bar{G}\,(dL^2 + d\bar{r}^2) , \qquad \bar{G}(r) = r^2 . \tag{2-149d}$$

Due to the presupposition made here for (2-143) the coordinate lines of (2-100) are preserved, i.e.,

$$\begin{aligned} &L\text{-lines} = \text{latitude or parallel circles} , \\ &\bar{r}\text{-lines} \equiv r\text{-lines} \equiv B\text{-lines} = \text{meridians} . \end{aligned} \tag{2-149e}$$

The following observations refer again to isothermal coordinates $\bar{u}^\alpha$ on arbitrary surfaces. Due to (2-68) the <u>geodesic curvatures of the coordinate lines</u> are given by

$$\kappa_{g1} = -(1/2)\, \bar{G}^{-3/2} \bar{G}_{,2} = (\bar{G}^{-1/2})_{,2} , \tag{2-150a}$$

$$\kappa_{g2} = (1/2)\, \bar{G}^{-3/2} \bar{G}_{,1} = -(\bar{G}^{-1/2})_{,1} . \tag{2-150b}$$

$\kappa_{g\alpha}$ is the geodesic curvature of the $\bar{u}^\alpha$-line. The partial dirivatives of κ_{g1} and κ_{g2} with respect to $\bar{u}^1$ and $\bar{u}^2$ thus become

$$\kappa_{g1,1} = -\kappa_{g2,2} = (\bar{G}^{-1/2})_{,12} , \tag{2-150c}$$

whereby

$$\kappa_{g1,1} + \kappa_{g2,2} = 0 \ . \tag{2-150d}$$

If the geodesic curvature of a coordinate line, e.g., a $\bar{u}^1$-line, is constant, then $\kappa_{g1,1} = 0$ and thus due to (2-150c,d)

$$\kappa_{g1,1} = \kappa_{g2,2} = (\bar{G}^{-1/2})_{,12} = 0 \tag{2-151a}$$

at all points along this line. For geodetic applications this frequently holds true for C_o according to (2-138d). In the case of (2-138e) κ_{g1} along C_o vanishes so that (2-150a) becomes

$$\kappa_{g1} = (\bar{G}^{-1/2})_{,2} = 0 \quad \rightarrow \quad \bar{G}_{,2} = 0 \ . \tag{2-151b}$$

If the geodesic curvature along all coordinate lines is constant, we speek of

<u>Circular isothermal surface coordinates</u> (2-152)

$$\kappa_{g\alpha} = \text{constant} \tag{2-152a}$$

for all $\bar{u}^\alpha$-lines, and thus according to (2-151a)

$$(\bar{G}^{-1/2})_{,12} = 0 \tag{2-152b}$$

at all points in the definition area of the $\bar{u}^\alpha$-systems.

The general solution of the partial differential equations of the second order (2-152b) has the form of

$$\bar{G}^{-1/2} = F_1(\bar{u}^1) + F_2(\bar{u}^2) \ . \tag{2-152c}$$

If a system of circular isothermal surface coordinates is to be constructed on a given surface, then the differential equations (2-134) or (2-141) must be solved together with (2-152b) or under the auxiliary condition of (2-152c). This is not possible for all surfaces. The latitude circles and meridians of (2-100) form circular isothermal coordinate lines on <u>surfaces of revolution</u> [see (2-148) and (2-149)].

2.2 FUNDAMENTALS OF COMPLEX ANALYSIS

2.2.1 PRELIMINARY REMARKS

As pointed out for (2-142), transformation equations between two systems of isothermal coordinates can be represented as complex analytic functions of a complex variable. The branch of mathematics dealing with functions of this type is the so-called complex analysis, which has been treated very thoroughly in numerous textbooks. In the following only the most important elements of complex analysis for transforming isothermal coordinates in the form required here will be summarized.

From the definition and construction of isothermal surface coordinates given in Chapter 2.1.13 we have learned that representing transformation equations between isothermal coordinates as functions of complex variables is possible, but by no means necessary. The main advantage of complex representations is the very clear formulation of initial value problems of the Cauchy-Riemannian differential equations of (2-141c,d).

2.2.2 FUNCTIONS OF A COMPLEX VARIABLE

Two real variables

$$u^{\alpha} , \quad \alpha \in \{1,2\} \tag{2-153a}$$

are combined into a <u>complex variable u</u> by prescribing that

$$\begin{aligned} u &= u^1 + i\, u^2 , & i &:= (-1)^{1/2} , \\ u^1 &= \text{real part} , & u^2 &= \text{imaginary part} . \end{aligned} \tag{2-153b}$$

It is helpful at times to give a geometrical explanation for complex numbers and variables, which is done, for example, in the form of a <u>Gauss plane</u>. This involves representing complex numbers u in a plane Cartesian coordinate system using points with the coordinates u^1 and u^2, corresponding to Fig. 2.7. Continuous ranges of a complex variable (2-153b) are represented by curves on Gauss planes. In addition to Cartesian coordinates (2-153a) <u>polar coordinates</u>

$$\varphi , \; r \tag{2-154a}$$

are also used for representing complex numbers [see Fig. 2.7]. The transformation equations between (2-153b) and (2-154a) read

$$u = u^1 + i\,u^2 = r\,(\cos\varphi + i\,\sin\varphi)\ , \qquad (2\text{-}154b)$$

$$\varphi = \arctan(u^2/u^1)\ , \qquad r = ((u^1)^2 + (u^2)^2)^{1/2}\ , \qquad (2\text{-}154c)$$

whereby r is the absolute value of u, i.e.,

$$r = |u|\ . \qquad (2\text{-}154d)$$

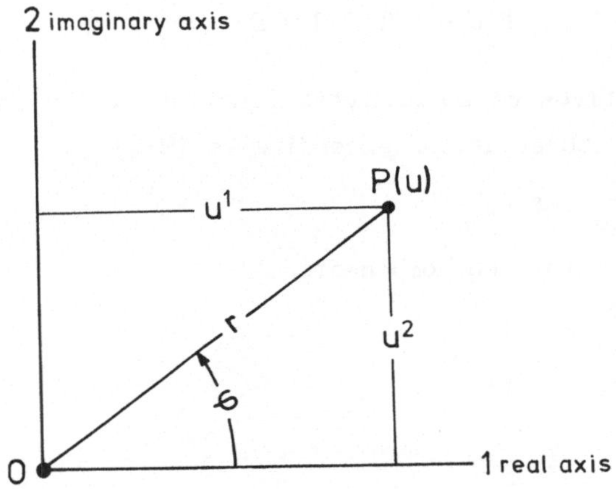

Fig. 2.7. Gauss plane

Functions of a complex variable

$$\bar{u} = f(u) \qquad (2\text{-}155a)$$

are generally complex valued:

$$\bar{u} = \bar{u}^1(u^\alpha) + i\,\bar{u}^2(u^\alpha)\ . \qquad (2\text{-}155b)$$

They are called analytic or regular in domains G of u^α when in these domains the derivatives

$$f'(u) := df/du \qquad (2\text{-}155c)$$

exist. Domains of u^α in which a function of a complex variable is not analytic are called singular. Special functions for complex variables are formally defined in the same way as for real variables [see mathematical literature]. Here, only the particularly important Eulerian formula is given, i.e.,

$$\exp(i\,\varphi) = \cos\varphi + i\,\sin\varphi\ , \qquad (2\text{-}156a)$$

$$\cos\varphi = (\exp(i\,\varphi) + \exp(-i\,\varphi))/2\ , \qquad (2\text{-}156b)$$

$$\sin\varphi = (\exp(i\,\varphi) - \exp(-i\,\varphi))/2\ .$$

Together with (2-156a), (2-154b) can be written as

$$u = r \exp(i\ \varphi)\ . \tag{2-157}$$

2.2.3 DIFFERENTIATION AND INTEGRATION OF ANALYTIC FUNCTIONS

The partial derivatives of an analytic function (2-155) with respect to u^α become with the abbreviations according to (2-2)

$$\bar{u}_{,\alpha} = \bar{u}^1_{,\alpha} + i\ \bar{u}^2_{,\alpha} = f'\ u_{,\alpha}$$

or are separated for the two components

$$\bar{u}_{,1} = \bar{u}^1_{,1} + i\ \bar{u}^2_{,1} = \quad f'\ , \tag{2-158a}$$

$$\bar{u}_{,2} = \bar{u}^1_{,2} + i\ \bar{u}^2_{,2} = i\ f'\ , \tag{2-158b}$$

resulting in

$$f' = f_{,1} = -i\ f_{,2}\ . \tag{2-159}$$

The second expression in (2-159) can thus be written as

$$\bar{u}^1_{,1} + i\ \bar{u}^2_{,1} = \bar{u}^2_{,2} - i\ \bar{u}^1_{,2}\ ,$$

from which the <u>Cauchy-Riemannian differential equations</u> are obtained by equating the real and imaginary parts, i.e.,

$$\bar{u}^1_{,1} = \bar{u}^2_{,2}\ , \qquad\qquad \bar{u}^1_{,2} = -\bar{u}^2_{,1}\ . \tag{2-160}$$

Differentiating (2-160) leads to the <u>Laplacian differential equation</u>, i.e.,

$$\bar{u}^\alpha_{,11} + \bar{u}^\alpha_{,22} = 0\ . \tag{2-161}$$

Satisfying the Cauchy-Riemannian differential equations is necessary and sufficient so that the two functions

$$\bar{u}^1(u^\alpha)\ , \qquad\qquad \bar{u}^2(u^\alpha) \tag{2-162}$$

are able to be the real and imaginary parts of an analytic function (2-155), respectively. The Laplacian differential equations are only necessary and not sufficient conditions [see also (2-136) and (2-137)].

The integration of functions of a complex variable is formally explained in the same manner as that of functions of a real variable. If $f(u)$ and

$g(u)$ are thus analytic functions in a domain G of u^α and

$$f'(u) = g(u) \ , \tag{2-163}$$

then $f(u)$ is referred to as the <u>indefinite integral</u> of the function $g(u)$. <u>Definite integrals</u> of functions of a complex variable $g(u)$, i.e.,

$$f_{ab} = f^1_{ab} + i\, f^2_{ab} = \int_{u_a}^{u_b} g(u)\, du \ , \tag{2-164a}$$

have the character of line integrals between the points

$$P_a(u_a) \ , \qquad P_b(u_b) \tag{2-164b}$$

on a Gauss plane. If C is an arbitrary curve through these points, and t is a curve coordinate of C, then (2-164a) together with

$$g(u) = \bar{u}^1(u^\alpha) + i\, \bar{u}^2(u^\alpha) \ , \tag{2-164c}$$

$$du = du^1 + i\, du^2 = ((du^1/dt) + i\, (du^2/dt))\, dt \tag{2-164d}$$

can be written as

$$f_{ab} = f^1_{ab} + i\, f^2_{ab} = \int_{a(C)}^{b} (\bar{u}^1 du^1 - \bar{u}^2 du^2) + i \int_{a(C)}^{b} (\bar{u}^1 du^2 + \bar{u}^2 du^1) \ . \tag{2-164e}$$

C is the integral path. The Cauchy-Riemannian differential equations (2-160) are valid for analytic functions (2-164c). However, they are also the conditions for the fact that both integrals in (2-164e) and thus f_{ab} are independent of the choice of the curve C. Consequently, the integral of (2-164a) has a complex value f_{ab} independent of the path. This statement is called <u>Cauchy's integral theorem</u>.

2.2.4 POWER SERIES OF ANALYTIC FUNCTIONS

Of validity here is the

<u>Expansion theorem</u> (2-165)

Let $f(u)$ be an analytic function in a domain G of u, $P_o(u_o)$ a point of G, and r_o the radius of the largest circle around P_o on the Gauss plane which still belongs completely to G. Then, within this circle, i.e., for

$$|u - u_o| < r_o \ , \tag{2-165a}$$

$f(u)$ can always be expanded into one and only one power series, i.e.,

$$f(u) = \sum_{n=0}^{\infty} \bar{a}_n (u - u_o)^n \quad , \quad \bar{a}_n = \bar{a}_n^1 + i\, \bar{a}_n^2 \; . \tag{2-165b}$$

For the complex coefficients we obtain

$$\bar{a}_n = (1/n!)(d^n f/du^n)_{u=u_o} =: (1/n!)\, f_o^{(n)} \; . \tag{2-165c}$$

By decomposing $f(u)$ into real and imaginary parts, i.e.,

$$f(u) = f^1(u^\alpha) + i\, f^2(u^\alpha) \; , \tag{2-166a}$$

the binomial theorem has to be applied to the powers of

$$u - u_o =: \Delta u = \Delta u^1 + i\, \Delta u^2 \tag{2-166b}$$

in (2-165b). Doing so, we obtain

$$\begin{aligned} f_n &= f_n^1 + i\, f_n^2 := \bar{a}_n\, \Delta u^n \\ &= (\bar{a}_n^1 + i\, \bar{a}_n^2) \sum_{\nu=1}^{n} (n;\nu)(\Delta u^1)^{n-\nu}(i\, \Delta u^2)^\nu \; , \end{aligned} \tag{2-166c}$$

$$(n;\nu) := n!/(\nu!(n-\nu)!) \tag{2-166d}$$

for the individual terms of the series (2-165b). Decomposing into the real part f_n^1 and the imaginary part f_n^2 yields

$$\begin{Bmatrix} f_n^1 \\ f_n^2 \end{Bmatrix} = \sum_{\nu=0}^{r} (-1)^\nu \left[\begin{Bmatrix} \bar{a}_n^1 \\ \bar{a}_n^2 \end{Bmatrix} (n;2\nu)\, (\Delta u^1)^{n-2\nu}\, (\Delta u^2)^{2\nu} \right. \\ \left. \begin{Bmatrix} -\bar{a}_n^2 \\ +\bar{a}_n^1 \end{Bmatrix} (n;2\nu+1)(\Delta u^1)^{n-2\nu-1}(\Delta u^2)^{2\nu+1} \right] \; , \tag{2-166e}$$

r = largest integer $\leq n/2$.

With the results of (2-166e) $f(u)$ [see (2-166a)] can be calculated as follows:

$$f(u) = \sum_{n=0}^{\infty} f_n^1(u^\alpha) + i \sum_{n=0}^{\infty} f_n^2(u^\alpha) \; . \tag{2-166f}$$

Let

$$P_o(u_o) \; , \qquad\qquad P_1(u_1) \tag{2-167a}$$

be two points in a domain G of u. For points $P(u)$ of G for which the two power series

$$f(u) = \sum_{m=0}^{\infty} \bar{a}_m (u - u_o)^m = \sum_{n=0}^{\infty} \bar{b}_n (u - u_1)^n \tag{2-167b}$$

converge, the following is true:

$$\bar{b}_n = \sum_{m=0}^{\infty} (m+n;n)\, \bar{a}_{m+n} (u_1 - u_o)^m , \quad (m+n;n) \text{ after } (2\text{-}166d) . \tag{2-167c}$$

We will now discuss the <u>inversion of a power series (2-165b)</u>. Correspon-ding to (2-155a,b), let

$$\bar{u} = \bar{u}^1 + i\, \bar{u}^2 = f(u) , \tag{2-168a}$$

and the accompanying inverse function reads

$$u = u^1 + i\, u^2 = g(\bar{u}) . \tag{2-168b}$$

Thus, for the reference point $P_o(u_o, \bar{u}_o)$

$$\bar{u}_o = f(u_o) , \qquad u_o = g(\bar{u}_o) \tag{2-168c}$$

are valid. Corresponding to (2-166b) the differences

$$\Delta u = u - u_o , \qquad \Delta\bar{u} = \bar{u} - \bar{u}_o \tag{2-168d}$$

arise, whereby the power series of (2-165b) for (2-168a) limited to the n' term can be written as

$$\Delta\bar{u} = \sum_{n=1}^{n'} \bar{a}_n \Delta u^n , \qquad \bar{a}_n = \bar{a}_n^1 + i\, \bar{a}_n^2 . \tag{2-169a}$$

For the inverse function (2-168b) we analogously obtain the following setup for the power series:

$$\Delta u = \sum_{n=1}^{n'} a_n \Delta\bar{u}^n , \qquad a_n = a_n^1 + i\, a_n^2 . \tag{2-169b}$$

Proceeding on the assumption that the coefficients $\bar{a}_n$ of (2-169a) are given, the coefficients a_n of (2-169b) must be determined. For this pur-pose (2-169a) is inserted into (2-169b), whereby we obtain the identity

$$\Delta u = \sum_{n=1}^{n'} a_n \Big(\sum_{m=1}^{p} \bar{a}_m \Delta u^m \Big)^n , \qquad p = n' - n + 1 . \tag{2-170a}$$

Using the polynomial theorem we have

$$\Big(\sum_{m=1}^{p} \bar{a}_m \Delta u^m \Big)^n = \sum_{r=n}^{n'} c_{nr} \Delta u^r \tag{2-170b}$$

with the coefficients

$$c_{nr} = n! \sum (\bar{a}_1^{q_1} \bar{a}_2^{q_2} \ldots \bar{a}_p^{q_p})/(q_1!q_2!\ldots q_p!) \ , \tag{2-170c}$$

$$q_m \in \{0,1,2\ldots n\} \ , \qquad \sum q_m = n \ , \qquad \sum m q_m = r \ . \tag{2-170d}$$

The expressions of c_{nr} are homogeneous polynomials of degree n in coefficients $\bar{a}_m$. In the special case of $n = r$ (2-170d) yields

$$q_m = \begin{cases} r & \text{for } m = 1 \\ 0 & \text{for } m > 1 \end{cases}$$

so that

$$c_{rr} = \bar{a}_1^r \ . \tag{2-170e}$$

By inserting (2-170b) into (2-170a) we have

$$\Delta u = \sum_{n=1}^{n'} a_n \left(\sum_{r=n}^{n'} c_{nr} \, \Delta u^r \right) = \sum_{r=1}^{n'} \left(\sum_{n=1}^{r} a_n \, c_{nr} \right) \Delta u^r \ ,$$

out of which

$$a_1 c_{11} = a_1 \bar{a}_1 = 1 \quad \text{for} \quad r = 1 \ , \qquad \sum_{n=1}^{r} a_n c_{nr} = 0 \quad \text{for} \quad r > 1 \tag{2-171a}$$

by comparing coefficients. Observing (2-170e) the recursion formulas $(X,4_3)$ by König & Weise (1951) for calculating the desired coefficients a_r of the inversion series (2-169b) are ultimately obtained, i.e.,

$$a_r = -(1/\bar{a}_1^r) \sum_{n=1}^{r-1} a_n \, c_{nr} \ . \tag{2-171b}$$

Up to $r = 4$ we have

$$\begin{aligned} &a_1 = 1/\bar{a}_1 \ , \qquad a_2 = -\bar{a}_2/\bar{a}_1^3 \ , \qquad a_3 = -(\bar{a}_3 - 2\,\bar{a}_2^2/\bar{a}_1)/\bar{a}_1^4 \ , \\ &a_4 = -(\bar{a}_4 - 5\,\bar{a}_2\bar{a}_3/\bar{a}_1 + 5\,\bar{a}_2^3/\bar{a}_1^2)/\bar{a}_1^5 \ . \end{aligned} \tag{2-171c}$$

According to (2-169a,b) both $\bar{a}_n$ and a_n are generally complex valued. If the coefficients a_n are calculated based on (2-171b,c), then (2-169b) can also be decomposed into real and imaginary parts analogous to (2-166).

3. REPRESENTING THE TRANSFORMATION EQUATIONS BETWEEN SURFACE COORDINATES BY POWER SERIES

3.1 CONSTRUCTING SURFACE COORDINATES

The construction of surface coordinates $\bar{u}^\alpha$ described for (1-12) assumes that in the case of "intrinsic coordinates $\bar{u}^\alpha$ " the intrinsic geometry as a function of the surface coordinates u^γ is given in the form of the metric tensor (1-12b), i.e.,

$$g_{\alpha\beta} = g_{\alpha\beta}(u^\gamma) \ . \tag{3-1a}$$

If $\bar{u}^\alpha$ involves "extrinsic coordinates", the extrinsic surface geometry must also be known in addition to (3-1a). In keeping with the remarks made regarding (2-50) knowing the second fundamental tensor as a function of u^γ suffices as well, i.e.,

$$L_{\alpha\beta} = L_{\alpha\beta}(u^\gamma) \ . \tag{3-1b}$$

The requirements of (3-1a,b) are always fulfilled when the position vector of the surface is given as a function of arbitrary surface coordinates u^γ from the very beginning, i.e.,

$$x_{i.} = x_{i.}(u^\gamma) \ . \tag{3-1c}$$

The objective of constructing surface coordinates $\bar{u}^\alpha$ is the determination of the <u>transformation equations of the coordinates and the metric tensor</u> according to (1-12c) and (2-8) as well as (2-27), respectively:

$$\bar{u}^\alpha = \bar{u}^\alpha(u^\beta) \ , \qquad u^\alpha = u^\alpha(\bar{u}^\beta) \ , \tag{3-2a}$$

$$\bar{g}_{\alpha\beta} = u^\gamma_{,\bar{\alpha}}\, u^\delta_{,\bar{\beta}}\, g_{\gamma\delta} \ , \qquad \bar{g}^{\alpha\beta} = \bar{u}^\alpha_{,\gamma}\, \bar{u}^\beta_{,\delta}\, g^{\gamma\delta} \ . \tag{3-2b}$$

This primarily involves the coordinate transformations of (3-2a). If these equations are given, the transformation matrices of (2-15) necessary for (3-2b) can be calculated in the form of partial derivatives.

Here, only the geodesic surface coordinates of (2-110) or isothermal surface coordinates of (2-132), which are defined based on the differential equations for (3-2a) in association with the initial or boundary conditions, will be considered for both u^α and $\bar{u}^\alpha$. Consequently, the determination of the transformation equations of (3-2a) consists essentially of integrating the accompanying differential equations whereby the real problem is to be seen in geodetic applications. For these purposes power series, as analytic solutions for (3-2a), still play an important role. Power series are particularly well suited for an initial introduction to the transformations of (3-2) as well. In the following the practical application of power series in the coordinates u^{α} and $\bar{u}^{\alpha}$ in (3-2) will be explained with the help of some examples.

3.2 REPRESENTING POWER SERIES

Let

$$(u^{\alpha})_o , \qquad (\bar{u}^{\alpha})_o \tag{3-3a}$$

be surface coordinates of a point P_o in the systems of u^{α} and $\bar{u}^{\alpha}$. For arbitrary test points P the coordinate differences relative to P_o are designated as

$$\Delta u^{\alpha} = u^{\alpha} - (u^{\alpha})_o , \qquad \Delta\bar{u}^{\alpha} = \bar{u}^{\alpha} - (\bar{u}^{\alpha})_o . \tag{3-3b}$$

A power series generated for the initial transformation equations of (3-2a), i.e.,

$$\bar{u}^{\alpha} = \bar{u}^{\alpha}(u^{\beta}) , \tag{3-4a}$$

at the point (3-3a) can be written together with (3-3b) as

$$\Delta\bar{u}^{\alpha} = \sum_{n=1}^{n'} \bar{a}^{\alpha}_{\beta_1\beta_2\ldots\beta_n} \Delta u^{\beta_1} \Delta u^{\beta_2} \ldots \Delta u^{\beta_n}, \quad \alpha, \beta_{\nu} \in \{1,2\} . \tag{3-4b}$$

For the coefficients of this series

$$\bar{a}^{\alpha}_{\beta_1\beta_2\ldots\beta_n} = (1/n!)(\partial^n \bar{u}^{\alpha}/(\partial u^{\beta_1} \partial u^{\beta_2} \ldots \partial u^{\beta_n}))_o . \tag{3-4c}$$

is true. An alternative form of (3-4b) is

$$\Delta\bar{u}^{\alpha} = \sum_{n=1}^{n'} \bar{b}^{\alpha}_{n_1 n_2} (\Delta u^1)^{n_1} (\Delta u^2)^{n_2}, \quad n_1, n_2 \in \{1,2\ldots n'\} , \; n_1 + n_2 = n \tag{3-4d}$$

with

$$\bar{b}^{\alpha}_{n_1 n_2} = (1/(n_1! n_2!))(\partial^n \bar{u}^{\alpha}/((\partial u^1)^{n_1} (\partial u^2)^{n_2}))_o . \tag{3-4e}$$

In the case of (3-4b,c)

$$N_n = 2(n+1) \tag{3-5a}$$

gives the number of independent series coefficients of the nth order when the coefficients in the lower indices are symmetrical. In the case of (3-4d,e) N_n indicates the total number of coefficients of the nth order. The relationship

$$\bar{b}^{\alpha}_{n_1 n_2} = \sum \bar{a}^{\alpha}_{\beta_1\beta_2\ldots\beta_n} \ , \quad n_1 + n_2 = n \ , \quad \beta_1\beta_2\ldots\beta_n = n_1{\times}1 \ n_2{\times}2 \ , \tag{3-5b}$$

which can be simplified to

$$\bar{b}^{\alpha}_{n_1 n_2} = (n!/(n_1!n_2!))\bar{a}^{\alpha}_{n_1\times 1 \ n_2\times 2} \ , \quad n_1 + n_2 = n \tag{3-5c}$$

for the symmetrical lower indices of the coefficients of (3-4c), generally exists between the coefficients of (3-4c) and (3-4e).

For the second transformation equations of (3-2a), i.e., the inverse transformation of (3-4a) of

$$u^{\alpha} = u^{\alpha}(\bar{u}^{\beta}) \ , \tag{3-6a}$$

we obtain, analogous to (3-4b,c), the power series

$$\Delta u^{\alpha} = \sum_{n=1}^{n'} a^{\alpha}_{\beta_1\beta_2\ldots\beta_n} \Delta\bar{u}^{\beta_1} \Delta\bar{u}^{\beta_2} \ldots \Delta\bar{u}^{\beta_n} \ , \tag{3-6b}$$

$$a^{\alpha}_{\beta_1\beta_2\ldots\beta_n} = (1/n!)(\partial^n u^{\alpha}/(\partial\bar{u}^{\beta_1}\partial\bar{u}^{\beta_2}\ldots\partial\bar{u}^{\beta_n}))_o \tag{3-6c}$$

and alternatively the form corresponding to (3-4d,e), i.e.,

$$\Delta u^{\alpha} = \sum_{n=1}^{n'} b^{\alpha}_{n_1 n_2} (\Delta\bar{u}^1)^{n_1} (\Delta\bar{u}^2)^{n_2} \ , \quad n_1 + n_2 = n \ , \tag{3-6d}$$

$$b^{\alpha}_{n_1 n_2} = (1/(n_1!n_2!))(\partial^n u^{\alpha}/((\partial\bar{u}^1)^{n_1}(\partial\bar{u}^2)^{n_2}))_o \ . \tag{3-6e}$$

(3-5) is also valid here.

If the coefficients of series (3-6b,d) are known, those of series (3-4b,d) can be calculated using a so-called series inversion, and naturally vice versa as well. For these purposes the recursive methods described for (2-170) and (2-171) can be applied here analogously, which will be explained briefly in the following for (3-6b) and (3-4b). Due to

$$a^{\alpha}_{\beta} = (u^{\alpha}_{,\bar{\beta}})_o \ , \qquad \bar{a}^{\alpha}_{\beta} = (\bar{u}^{\alpha}_{,\beta})_o$$

(2-15b) directly yields the following four linear conditional equations for $\bar{a}^{\alpha}_{\beta}$:

$$a^{\alpha}_{\gamma}\,\bar{a}^{\gamma}_{\beta} = \delta^{\alpha}_{\beta}\;. \tag{3-7a}$$

If they are calculated, we thus obtain

$$\Delta\bar{u}^{\alpha} = \bar{a}^{\alpha}_{\gamma}\,\Delta u^{\gamma} + \ldots =: w^{\alpha} \tag{3-7b}$$

as a first approximation. Herewith, the inner multiplication of series (3-6b) up to $n = 2$ with $\bar{a}^{\alpha}_{\gamma}$ after exchanging the indices yields

$$\Delta\bar{u}^{\alpha} = w^{\alpha} - \bar{a}^{\alpha}_{\gamma}\,a^{\gamma}_{\mu_1\mu_2}\,w^{\mu_1}w^{\mu_2} + \ldots \tag{3-7c}$$

as a second approximation. If in turn (3-7c) is inserted into series (3-6b) after inner multiplication with $\bar{a}^{\gamma}_{\alpha}$, we then obtain the third approximation, i.e.,

$$\begin{aligned}\Delta\bar{u}^{\alpha} = {} & w^{\alpha} - \bar{a}^{\alpha}_{\gamma}\,a^{\gamma}_{\mu_1\mu_2}\,w^{\mu_1}w^{\mu_2}\\ & - \bar{a}^{\alpha}_{\gamma}\,(a^{\gamma}_{\mu_1\mu_2\mu_3} - \bar{a}^{\delta}_{\epsilon}\,(a^{\gamma}_{\mu_1\delta} + a^{\gamma}_{\delta\mu_1})\,a^{\epsilon}_{\mu_2\mu_3})\,w^{\mu_1}w^{\mu_2}w^{\mu_3} + \ldots\;.\end{aligned} \tag{3-7d}$$

This process of recursion can be continued as long as desired. Based on (3-7a-d) the following results are obtained for the coefficients of (3-4c) and (3-6c) up to $n = 4$:

$$\begin{aligned}&\bar{a}^{\alpha}_{\beta_1} = \text{solutions to (3-7a)}\;, \qquad \bar{a}^{\alpha}_{\beta_1\beta_2} = -\,\bar{a}^{\alpha}_{\gamma}\,\bar{a}^{\mu_1}_{\beta_1}\,\bar{a}^{\mu_2}_{\beta_2}\,a^{\gamma}_{\mu_1\mu_2}\;,\\ &\bar{a}^{\alpha}_{\beta_1\beta_2\beta_3} = -\,\bar{a}^{\alpha}_{\gamma}\,\bar{a}^{\mu_1}_{\beta_1}\,\bar{a}^{\mu_2}_{\beta_2}\,\bar{a}^{\mu_3}_{\beta_3}\,(a^{\gamma}_{\mu_1\mu_2\mu_3} - \bar{a}^{\delta}_{\epsilon}\,(a^{\gamma}_{\mu_1\delta} + a^{\gamma}_{\delta\mu_1})\,a^{\epsilon}_{\mu_2\mu_3})\;,\\ &\bar{a}^{\alpha}_{\beta_1\beta_2\beta_3\beta_4} = -\,\bar{a}^{\alpha}_{\gamma}\,\bar{a}^{\mu_1}_{\beta_1}\,\bar{a}^{\mu_2}_{\beta_2}\,\bar{a}^{\mu_3}_{\beta_3}\,\bar{a}^{\mu_4}_{\beta_4}\Big\{a^{\gamma}_{\mu_1\mu_2\mu_3\mu_4}\\ &\quad - \bar{a}^{\delta_1}_{\epsilon_1}\,[(a^{\gamma}_{\mu_1\delta_1} + a^{\gamma}_{\delta_1\mu_1})\,a^{\epsilon_1}_{\mu_2\mu_3\mu_4} + (a^{\gamma}_{\mu_1\mu_2\delta_1} + a^{\gamma}_{\mu_1\delta_1\mu_2} + a^{\gamma}_{\delta_1\mu_1\mu_2})\,a^{\epsilon_1}_{\mu_3\mu_4}\\ &\qquad - \bar{a}^{\delta_2}_{\epsilon_2}\,(a^{\gamma}_{\delta_1\delta_2}\,a^{\epsilon_2}_{\mu_1\mu_2} + (a^{\gamma}_{\mu_1\delta_2} + a^{\gamma}_{\delta_2\mu_1})(a^{\epsilon_2}_{\mu_2\delta_1} + a^{\epsilon_2}_{\delta_1\mu_2}))\,a^{\epsilon_1}_{\mu_3\mu_4}]\Big\}\;.\end{aligned} \tag{3-7e}$$

In the most frequent case of

$$a^{\alpha}_{\beta} = \delta^{\alpha}_{\beta} \tag{3-7f}$$

the results of (3-7e) can be simplified as follows:

$$\begin{aligned}&\bar{a}^{\alpha}_{\beta_1} = \delta^{\alpha}_{\beta_1}\;, \qquad \bar{a}^{\alpha}_{\beta_1\beta_2} = -\,a^{\alpha}_{\beta_1\beta_2}\;,\\ &\bar{a}^{\alpha}_{\beta_1\beta_2\beta_3} = -\,a^{\alpha}_{\beta_1\beta_2\beta_3} + (a^{\alpha}_{\beta_1\gamma} + a^{\alpha}_{\gamma\beta_1})\,a^{\gamma}_{\beta_2\beta_3}\;,\end{aligned} \tag{3-7g}$$

$$\bar{a}^{\alpha}_{\beta_1\beta_2\beta_3\beta_4} = - a^{\alpha}_{\beta_1\beta_2\beta_3\beta_4} + (a^{\alpha}_{\beta_1\gamma} + a^{\alpha}_{\gamma\beta_1})\, a^{\gamma}_{\beta_2\beta_3\beta_4} \tag{3-7g}$$

$$+ [a^{\alpha}_{\beta_1\beta_2\gamma} + a^{\alpha}_{\beta_1\gamma\beta_2} + a^{\alpha}_{\gamma\beta_1\beta_2} - a^{\alpha}_{\gamma\delta}\, a^{\delta}_{\beta_1\beta_2} - (a^{\alpha}_{\beta_1\delta} + a^{\alpha}_{\delta\beta_1})(a^{\delta}_{\beta_2\gamma} + a^{\delta}_{\gamma\beta_2})]\, a^{\gamma}_{\beta_3\beta_4} .$$

The recursion method for inverting power series of the types in (3-6b) and (3-4b) described above can also be applied to the series types of (3-6d) and (3-4d) in the same way. In this context, König & Weise (1951, X, 11) gave general recursion formulas for arbitrary orders, whose representation will not be given here [regarding the problem of series inversion, see also Glasmacher & Krack 1984].

Solving a transformation problem (3-2) by means of power series is generally done in the following manner. Power series in the form of (3-4b) and (3-6b) or (3-4d) and (3-6d) are chosen for transforming (3-4a) or (3-6b), respectively. The accompanying coefficients must be determined so that the transformation equations approximated using the power series satisfy

- the accompanying differential equations and (3-8a)
- the initial or boundary conditions resulting from the special definition of $\bar{u}^{\alpha}$ in u^{α}. (3-8b)

After this problem has been solved, a power series for inverse transformation [see (3-6) and (3-4)] can then be calculated using, for example, series inversion according to (3-7). The transformation matrices

$$u^{\alpha}_{,\bar{\beta}} \qquad \text{and} \qquad \bar{u}^{\alpha}_{,\beta} \tag{3-8c}$$

necessary for (3-2b) ensue as partial derivatives of power series (3-6b,d) or (3-4b,d), respectively.

This method can be simplified for transformations between two isothermal coordinate systems u^{α} and $\bar{u}^{\alpha}$ when a complex setup of the power series corresponding to theorem (2-165) is selected in place of (3-4b,d) or (3-6b,d). For the transformations of (3-4) this reads

$$\Delta\bar{u} = \Delta\bar{u}^1 + i\,\Delta\bar{u}^2 = \sum_{n=1}^{n'} \bar{a}_n (\Delta u)^n ,$$
$$\bar{a}_n = \bar{a}^1_n + i\,\bar{a}^2_n , \quad \Delta u = \Delta u^1 + i\,\Delta u^2 . \tag{3-9}$$

In this case the equations of (3-8a) are the Cauchy-Riemannian differential equations (2-141b) and (2-160), which are satisfied by (3-9) from the very beginning. In this context this is the main advantage of these complex considerations. In determining the coefficients of (3-9) only (3-8b) must be observed. The inverse transformation of (3-9) can be done in the same way proceeding from the Cauchy-Riemannian differential equations of (2-141c), instead of (2-141b), and the initial conditions. However, series (3-9) can also be inverted according to the method described for (2-168) to (2-171).

3.3 TRANSFORMATIONS BETWEEN GEODESIC POLAR COORDINATES AND ARBITRARY SURFACE COORDINATES

3.3.1 GENERAL TRANSFORMATION EQUATIONS

Solutions to the transformation problems of (2-114) and (2-115) are given here in the form of power series. We will begin with the direct or first geodetic problem (2-114), i.e., the transformation of

$$\bar{u}^{\alpha} = (\beta, r) = \text{geodesic polar coordinates} \tag{3-10a}$$

into

$$u^{\alpha} = \text{arbitrary surface coordinates} \tag{3-10b}$$

or

$$\Delta u^{\alpha} = u^{\alpha} - (u^{\alpha})_o \;, \quad (u^{\alpha})_o = \text{coordinates of the pole } P_o \text{ of } \bar{u}^{\alpha} \;. \tag{3-10c}$$

The presuppositions for (3-1) are assumed to be given for u^{α}. Since (3-10a) involves "intrinsic coordinates", the knowledge of the metric tensor (3-1a) is sufficient. Solutions of the initial value problems of the geodesics of (2-91) for the r-lines through P_o with $s =: r$ and the initial values

$$(\Delta u^{\alpha})_o = 0 \;, \quad (\Delta u^{\alpha'})_o = (u^{\alpha'})_o \quad \text{according to (2-113f,g)} \tag{3-11}$$

correspond to the transformations of problem (2-114). We also usually begin with a power series in r, i.e.,

$$\Delta u^{\alpha} = (u^{\alpha'})_o r + \sum_{n=2}^{n'} (1/n!)(u^{\alpha(n)})_o r^n \ , \tag{3-12a}$$

$$(u^{\alpha(n)})_o := (\partial^n u^{\alpha}/\partial r^n)_{P_o} \ , \quad u^{\alpha(1)} =: u^{\alpha'} . \tag{3-12b}$$

The first term of series (3-12a) is determined directly with the initial values of (3-11) and r, i.e., with the given polar coordinates of (3-10a). The terms of the orders $n > 1$ must be determined step by step proceeding from the differential equation of the geodesics of (2-90a), i.e.,

$$u^{\alpha''} = u^{\alpha(2)} = -\Gamma^{\alpha}_{\beta_1\beta_2} u^{\beta_1'} u^{\beta_2'} \ , \tag{3-13}$$

from which

$$(u^{\alpha(2)})_o = -(\Gamma^{\alpha}_{\beta_1\beta_2})_o (u^{\beta_1'})_o (u^{\beta_2'})_o \tag{3-14a}$$

directly results. Differentiating (3-13) with respect to r and observing (3-14a) we have

$$(u^{\alpha(3)})_o = -(\Gamma^{\alpha}_{\beta_1\beta_2,\beta_3} - 2\Gamma^{\alpha}_{\beta_1\gamma}\Gamma^{\gamma}_{\beta_2\beta_3})_o (u^{\beta_1'})_o (u^{\beta_2'})_o (u^{\beta_3'})_o \ . \tag{3-14b}$$

The higher derivatives of (3-12b) are obtained in the same way. The results can generally be written as follows:

$$(u^{\alpha(n)})_o = n! a^{\alpha}_{\beta_1\beta_2\ldots\beta_n} (u^{\beta_1'})_o (u^{\beta_2'})_o \ldots (u^{\beta_n'})_o \ . \tag{3-14c}$$

With (3-14c) and the introduction of normal coordinates (2-131j), i.e.,

$$\bar{v}^{\alpha} = (u^{\alpha'})_o r \ , \tag{3-15a}$$

(3-12a) becomes a so-called <u>Legendrean series</u>, i.e.,

$$\Delta u^{\alpha} = \bar{v}^{\alpha} + \sum_{n=2}^{n'} a^{\alpha}_{\beta_1\beta_2\ldots\beta_n} \bar{v}^{\beta_1} \bar{v}^{\beta_2} \ldots \bar{v}^{\beta_n} \ , \tag{3-15b}$$

with coefficients up to $n = 4$, i.e.,

$$a^{\alpha}_{\beta_1\beta_2} = -(\Gamma^{\alpha}_{\beta_1\beta_2})_o/2 \ , \quad a^{\alpha}_{\beta_1\beta_2\beta_3} = -(\Gamma^{\alpha}_{\beta_1\beta_2,\beta_3} - 2\Gamma^{\alpha}_{\beta_1\gamma}\Gamma^{\gamma}_{\beta_2\beta_3})_o/6 \ ,$$

$$a^{\alpha}_{\beta_1\beta_2\beta_3\beta_4} = -[\, \Gamma^{\alpha}_{\beta_1\beta_2,\beta_3\beta_4} - 2\Gamma^{\alpha}_{\beta_1\gamma}\Gamma^{\gamma}_{\beta_2\beta_3,\beta_4} \tag{3-15c}$$

$$- (\Gamma^{\alpha}_{\beta_1\beta_2,\delta} + 4\Gamma^{\alpha}_{\beta_1\delta,\beta_2} - 2\Gamma^{\alpha}_{\gamma\delta}\Gamma^{\gamma}_{\beta_1\beta_2} - 4\Gamma^{\alpha}_{\beta_1\gamma}\Gamma^{\gamma}_{\beta_2\delta})\Gamma^{\delta}_{\beta_3\beta_4}]_o/24 \ .$$

The coefficients of the third and fourth orders are generally nonsymmetrical in the lower indices so that (3-5b) has to be used for converting (3-15b) into the form of (3-6d), i.e.,

$$\Delta u^{\alpha} = \bar{v}^{\alpha} + \sum_{n=2}^{n'} b^{\alpha}_{n_1 n_2} (\bar{v}^1)^{n_1} (\bar{v}^2)^{n_2} , \quad n_1 + n_2 = n . \tag{3-15d}$$

For solving the direct geodetic problem by means of the Legendrean series of (3-15b,d) the polar coordinates $\bar{u}^{\alpha}$ [see (3-10a)] are first transformed into normal coordinates $\bar{v}^{\alpha}$ according to (3-15a) which are then inserted into the right side of (3-15b) or (3-15d). For the inverse transformation of (3-15b), i.e., the solution to the inverse geodetic problem, we obtain a series in the form of (3-4b), i.e.,

$$\bar{v}^{\alpha} = \Delta u^{\alpha} + \sum_{n=2}^{n'} \bar{a}^{\alpha}_{\beta_1 \beta_2 \ldots \beta_n} \Delta u^{\beta_1} \Delta u^{\beta_2} \ldots \Delta u^{\beta_n} . \tag{3-15e}$$

Based on (3-7g) and up to $n = 4$ the coefficients become

$$\bar{a}^{\alpha}_{\beta_1\beta_2} = (\Gamma^{\alpha}_{\beta_1\beta_2})_o/2 , \qquad \bar{a}^{\alpha}_{\beta_1\beta_2\beta_3} = (\Gamma^{\alpha}_{\beta_1\beta_2,\beta_3} + \Gamma^{\alpha}_{\beta_1\gamma}\, \Gamma^{\gamma}_{\beta_2\beta_3})_o/6 ,$$

$$\bar{a}^{\alpha}_{\beta_1\beta_2\beta_3\beta_4} = [\, \Gamma^{\alpha}_{\beta_1\beta_2,\beta_3\beta_4} + 2\Gamma^{\alpha}_{\beta_1\gamma}\, \Gamma^{\gamma}_{\beta_2\beta_3,\beta_4} + (\Gamma^{\alpha}_{\beta_1\beta_2,\delta} + \Gamma^{\alpha}_{\gamma\delta}\, \Gamma^{\gamma}_{\beta_1\beta_2})\, \Gamma^{\delta}_{\beta_3\beta_4}]_o/24 . \tag{3-15f}$$

Using (3-5b), (3-15e) can be converted into the same form as that of (3-4d), i.e.,

$$\bar{v}^{\alpha} = \Delta u^{\alpha} + \sum_{n=2}^{n'} \bar{b}^{\alpha}_{n_1 n_2} (\Delta u^1)^{n_1} (\Delta u^2)^{n_2} , \quad n_1 + n_2 = n . \tag{3-15g}$$

The polar coordinates are obtained with (2-131j).

Proceeding from (3-15d) and observing (3-15a) the tangent vector on the geodesic r-line at point $P(\beta, r)$ becomes

$$u^{\alpha'} = (u^{\alpha'})_o + (1/r) \sum_{n=2}^{n'} n\, b^{\alpha}_{n_1 n_2} (\bar{v}^1)^{n_1} (\bar{v}^2)^{n_2} . \tag{3-16a}$$

Due to the relationships given for (2-113) and using (3-16a) the direction angle θ between the r-line and the u^2-line can be calculated, whereby

$$P_o =: P , \qquad \beta =: \theta . \tag{3-16b}$$

In the case of rectangular coordinates u^{α},

$$g_{12} = 0 \tag{3-16c}$$

follows from (2-113g):

$$\theta = \arctan((g_{11}/g_{22})^{1/2} u^{1'}/u^{2'}) \ . \tag{3-16d}$$

Especially at P_o,

$$\theta_o = \beta \tag{3-16e}$$

is true. However, directly analogous to (3-12) a power series in r can also be set up for θ, i.e.,

$$\theta = \beta + \sum_{n=1}^{n'} (1/n!)\theta_o^{(n)} r^n \ , \quad \theta_o^{(n)} = (\partial^n\theta/\partial r^n)_{P_o} \ . \tag{3-17a}$$

Under the condition of (3-16c) we obtain the derivatives of θ by stepwise differentiating (2-97a), i.e.,

$$\theta^{(1)} = \theta' = -(1/2)g^{-1/2}(g_{11,2}u^{1'} - g_{22,1}u^{2'}) \ , \tag{3-17b}$$

and observing (3-13) analogous to the method given for (3-14). The results have the same form as those of (3-14c), i.e.,

$$\theta_o^{(n)} = n!\ d_{\beta_1\beta_2\ldots\beta_n}(u^{\beta_1'})_o(u^{\beta_2'})_o\ldots(u^{\beta_n'})_o \ . \tag{3-17c}$$

Hereby, (3-17a) changes into the following forms corresponding to (3-15b,d):

$$\theta = \beta + \sum_{n=1}^{n'} d_{\beta_1\beta_2\ldots\beta_n} \bar{v}^{\beta_1}\bar{v}^{\beta_2}\ldots\bar{v}^{\beta_n} \ , \tag{3-18a}$$

$$\theta = \beta + \sum_{n=1}^{n'} c_{n_1n_2}(\bar{v}^1)^{n_1}(\bar{v}^2)^{n_2} \ , \quad n_1 + n_2 = n \ . \tag{3-18b}$$

(3-15) and (3-16) involve Taylor series with reference to the coordinates and tangent vector at an endpoint P_o of the geodesic r-line. In the following the expressions

$$P_a := P_o \ , \qquad P_b := P \tag{3-19}$$

will be used for the endpoints P_o and P so that the previous expansions will refer to the initial values of

$$(u^{\alpha})_a \ , \qquad (u^{\alpha'})_a \ . \tag{3-20}$$

Essentially better converging series are obtained when the mean values of

$$(u^\alpha)_m = ((u^\alpha)_a + (u^\alpha)_b)/2 = \text{mid-coordinates} , \tag{3-21a}$$

$$(u^{\alpha'})_m = ((u^{\alpha'})_a + (u^{\alpha'})_b)/2 = \text{mid-tangents} \tag{3-21b}$$

are used as initial quantities of the expansions instead of (3-20). In (3-21b) both tangent vectors at P_a and P_b are to have the same sense of direction $P_a \rightarrow P_b$ so that

$$(u^{\alpha'})_a \rightarrow (u^{\alpha'})_b \qquad \text{for} \qquad r \rightarrow 0 . \tag{3-21c}$$

The power series referring to (3-21a,b) are so-called mid-coordinate formulas, which can be traced back to C.F. Gauss. They are suited above all for solving the transformations of (2-115), i.e., for solving the inverse geodetic problems of (2-92), and will be considered in the following only for this purpose. Here, the method of derivation first described by Hubeny (1954) has been selected. For this purpose, normal coordinates (3-15a) related to P_a and P_b are introduced, i.e.,

$$(\bar{v}^\alpha)_p = (u^{\alpha'})_p r , \qquad p \in \{a,b\} , \tag{3-22a}$$

and

$$(\bar{v}^\alpha)_m = ((\bar{v}^\alpha)_a + (\bar{v}^\alpha)_b)/2 = (u^{\alpha'})_m r \tag{3-22b}$$

is assumed. The basis of further calculations is formed by series (3-15g) for $\bar{v}^\alpha$ obtained from (3-15b,d) through series inversion. If these series are calculated once relative to P_a and once relative to P_b, the following results are obtained observing (3-21c) and (3-22a):

$$(\bar{v}^\alpha)_a = \Delta u^\alpha + \sum_{n=2}^{n'} \bar{a}^\alpha_{n_1 n_2 a} (\Delta u^1)^{n_1} (\Delta u^2)^{n_2} , \tag{3-23a}$$

$$(\bar{v}^\alpha)_b = \Delta u^\alpha + \sum_{n=2}^{n'} \bar{a}^\alpha_{n_1 n_2 b} (\Delta u^1)^{n_1} (\Delta u^2)^{n_2} (-1)^{n-1} , \tag{3-23b}$$

$$\Delta u^\alpha = (u^\alpha)_b - (u^\alpha)_a , \qquad n_1 + n_2 = n . \tag{3-23c}$$

The indices a and b in the coefficients of (3-23a,b) indicate that they are to be formed at P_a and P_b, respectively. If half of the sum of (3-23a,b) and half of the difference are formed, then the following series are obtained with (3-21) and (3-22):

$$\begin{aligned}(\bar{v}^\alpha)_m &= ((\bar{v}^\alpha)_a + (\bar{v}^\alpha)_b)/2 = (u^{\alpha'})_m r \\ &= \Delta u^\alpha + (1/2) \sum_{n=2}^{n'} (\bar{a}^\alpha_{n_1 n_2 a} - (-1)^n \bar{a}^\alpha_{n_1 n_2 b}) (\Delta u^1)^{n_1} (\Delta u^2)^{n_2} ,\end{aligned} \tag{3-24a}$$

$$\delta\bar{v}^{\alpha}_{ab} = ((\bar{v}^{\alpha})_b - (\bar{v}^{\alpha})_a)/2$$

$$= -(1/2)\sum_{n=2}^{n'} (\bar{a}^{\alpha}_{n_1n_2a} + (-1)^n \bar{a}^{\alpha}_{n_1n_2b})(\Delta u^1)^{n_1}(\Delta u^2)^{n_2} \; . \qquad (3\text{-}24b)$$

Proceeding from the values of $\bar{a}^{\alpha}_{n_1n_2m}$ at the point with the mid-coordinates (3-21a) the coefficients $\bar{a}^{\alpha}_{n_1n_2p}$ can be represented as power series as follows:

$$\left.\begin{matrix}\bar{a}^{\alpha}_{n_1n_2a}\\ \bar{a}^{\alpha}_{n_1n_2b}\end{matrix}\right\} = \bar{a}^{\alpha}_{n_1n_2m} + \sum_{q=1}^{q'} \left\{\begin{matrix}(-1/2)^q\\ (1/2)^q\end{matrix}\right\} c^{\alpha}_{n_1n_2q_1q_2}(\Delta u^1)^{q_1}(\Delta u^2)^{q_2} \; , \qquad (3\text{-}25)$$

$$q_1 + q_2 = q \; .$$

Inserting (3-25) into (3-24a,b) ultimately yields the desired <u>mid-coordinate formulas</u> of

$$(\bar{v}^{\alpha})_m = \Delta u^{\alpha} + \sum_{n=3}^{n'} ((1-(-1)^n)/2) A^{\alpha}_{n_1n_2m}(\Delta u^1)^{n_1}(\Delta u^2)^{n_2} \; , \qquad (3\text{-}26a)$$

$$\delta\bar{v}^{\alpha}_{ab} = \sum_{n=2}^{n'} ((1+(-1)^n)/2) B^{\alpha}_{n_1n_2m}(\Delta u^1)^{n_1}(\Delta u^2)^{n_2} \; , \qquad (3\text{-}26b)$$

whose coefficients $A^{\alpha}_{n_1n_2m}$ and $B^{\alpha}_{n_1n_2m}$ are functions of

$$\bar{a}^{\alpha}_{n_1n_2m}, \; c^{\alpha}_{n_1n_2q_1q_2} \quad \text{for} \quad 1 \leq q \leq n'-n \; . \qquad (3\text{-}26c)$$

(3-26a) and (3-26b) are odd and even series in n, respectively, so that (3-26a) has less terms than the initial series of (3-23a,b). Yet, the fact that (3-26a) converges significantly faster than (3-23a,b) is more important. The calculation of the coefficients of (3-26a,b) involve, however, considerably more development, which for higher orders is only possible with the help of a computer [see Krack 1982b].

The mid-coordinate formulas of (3-26) can be used directly for solving the transformation (2-115) of coordinates (3-10b) into polar coordinates (3-10a), i.e.,

$$u^{\alpha} \rightarrow \bar{u}^{\alpha} = (\beta, r) \; , \qquad (3\text{-}27a)$$

which is identical to the solution for the first boundary value problem of the geodesic line (2-92), also called the inverse geodetic problem. Based on the given coordinates u^{α} the differences of (3-23c) and the mid-coordinates (3-21a) can be calculated. Thus, series (3-26a,b) can be evaluated with the results of $(\bar{v}^\alpha)_m$ and $\delta\bar{v}^\alpha_{ab}$, for which equations

$$(\bar{v}^\alpha)_b + (\bar{v}^\alpha)_a = 2(\bar{v}^\alpha)_m \;, \qquad (\bar{v}^\alpha)_b - (\bar{v}^\alpha)_a = 2\delta\bar{v}^\alpha_{ab} \tag{3-27b}$$

are valid corresponding to (3-24a,b). Observing (3-22a), the inversion of these equations yields

$$\begin{aligned} (\bar{v}^\alpha)_a &= (u^{\alpha'})_a r = (\bar{v}^\alpha)_m - \delta\bar{v}^\alpha_{ab} \;, \\ (\bar{v}^\alpha)_b &= (u^{\alpha'})_b r = (\bar{v}^\alpha)_m + \delta\bar{v}^\alpha_{ab} \;. \end{aligned} \tag{3-27c}$$

Then, observing (2-30) we obtain

$$r^2 = (g_{\alpha\beta})_p(\bar{v}^\alpha)_p(\bar{v}^\beta)_p \;, \qquad p \in \{a,b\} \;. \tag{3-27d}$$

If r is calculated in the same manner, the tangent vectors in P_a and P_b then become

$$(u^{\alpha'})_p = (\bar{v}^\alpha)_p/r \tag{3-27e}$$

based on (3-27c). When dealing with rectangular coordinates u^α, β_p can be calculated as

$$\beta_p = \arctan((g_{11}/g_{22})_p^{1/2}(\bar{v}^1)_p/(\bar{v}^2)_p) \;, \qquad p \in \{a,b\} \tag{3-27f}$$

based on (2-113g) or (3-16d).

The previous considerations of the transformation between geodesic polar coordinates $\bar{u}^\alpha$ [see (3-10a)] and arbitrary surface coordinates u^α [see (3-10b)] were aimed only at determining the transformation equations of the coordinates of (3-2a). Proceeding from the results of (3-15d) and (3-16) we will now go into the <u>transformation of the first fundamental tensors</u> according to the first equation of (3-2b). For this we need the transformation matrices

$$u^\gamma_{,\bar{\alpha}} = \partial u^\gamma/\partial\bar{u}^\alpha = (\partial u^\gamma/\partial\beta,\ \partial u^\gamma/\partial r)_\alpha \;, \tag{3-28a}$$

whose calculation in (3-15d) necessitates the partial derivatives of normal coordinates (3-15a) with respect to β and r. Observing (2-113f) we have

$$\begin{aligned} \bar{v}^\beta_{,\bar{1}} &= \partial\bar{v}^\beta/\partial\beta = \epsilon^\beta_\gamma(u^{\gamma'})_o r = \epsilon^\beta_\gamma\,\bar{v}^\gamma \;, \\ \bar{v}^\beta_{,2} &= \partial\bar{v}^\beta/\partial r = \quad (u^{\beta'})_o \quad = r^{-1}\bar{v}^\beta \;. \end{aligned} \tag{3-28b}$$

Thereby, for the orthogonal surface coordinates u^α, i.e.,

$$g_{12} = 0 , \tag{3-29a}$$

with the accompanying ϵ-tensor according to (2-44c), i.e.,

$$\epsilon^{\beta}_{\gamma} = -(g_{22}/g_{11})^{1/2}\delta^{\beta}_{1}\,\delta^{2}_{\gamma} + (g_{11}/g_{22})^{1/2}\delta^{\beta}_{2}\,\delta^{1}_{\gamma} , \tag{3-29b}$$

the transformation matrices of (3-28a) are written as follows based on (3-15c):

$$\begin{aligned} u^{\alpha}_{,\bar{1}} = {} & -(g_{22}/g_{11})^{1/2}_{o}\,\bar{v}^{2}\delta^{\alpha}_{1} + (g_{11}/g_{22})^{1/2}_{o}\,\bar{v}^{1}\delta^{\alpha}_{2} \\ & + \sum_{n=2}^{n'} a^{\alpha}_{n_1 n_2}\Big[-n_1(g_{22}/g_{11})^{1/2}_{o}(\bar{v}^{1})^{n_1-1}(\bar{v}^{2})^{n_2+1} \\ & \qquad +n_2(g_{11}/g_{22})^{1/2}_{o}(\bar{v}^{1})^{n_1+1}(\bar{v}^{2})^{n_2-1}\Big] , \end{aligned} \tag{3-30a}$$

$$u^{\alpha}_{,\bar{2}} = (1/r)(\bar{v}^{\alpha} + \sum_{n=2}^{n'} n\, a^{\alpha}_{n_1 n_2}(\bar{v}^{1})^{n_1}(\bar{v}^{2})^{n_2}) . \tag{3-30b}$$

The first fundamental tensor $\bar{g}_{\alpha\beta}$ of the geodesic polar coordinates have the components

$$\bar{g}_{11} = h^2,\quad \bar{g}_{22} = 1 ,\quad \bar{g}_{12} = 0 \tag{3-31}$$

according to (2-122d) so that only h needs to be calculated using (3-2b). With the assumption of (3-29b) we obtain

$$h^2 = (u^{1}_{,\bar{1}})^2 g_{11} + (u^{2}_{,\bar{1}})^2 g_{22} ,\quad u^{\alpha}_{,\bar{1}} \text{ after (3-30a)} . \tag{3-32}$$

In (3-31b) g_{11} and g_{22} are to be formed at the test point for which h^2 should be calculated. Thus, the complete expansion of the power series of h^2 also requires power series for g_{11} and g_{22} in $\bar{v}^{\alpha}$ corresponding to (3-30a). Then, the <u>first fundamental tensor of the Riemannian normal coordinates</u> of (2-129g) is determined with h as well.

3.3.2 CALCULATING SMALL GEODESIC TRIANGLES

Geodesic triangles are areas of a surface bound by three geodesics whose corners, sides, and angles are designated in the following corresponding to Fig. 3.1. The objective here is the derivation of equations for the sides and angles of geodesic triangles, analogous to the laws of sines, projection, and cosines in plane trigonometry. For this we will first introduce Riemannian normal coordinates v^α [see (2-129)] assuming

$$\begin{aligned}&\text{pole } P_o =: P_c ,\\&\text{reference direction for } \beta = \text{tangent on } P_c - P_a \text{ at } P_c\end{aligned} \tag{3-33a}$$

(cf. Fig. 2.6). Thereby, the three corners of the geodesic triangle have the following Riemannian normal coordinates:

$$(v^\alpha)_a = \delta_2^\alpha s_b , \quad (v^\alpha)_b = (\sin\alpha_c , \cos\alpha_c)^\alpha s_a , \quad (v^\alpha)_c = 0 . \tag{3-33b}$$

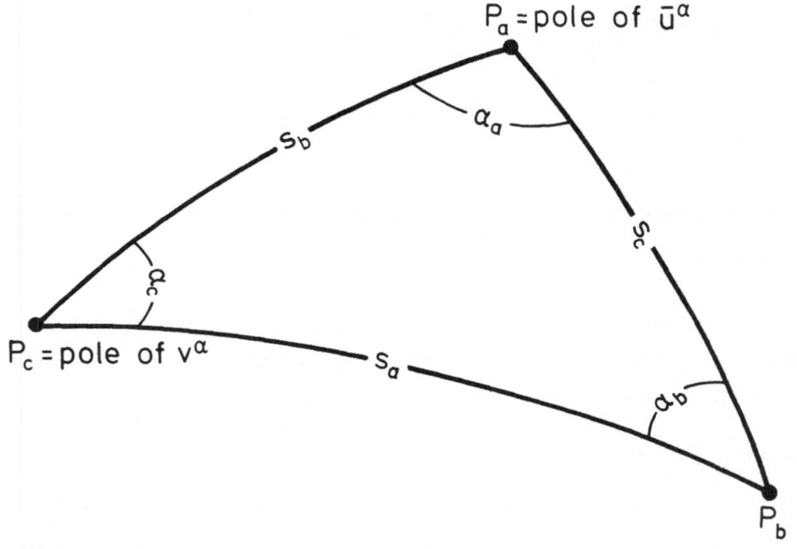

Fig. 3.1. Geodesic triangle

The metric tensor $g_{\alpha\beta}$ is determined by

$$h = h(\beta, r) \tag{3-33c}$$

according to (2-129g). A presupposition here is that this function is given. Assuming arbitrary surface coordinates u^α, h can be calculated based on (3-30) and (3-32).

In addition to Riemannian normal coordinates v^α, geodesic polar coordinates $\bar{u}^\alpha$ [see (2-112)] are also introduced here as follows:

$$\begin{aligned}&\text{pole } P_a , \quad \bar{u}^\alpha = (\gamma, \bar{r}) ,\\&\gamma = \text{direction angle relative to the } v^2\text{-line through } P_a .\end{aligned} \tag{3-34a}$$

Consequently, P_b has the polar coordinates

$$(\bar{u}^\alpha)_b = (\pi - \alpha_a, s_c) \qquad (3\text{-}34b)$$

and with reference to P_a the normal coordinates

$$(\bar{v}^\alpha)_b = (v^{\alpha'})_{ab} s_c , \qquad (v^{\alpha'})_{ab} = (\partial v^\alpha / \partial s_c)_a \qquad (3\text{-}34c)$$

according to (2-131j). At P_a, $\beta = 0$ so that based on (2-129g) the metric tensor of v^α has the following components:

$$(g_{11})_a = (h_a/s_b)^2 , \qquad (g_{22})_a = 1 , \qquad (g_{12})_a = 0 . \qquad (3\text{-}34d)$$

According to (2-113g), (3-34c) then becomes

$$(v^{\alpha'})_{ab} = ((s_b/h_a)\sin\alpha_a, -\cos\alpha_a)^\alpha . \qquad (3\text{-}34e)$$

We obtain the desired equations for calculating the sides and angles of small geodesic triangles by transforming the geodesic polar coordinates of (3-34b) into Riemannian normal coordinates based on the power series of (3-15a-g). Proceeding from (3-15b) and observing (3-33b) and (3-34c,e) up to $n' = 3$ we obtain

$$\begin{aligned} \Delta v^\alpha &= (v^\alpha)_b - (v^\alpha)_a = (s_a \sin\alpha_c, s_a\cos\alpha_c - s_b)^\alpha \\ &= (s_c(s_b/h_a)\sin\alpha_a, -s_c\cos\alpha_a)^\alpha + \sigma^\alpha \end{aligned} \qquad (3\text{-}35a)$$

with

$$\sigma^\alpha = a^\alpha_{\beta\gamma} (\bar{v}^\beta)_b(\bar{v}^\gamma)_b + a^\alpha_{\beta\gamma\delta} (\bar{v}^\beta)_b(\bar{v}^\gamma)_b(\bar{v}^\delta)_b + \ldots . \qquad (3\text{-}35b)$$

Series (3-35b) can basically be continued up to any order of $n' > 3$, in line with (3-15b).

The $\alpha = 1$ component of (3-35a) directly yields the <u>generalized law of sines</u>, i.e.,

$$s_a \sin\alpha_c = s_c \sin\alpha_a + s_c((s_b - h_a)/h_a)\sin\alpha_a + \sigma^1 , \qquad (3\text{-}36a)$$

and the $\alpha = 2$ component of (3-35a) provides the <u>generalized law of projection</u>, i.e.,

$$s_b = s_a\cos\alpha_c + s_c\cos\alpha_a - \sigma^2 . \qquad (3\text{-}36b)$$

Finally, taking both components of (3-35a) into account we obtain the <u>generalized law of cosines</u>, i.e.,

$$\begin{aligned} s_a^2 &= (\Delta v^1)^2 + (\Delta v^2 + s_b)^2 \\ &= s_b^2 + s_c^2 - 2s_b s_c\cos\alpha_a + (s_b^2 - h_a^2)(s_c/h_a)^2\sin^2\alpha_a \\ &\quad + \sigma^1(2s_c(s_b/h_a)\sin\alpha_a + \sigma^1) + \sigma^2(2(s_b - s_c\cos\alpha_a) + \sigma^2) . \end{aligned} \qquad (3\text{-}36c)$$

For plane surfaces

$$h_a = s_b \, , \qquad\qquad \sigma^\alpha = 0 \, ,$$

whereby the equations of (3-36a-c) reduce to the laws of sines, projection, and cosines of plane trigonometry.

For calculating σ^α according to (3-35b) we first need to calculate the $a^\alpha_{\beta\gamma}$ and $a^\alpha_{\beta\gamma\delta}$ coefficients in the same manner as that given for (3-14) and (3-15c). The Christoffel's symbols refer to the Riemannian normal coordinates v^α and thus to their metric tensor (2-129g), which is designated here by $g_{\alpha\beta}$. The necessary first and second derivatives of (2-129g) with respect to v^α are obtained taking (2-129e) into account. Here, these calculations will only be made for the first term in (3-35b), and for h [see (3-33c)] (2-125d) is taken, i.e.,

$$h = r - (1/6)\, K\, r^3 \, . \qquad (3\text{-}37a)$$

K is the Gaussian curvature, which is constant within the geodesic triangle so that the area of the triangle is approximated by that of a sphere. Thus, based on (2-129g) we obtain the following for P_a:

$$(g_{11})_a = 1 - (K/3)\, s_b^2 \, , \qquad (g_{22})_a = 1 \, , \qquad (g_{12})_a = 0 \; ; \qquad (3\text{-}37b)$$

$$\begin{aligned} (g_{11,1})_a &= 0 \, , & (g_{11,2})_a &= -2(K/3)\, s_b \, , \\ (g_{22,1})_a &= 0 \, , & (g_{22,2})_a &= 0 \, , \\ (g_{12,1})_a &= (K/3)\, s_b \, , & (g_{12,2})_a &= 0 \, . \end{aligned} \qquad (3\text{-}37c)$$

From (3-15c)

$$a^\alpha_{\beta\gamma} = -(1/2)(\Gamma^\alpha_{\beta\gamma})_a \qquad (3\text{-}37d)$$

follows. With (3-37b,c) we obtain

$$a^1_{12} = a^1_{21} = (K/6)\, s_b \, , \qquad a^2_{11} = -(K/3)\, s_b \, , \qquad (3\text{-}37e)$$

and all remaining components of (3-37d) vanish. In (3-35b), (3-34c) together with (3-34e) can just as well be replaced by

$$(\bar{v}^\beta)_b = (\sin\alpha_a , -\cos\alpha_a)\, s_c \, . \qquad (3\text{-}37f)$$

With (3-37e,f) we ultimately obtain

$$\sigma^\alpha = -(K/3)\, s_b s_c^2 (\sin\alpha_a \cos\alpha_a , \sin^2\alpha_a)^\alpha \qquad (3\text{-}37g)$$

according to (3-35b).

$$(s_b - h_a)/h_a = (K/6)\, s_b^2 \, , \qquad (s_b^2 - h_a^2)/h_a^2 = (K/3)\, s_b^2$$

are true with the same accuracy. If these expressions and (3-37g) are inserted in (3-36a,c), the laws of sines, projection, and cosines take on the following forms:

$$s_a \sin\alpha_c - s_c \sin\alpha_a = (K/6)\, s_b s_c (s_b - 2s_c \cos\alpha_a) \sin\alpha_a \,, \tag{3-38a}$$

$$s_b - s_a \cos\alpha_c - s_c \cos\alpha_a = (K/3)\, s_b s_c^2 \sin^2\alpha_a \,, \tag{3-38b}$$

$$s_a^2 - s_b^2 - s_c^2 + 2 s_b s_c \cos\alpha_a = -(K/3)\, s_b^2 s_c^2 \sin^2\alpha_a \,. \tag{3-38c}$$

These equations for calculating small geodesic triangles in spherical approximation are to be supplemented further by the angular sum equation (2-128e), i.e.,

$$\alpha_a + \alpha_b + \alpha_c - \pi = \epsilon = K\, \Delta F \,, \tag{3-39a}$$

in which

$$\Delta F = (1/2)\, s_b s_c \sin\alpha_a \tag{3-39b}$$

is accurate enough and ϵ is the spherical excess. The Gaussian curvature K in (3-38) can be eliminated by the spherical excess using (3-39). Furthermore,

$$s_b - 2s_c \cos\alpha_a = s_a \cos\alpha_c - s_c \cos\alpha_a \,,$$
$$s_c \sin\alpha_a = (s_a \sin\alpha_c + s_c \sin\alpha_a)/2$$

can be inserted in the right sides of (3-38a,b). Based on the aforementioned substitutions the equations of (3-38) become

$$s_a \sin\alpha_c = s_c \sin\alpha_a + (\epsilon/3)(s_a \cos\alpha_c - s_c \cos\alpha_a) \,, \tag{3-40a}$$

$$s_b = s_a \cos\alpha_c + s_c \cos\alpha_a + (\epsilon/3)(s_a \sin\alpha_c + s_c \sin\alpha_a) \,, \tag{3-40b}$$

$$s_a^2 = s_b^2 + s_c^2 - 2 s_b s_c \cos\alpha_a - 2(\epsilon/3)\, s_b s_c \sin\alpha_a \,. \tag{3-40c}$$

Within the limits of the accuracy of (3-38) and (3-40) the second and higher powers of ϵ can be neglected so that the equations of (3-40) are equivalent to

$$s_a \sin(\alpha_c - \epsilon/3) = s_c \sin(\alpha_a - \epsilon/3) \,, \tag{3-41a}$$

$$s_b = s_a \cos(\alpha_c - \epsilon/3) + s_c \cos(\alpha_a - \epsilon/3) \,, \tag{3-41b}$$

$$s_a^2 = s_b^2 + s_c^2 - 2 s_b s_c \cos(\alpha_a - \epsilon/3) \,. \tag{3-41c}$$

The laws of sines, projection, and cosines in the form of (3-41) contain the proof of the

Theorem of Legendre (3-42)

Small geodesic triangles can be calculated like plane triangles with the same sides, i.e.,

$s_a \,,\; s_b \,,\; s_c$

and angles decreased by $\epsilon/3$, i.e.,

$\alpha_a - \epsilon/3 \,,\; \alpha_b - \epsilon/3 \,,\; \alpha_c - \epsilon/3$.

3.4 TRANSFORMATIONS BETWEEN GEODESIC PARALLEL COORDINATES AND ARBITRARY SURFACE COORDINATES

3.4.1 INDIRECT REPRESENTATION BY POWER SERIES

We will proceed here from the definitions of

$$\bar{u}^{\alpha} = \text{geodesic parallel coordinates} \tag{3-43a}$$

given for (2-116a-e) [see Fig. 2.5]. This involves intrinsic coordinates so that for them to be constructed according to Chapter 3.1 the metric tensor (3-1a) as a function of

$$u^{\alpha} = \text{arbitrary surface coordinates} \tag{3-43b}$$

must be given, which is always a presupposition in the following.

We will begin with the transformation of $\bar{u}^{\alpha}$ in u^{α} corresponding to the method of calculation described for (2-120) assuming that

$$C_o = \text{geodesic line} \,, \qquad m = 1 \,. \tag{3-44a}$$

Relative to the origin P_o of $\bar{u}^{\alpha}$ the normal coordinates

$$(\bar{v}^{\alpha})_o = \upsilon^{\alpha}_{Co}\, \bar{u}^{1} \tag{3-44b}$$

are introduced for the foot point Q. With these normal coordinates the coordinate differences $P_o - Q$ can be calculated using a Legendrean series (3-15d) as follows:

$$(\Delta u^{\alpha})_{oQ} = (u^{\alpha})_Q - (u^{\alpha})_o = (\bar{v}^{\alpha})_o + \sum_{n=2}^{n'} a^{\alpha}_{n_1 n_2 o} (\bar{v}^1)_o^{n_1} (\bar{v}^2)_o^{n_2} \,, \tag{3-44c}$$

$n_1 + n_2 = n$.

The tangent vector on the $\bar{u}^2$-line in Q [see (2-120f)] together with (3-16a) becomes

$$(u^{\alpha'})_{QP} = \pm(\epsilon^{\alpha}_{\beta})_Q(\nu^{\beta}_{Co} \pm (1/\bar{u}^1) \sum_{n=2}^{n'} n\, a^{\beta}_{n_1 n_2 o}(\bar{v}^1)_o^{n_1}(\bar{v}^2)_o^{n_2}) . \qquad (3\text{-}44d)$$

With the normal coordinates of P with respect to Q, i.e.,

$$(\bar{v}^{\alpha})_Q = (u^{\alpha'})_{QP}\, \bar{u}^2 , \qquad (3\text{-}44e)$$

the coordinate differences Q - P can finally be calculated using the Legendrean series

$$(\Delta u^{\alpha})_{QP} = u^{\alpha} - (u^{\alpha})_Q = (\bar{v}^{\alpha})_Q + \sum_{n=2}^{n'} a^{\alpha}_{n_1 n_2 Q}(\bar{v}^1)_Q^{n_1}(\bar{v}^2)_Q^{n_2} , \qquad (3\text{-}44f)$$

$$n_1 + n_2 = n$$

according to (3-15d). Based on the intermediate results of (3-44c,f) the desired coordinates u^{α} of P become

$$u^{\alpha} = (u^{\alpha})_o + (\Delta u^{\alpha})_{oQ} + (\Delta u^{\alpha})_{QP} . \qquad (3\text{-}44g)$$

Now the method of <u>transforming u^{α} into $\bar{u}^{\alpha}$</u> [see (2-121)] will be given. Generally seen, this involves solving a second boundary value problem (2-92) for the geodesic $\bar{u}^1$-line. Yet, this transformation problem is generally solved in a slightly different way. First, an inverse geodetic problem (i.e., first boundary value problem) for the geodesic P_o - P will be solved with the help of the inverse Legendrean series (3-15e-g) or the mid-coordinate formulas (3-26) and (3-27). The results are

s_{oP} = length of the geodesic line P_o-P ,
β = direction angle of the geodesic line P_o - P at P_o relative to C_o . (3-45a)

The rectangular geodesic triangle P_o - Q - P is thus determined. Based on the angle sum equation (2-39) the angle at P becomes

$$\alpha = \pi/2 - \beta + \epsilon , \qquad (3\text{-}45b)$$

providing that the geodesic triangle is small enough. The law of sines (3-40a) and (3-41a) finally yields, with

$P_c =: Q$, $\alpha_c = \pi/2$ the following:

$$\bar{u}^1 = s_{oP}(\sin\alpha - (\epsilon/3)\cos\alpha) = s_{oP} \sin(\alpha - \epsilon/3) ,$$
$$\bar{u}^2 = s_{oP}(\sin\beta - (\epsilon/3)\cos\beta) = s_{oP} \sin(\beta - \epsilon/3) . \qquad (3\text{-}45c)$$

With greater $\bar{u}^1$-values it is advisable to introduce an auxiliary point P_1

on the abscissa line C_o near Q with the given coordinates

$$(\bar{u}^{\alpha})_1 = ((\bar{u}^1)_1, 0) . \tag{3-45d}$$

The accompanying coordinates $(u^{\alpha})_1$ can be calculated with $Q =: P_1$ after (3-44c), and then the way given by (3-45a-c) will be followed with P_1 instead of P_o. In the same manner (3-45c) furnishes the coordinates

$$\bar{u}^1 - (\bar{u}^1)_1 , \qquad \bar{u}^2 , \tag{3-45e}$$

and observing (3-45d) we obtain

$$\bar{u}^1 = (\bar{u}^1 - (\bar{u}^1)_1) + (\bar{u}^1)_1 . \tag{3-45f}$$

3.4.2 DIRECT REPRESENTATION BY POWER SERIES

The <u>transformation of $\bar{u}^{\alpha}$ into u^{α}</u> as described for (3-44) in Chapter 3.4.1 was done by twice applying Legendrean series with the starting points P_o and Q. Proceeding from this, indirect representation by power series the method of expanding direct power series for the desired transformation equations in the form of (3-5) will be explained in the following. For (3-44g) we will begin here with the setup

$$\Delta u^{\alpha} = u^{\alpha} - (u^{\alpha})_o = \sum_{n=1}^{n'} b^{\alpha}_{n_1 n_2} (\bar{u}^1)^{n_1} (\bar{u}^2)^{n_2} , \qquad n_1 + n_2 = n \tag{3-46a}$$

in which the coefficients are calculated in line with (3-5c), i.e.,

$$b^{\alpha}_{n_1 n_2} = (1/(n_1! n_2!))(\partial^n u^{\alpha}/((\partial \bar{u}^1)^{n_1} (\partial \bar{u}^2)^{n_2}))_{P_o} . \tag{3-46b}$$

Proceeding from (3-44g), i.e.,

$$u^{\alpha} = (u^{\alpha})_o + (\Delta u^{\alpha})_{oQ} + (\Delta u^{\alpha})_{QP} , \tag{3-46c}$$

with

$$(\Delta u^{\alpha})_{oQ} \text{ after (3-44c)}, \qquad (\Delta u^{\alpha})_{QP} \text{ after (3-44f)} \tag{3-46d}$$

the partial derivatives of u^{α} are to be formed at

$$P_o : \ (\bar{u}^{\alpha})_o = 0 . \tag{3-46e}$$

Hereby, it must be noted that (3-44c) is solely a function of $\bar{u}^1$, but (3-44f) is a function of both coordinates of $\bar{u}^{\alpha}$.

Proceeding from the coefficients of series (3-44c), those of series (3-44f) can be represented by a power series in $\bar{u}^1$ or the normal coordinates (3-44b) as follows:

$$a^{\alpha}_{n_1n_2Q} = a^{\alpha}_{n_1n_2o} + \sum_{m=1}^{n'-n} a^{\alpha}_{n_1n_2m_1m_2}(\bar{v}^1)_o^{m_1}(\bar{v}^2)_o^{m_2}\ , \quad m_1+m_2 = m\ , \tag{3-47a}$$

$$a^{\alpha}_{n_1n_2m_1m_2} = (1/(m_1!m_2!))(\partial^m a^{\alpha}_{n_1n_2}/((\partial u^1)^{m_1}(\partial u^2)^{m_2}))_o\ . \tag{3-47b}$$

In the same manner we obtain the following power series for the ϵ-tensor in (3-44d):

$$(\epsilon^{\alpha}_{\beta})_Q = (\epsilon^{\alpha}_{\beta})_o + \sum_{m=1}^{m'} e^{\alpha}_{\beta\ m_1m_2}(\bar{v}^1)_o^{m_1}(\bar{v}^2)_o^{m_2}\ , \quad m_1+m_2 = m\ , \tag{3-47c}$$

$$e^{\alpha}_{\beta\ m_1m_2} = (1/(m_1!m_2!))(\partial^m \epsilon^{\alpha}_{\beta}/((\partial u^1)^{m_1}(\partial u^2)^{m_2}))_o\ . \tag{3-47d}$$

When the coefficients of (3-47b,d) have been calculated, the necessary foundation for determining the coefficients of the desired power series (3-46a) is thus given. According to (3-46b-e) the coefficients can then be calculated whereby the representations of (3-47a,c) must be regarded for the derivatives of $(\Delta u^{\alpha})_{QP}$ with respect to $\bar{u}^1$ in (3-44d-f). A similar possibility of deriving series (3-46a) consists of first inserting (3-47) into (3-44d) and multiplying out the resulting product of the power series in $\bar{u}^1$. The resulting power series in $\bar{u}^1$ and (3-47a) are then inserted into (3-44e,f). We then obtain the desired result (3-46a) by multiplying out the power series products in (3-44f) and after the addition to series (3-44c) and sorting according to the powers of $\bar{u}^1$ and $\bar{u}^2$. Another variation is described for (3-50).

For <u>transforming u^{α} into $\bar{u}^{\alpha}$</u> we begin with the inverse power series of (3-46a), i.e.,

$$\bar{u}^{\alpha} = \sum_{n=1}^{n'} \bar{b}^{\alpha}_{n_1n_2}(\Delta u^1)^{n_1}(\Delta u^2)^{n_2}\ , \quad n_1+n_2 = n\ . \tag{3-48}$$

The coefficients can be obtained by inverting series (3-46a) according to (3-7). However, we can also use the method described for (3-45). Then the mid-coordinate formulas (3-26) to (3-27c) yield power series in Δu^{α} for the Riemannian normal coordinates (2-129) of P relative to P_o as follows:

$$(v^{\alpha})_o = (\sin\beta, \cos\beta)\, s_{oP}, \quad s_{oP}\ , \ \beta \text{ after (3-45a)} \ . \tag{3-49a}$$

These power series are inserted into the equations of (3-45c), which can be written together with (3-45b), (3-49a), and (3-39) for ϵ as follows:

$$\begin{aligned} \bar{u}^1 &= (v^2)_o + (K/3)(v^1)_o^2\,(v^2)_o + \dots\ , \\ \bar{u}^2 &= (v^1)_o + (K/6)(v^1)_o\,(v^2)_o^2 + \dots\ . \end{aligned} \tag{3-49b}$$

In many geodetic applications the accuracy of the calculation of rectangular geodesic triangles $P_o - Q - P$, which is based solely on the intrinsic coordinates $(v^{\alpha})_o$ and $\bar{u}^{\alpha}$ is sufficient [cf.Schödlbauer 1963, II.2]. We obtain the desired answer to (3-48) by inserting the power series mentioned for (3-49b) into (3-49b) and by sorting the resulting expressions according to the powers of Δu^{α}. Here, if need be, $n' > 3$ may be chosen in dependency on the properties of the coordinates u^{α}, in contrast to (3-49b). If, for example, the u^{α} are geographic coordinates (1-9) on an ellipsoid of revolution as a reference surface, then an expansion of (3-48) with $n' = 5$ must have an accuracy equivalent to that of (3-49b). In this case this is due to the fact that the geographic coordinates u^{α} are extrinsic coordinates, for which the components of the transformation matrices, i.e.,

$$u^{\alpha}_{,\bar{\beta}} = \partial u^{\alpha}/\partial \bar{u}^{\beta}\ , \tag{3-49c}$$

attain relatively great values as a result of the curvature of the ellipsoid of revolution. The inverse series for (3-48), i.e., (3-46a), must of course be expanded up to $n' = 5$ as well.

A method for deriving (3-46a) corresponding to (3-49) begins with the inverse series for (3-49b), i.e.,

$$\begin{aligned} (v^1)_o &= \bar{u}^2 - (K/6)(\bar{u}^1)^2(\bar{u}^2) + \dots\ , \\ (v^2)_o &= \bar{u}^1 - (K/3)(\bar{u}^1)\,(\bar{u}^2)^2 + \dots\ . \end{aligned} \tag{3-50a}$$

In the following, the simplified assumptions of

$$g_{12} = 0\ , \qquad C_o = u^2\text{-line} \tag{3-50b}$$

for (3-43b) and (3-44a) are made so that

$$(\bar{v}^{\alpha})_{oP} = (v^{\alpha})_o \tag{3-50c}$$

is true for the normal coordinates of P with respect to P_o. Thus, in place of (3-44c) the Legendrean series for P up to n' = 5 will be applied directly as follows:

$$\Delta u^{\alpha} = u^{\alpha} - (u^{\alpha})_o = (v^{\alpha})_o + \sum_{n=2}^{5} a^{\alpha}_{n_1 n_2 o} (v^1)_o^{n_1} (v^2)_o^{n_2} , \quad (3\text{-}50d)$$

$$n_1 + n_2 = n .$$

Inserting (3-50a) into (3-50d) and sorting according to the powers of $\bar{u}^{\alpha}$ yield the result of (3-46a) for the special assumptions of (3-50b) [see Schödlbauer 1963].

For <u>transforming the first fundamental tensors</u> according to the first equation of (3-2b) and proceeding from (3-46a) the transformation matrices are to be formed as follows:

$$u^{\alpha}_{,\bar{\beta}} = \sum_{n=1}^{n'} b^{\alpha}_{n_1 n_2} (\bar{u}^1)^{n_1} (\bar{u}^2)^{n_2} \left(n_1 (\bar{u}^1)^{-1}, \; n_2 (\bar{u}^2)^{-1}\right)_{\beta} . \quad (3\text{-}51a)$$

Terms with the power −1 in $\bar{u}^1$ or $\bar{u}^2$ vanish. According to (2-110) the first fundamental tensor or metric tensor $\bar{g}_{\alpha\beta}$ of the geodesic parallel coordinates has the form of

$$\bar{g}_{11} , \quad \bar{g}_{22} = 1 , \quad \bar{g}_{12} = 0 \quad (3\text{-}51b)$$

so that only $\bar{g}_{11}$ is to be calculated based on (3-2b). We generally obtain

$$\bar{g}_{11} = u^{\gamma}_{,\bar{1}} \, u^{\delta}_{,\bar{1}} \, g_{\gamma\delta} , \quad (3\text{-}51c)$$

and in the case of rectangular coordinates u^{α}

$$\bar{g}_{11} = (u^1_{,\bar{1}})^2 g_{11} + (u^2_{,\bar{1}})^2 g_{22} , \quad g_{12} = 0 . \quad (3\text{-}51d)$$

In (3-51d) g_{11} and g_{22} are to be formed at the test point, for which $\bar{g}_{11}$ should be calculated. A complete power series expansion of $\bar{g}_{11}$ consequently requires power series expansions in $\bar{u}^{\alpha}$ for g_{11} and g_{22} corresponding to (3-51a) as well.

3.5 TRANSFORMATIONS BETWEEN ISOTHERMAL SURFACE COORDINATES AND ARBITRARY SURFACE COORDINATES

3.5.1 GENERAL TRANSFORMATION EQUATIONS

Let

$$\bar{u}^{\alpha} = \text{isothermal surface coordinates} \tag{3-52a}$$

according to definition (2-132), which is founded solely on the first fundamental form, i.e.,

$$ds^2 = \bar{G}(\bar{u}^{\alpha})((d\bar{u}^1)^2 + (d\bar{u}^2)^2) \ . \tag{3-52b}$$

Accordingly, to construct isothermal surface coordinates complying with Chapter 3.1 the metric tensor $g_{\alpha\beta}$ must be given as a function of

$$u^{\alpha} = \text{arbitrary surface coordinates} \ , \tag{3-52c}$$

which is always presumed in the following. The isothermal coordinates (3-52a) must satisfy the differential equations of (2-134), which are generally written as

$$\bar{u}^1_{,\beta} = \epsilon^{\gamma}_{\beta}\,\bar{u}^2_{,\gamma} \qquad \text{or} \qquad u^{\beta}_{,\bar{1}} = \epsilon^{\beta}_{\gamma}\,u^{\gamma}_{,\bar{2}} \tag{3-53a}$$

and for rectangular coordinates become

$$g_{22}^{1/2}\,\bar{u}^1_{,1} = g_{11}^{1/2}\,\bar{u}^2_{,2} \ , \qquad g_{11}^{1/2}\,\bar{u}^1_{,2} = -g_{22}^{1/2}\,\bar{u}^2_{,1} \ ,$$

or (3-53b)

$$g_{11}^{1/2}\,u^1_{,\bar{1}} = g_{22}^{1/2}\,u^2_{,\bar{2}} \ , \qquad g_{11}^{1/2}\,u^1_{,\bar{2}} = -g_{22}^{1/2}\,u^2_{,\bar{1}} \ .$$

If the coordinates of (3-52c) are also isothermal, we ultimately obtain the Cauchy-Riemannian differential equations (2-141b,c).

Transformation problems (2-139) involve initial value problems of the differential equations of (3-53) with the initial conditions of (2-138b) for a surface curve C_o, i.e.,

$$u^{\alpha}_o = u^{\alpha}(t) \ , \qquad \bar{u}^{\alpha}_o = \bar{u}^{\alpha}(t) \ , \qquad t = \text{curve coordinate} \ . \tag{3-54a}$$

These functions should be differentiable with respect to t as often as desired so that together with (3-54a) the derivatives

$$u^{\alpha(m)}_o := (d^m u^{\alpha}_o/dt^m)_{P_o} \ , \qquad \bar{u}^{\alpha(m)}_o := (d^m \bar{u}^{\alpha}_o/dt^m)_{P_o} \tag{3-54b}$$

with $P_o \in C_o$ can be assumed to be given, too.

For solving the first differential equations of (3-53a) or (3-53b) with (3-54) it is expedient in the following to begin with power series in the form of (3-4b) with (3-3), i.e.,

$$\Delta \bar{u}^{\alpha} = \sum_{n=1}^{n'} \bar{b}^{\alpha}_{\beta_1\beta_2\ldots\beta_n} \Delta u^{\beta_1} \Delta u^{\beta_2} \ldots \Delta u^{\beta_n} , \tag{3-55a}$$

in which $P_o \in C_o$ is to be chosen. Every order n has 2^{n+1} coefficients (3-4c)

$$\bar{b}^{\alpha}_{\beta_1\beta_2\ldots\beta_n} = (1/n!)\, \bar{u}^{\alpha}_{,\beta_1\beta_2\ldots\beta_n} \tag{3-55b}$$

with

$$\bar{u}^{\alpha}_{,\beta_1\beta_2\ldots\beta_n} := (\partial^n \bar{u}^{\alpha}/(\partial u^{\beta_1} \partial u^{\beta_2} \ldots \partial u^{\beta_n}))_{P_o} , \tag{3-55c}$$

which are symmetrical in all indices. The result of this is that corresponding to (3-5a) only

$$N_n = 2(n+1) \tag{3-55d}$$

of the coefficients of rank n of (3-55b,c) are linearly independent. Conforming with (3-5b,c) a series like (3-55a) can be modified into the form of (3-4d), which is more common for practical calculations.

The coefficients of (3-55b) are to be determined in a way that (3-55a) satisfies the differential equations of (3-53a) or (3-53b) and meets the initial conditions (3-54). Moreover, the partial derivatives (3-55c) are essential to the calculation for which we first obtain the four conditional equations of

$$\bar{u}^{1}_{,\beta_1} = \epsilon^{\gamma}_{\beta_1} \bar{u}^{2}_{,\gamma} , \qquad \beta_1 = 1,2 , \tag{3-56a}$$

$$\bar{u}^{\alpha(1)} = \bar{u}^{\alpha}_{,\beta_1} u^{\beta_1(1)} , \qquad \alpha = 1,2 \tag{3-56b}$$

for $n = 1$ due to (3-53a) and (3-54). In these equations all functions are to be formed at P_o, which is also true for the following considerations. In (3-56a,b) there are four linear equations which generally uniquely determine the $N_1 = 4$ partial derivatives at P_o, i.e.,

$$\bar{u}^{\alpha}_{,\beta_1} . \tag{3-56c}$$

The derivatives of (3-56a) with respect to u^{β_2} and of (3-56b) with respect to t, i.e.,

$$\bar{u}^1_{,\beta_1\beta_2} = \epsilon^\gamma_{\beta_1}\bar{u}^2_{,\gamma\beta_2} + \epsilon^\gamma_{\beta_1,\beta_2}\bar{u}^2_{,\gamma} , \tag{3-57a}$$

$$\bar{u}^{\alpha(2)} = \bar{u}^\alpha_{,\beta_1\beta_2} u^{\beta_1(1)} u^{\beta_2(1)} + \bar{u}^\alpha_{,\beta_1} u^{\beta_1(2)} , \tag{3-57b}$$

yield in view of (3-56c) six linear equations which generally uniquely determine the $N_2 = 6$ independent partial derivatives of the second order at P_o, i.e.,

$$\bar{u}^\alpha_{,\beta_1\beta_2} = \bar{u}^\alpha_{,\beta_2\beta_1} . \tag{3-57c}$$

This procedure can be continued stepwise, whereby

$$\bar{u}^1_{,\beta_1\beta_2\ldots\beta_n} = \epsilon^\gamma_{\beta_1}\bar{u}^2_{,\gamma\beta_2\ldots\beta_n} + (\ldots) , \tag{3-58a}$$

$$\bar{u}^{\alpha(n)} = \bar{u}^\alpha_{,\beta_1\beta_2\ldots\beta_n} u^{\beta_1(1)} u^{\beta_2(1)} \ldots u^{\beta_n(1)} + (\ldots) \tag{3-58b}$$

are obtained for the partial derivative of order n-1 of (3-56a) with respect to u^β and of (3-56b) with respect to t at points $P_o \in C_o$. (...) means that even more summands follow with partial derivatives of orders smaller than n which have already been calculated in previous steps. In (3-58) the equations produced by exchanging the indices β_1 and β_n are identical so that only n independent equations remain. Since $\beta_1 = 1,2$, (3-58a) consequently yields 2n independent equations, which together with the two equations of (3-58b) represent a system of collectively N_n linear equations [see (3-55d)], through which the same number of partial derivatives (3-55c) or coefficients (3-55b) is generally uniquely determined.

The method for solving transformation problems (2-139) by means of power series (3-55a) as analytic transformation equations was described here in a general way. Proceeding from the second differential equations of (3-53a) or (3-53b) the <u>inverse transformation (2-140)</u> can be solved with series inversion (3-7).

A few special cases which are of practical importance will now be elucidated briefly. The choice of

$$C_o = u^1\text{-line} \quad \rightarrow \quad u^2 = \text{constant for } P \in C_o \tag{3-59a}$$

together with the assumption

$$u^1 := t = \text{curve coordinate of } C_o \tag{3-59b}$$

results in the following for points $P \in C_o$:

$$u^{1(n)} = \begin{cases} 1 & \text{for } n = 1 \\ 0 & \text{for } n > 1 \end{cases}, \tag{3-59c}$$

$$u^{2(n)} = 0 , \tag{3-59d}$$

$$\bar{u}^{\alpha(n)} = \bar{u}^{\alpha}_{,11\ldots1} . \tag{3-59e}$$

These results facilitate the evaluation of the equations of (3-58b). Particularly extreme simplifications of (3-58a,b) are obtained for transformations which are defined by the special assumptions of

$$C_o = u^1\text{-line} = \bar{u}^1\text{-line} , \tag{3-60a}$$

$$u^1 = \bar{u}^1 := t , \qquad u^2, \bar{u}^2 = \text{constant for } P \in C_o , \tag{3-60b}$$

and are frequently used for the primary construction of isothermal coordinates $\bar{u}^{\alpha}$. The results of (3-59c,d) remain unchanged, and instead of (3-59e), we obtain the extended statements for $P_o \in C_o$, i.e.,

$$\bar{u}^{1(n)} = \bar{u}^1_{,11\ldots1} = \begin{cases} 1 & \text{for } n = 1 \\ 0 & \text{for } n > 1 \end{cases}, \tag{3-60c}$$

$$\bar{u}^{2(n)} = \bar{u}^2_{,11\ldots1} = 0 . \tag{3-60d}$$

Finally, the differential equations of (3-58a) for rectangular coordinates u^{α} are simplified since then, in place of (3-53a), the equations of (3-53b) are valid. If u^{α} are also isothermal coordinates, the Cauchy-Riemannian differential equations (2-141b,c) are thus valid. This case will be treated in more detail in Chapter 3.5.2.

The <u>metric tensor of the $\bar{u}^{\alpha}$-coordinates</u> is calculated according to (3-2b). Observing (2-132b,c) we obtain

$$1/\bar{G} = \bar{u}^1_{,\gamma}\bar{u}^1_{,\delta}\, g^{\gamma\delta} = \bar{u}^2_{,\gamma}\bar{u}^2_{,\delta}\, g^{\gamma\delta} . \tag{3-61a}$$

For rectangular coordinates u^{α} we have

$$1/\bar{G} = (\bar{u}^1_{,1})^2/g_{11} + (\bar{u}^1_{,2})^2/g_{22} = (\bar{u}^2_{,1})^2/g_{11} + (\bar{u}^2_{,2})^2/g_{22} . \tag{3-61b}$$

3.5.2 TRANSFORMATIONS BETWEEN TWO ISOTHERMAL COORDINATE SYSTEMS

The following considerations begin with the special case of (3-52) in which both

$$\bar{u}^{\alpha} = \text{isothermal surface coordinates} \tag{3-62a}$$

and

$$u^{\alpha} = \text{isothermal surface coordinates} . \tag{3-62b}$$

In conformity with (2-132b,c) the accompanying metric tensors are determined by

$$\bar{G}(\bar{u}^{\alpha}) \qquad \text{and} \qquad G(u^{\alpha}) . \tag{3-62c}$$

The transformation equations

$$\bar{u}^{\alpha} = \bar{u}^{\alpha}(u^{\beta}) \qquad \text{and} \qquad u^{\alpha} = u^{\alpha}(\bar{u}^{\beta}) \tag{3-63a}$$

satisfy the Cauchy-Riemannian differential equations for (3-62a,b) ensuing from (3-53b), i.e.,

$$\bar{u}^1_{,1} = \bar{u}^2_{,2} , \quad \bar{u}^1_{,2} = -\bar{u}^2_{,1} \qquad \text{or} \qquad u^1_{,\bar{1}} = u^2_{,\bar{2}} , \quad u^1_{,\bar{2}} = -u^2_{,\bar{1}} . \tag{3-63b}$$

For solving the <u>transformation problems (2-139)</u> by means of power series we can proceed in the same manner as that described for the general case of (3-52) for (3-54) to (3-58). (3-58a) is greatly simplified, and (3-58b) remains unchanged, i.e.,

$$\bar{u}^1_{,1\beta_2\ldots\beta_n} = \bar{u}^2_{,2\beta_2\ldots\beta_n} , \quad \bar{u}^1_{,2\beta_2\ldots\beta_n} = -\bar{u}^2_{,1\beta_2\ldots\beta_n} , \tag{3-64a}$$

$$\bar{u}^{\alpha(n)} = \bar{u}^{\alpha}_{,\beta_1\beta_2\ldots\beta_n} u^{\beta_1(1)} u^{\beta_2(1)} \ldots u^{\beta_n(1)} + (\ldots) . \tag{3-64b}$$

The inverse <u>transformation problems (2-140)</u> can be solved in the same way in this case: we then begin with the latter equations of (3-63b), and only exchange the coordinates u^{α} and $\bar{u}^{\alpha}$ in (3-64).

In addition to the method described above for determining the transformation equations (3-63a) we also make use of the fact that due to (3-63b) these equations can be represented as complex functions of a complex variable corresponding to (2-155). Accordingly, for the first function of (3-63a)

$$\bar{u} = \bar{u}^1(u^{\alpha}) + i\,\bar{u}^2(u^{\alpha}) = f(u) , \qquad u = u^1 + i\,u^2 \tag{3-65}$$

is valid. Thereby, a power series in the form of (2-165) replaces the power series of (3-55) and is written in the following way:

$$\Delta\bar{u} = \sum_{n=1}^{n'} \bar{a}_n \, \Delta u^n \,, \tag{3-66a}$$

$$\Delta\bar{u} = \bar{u} - \bar{u}_o \,, \qquad \Delta u = u - u_o \,, \tag{3-66b}$$

$$\bar{a}_n = \bar{a}^1_n + i\,\bar{a}^2_n = \text{complex coefficients} \,. \tag{3-66c}$$

The number of coefficients of rank n is constant, i.e.,

$$\bar{N}_n = 2 \quad \text{for} \quad n \in \{0,1,2...\} \,. \tag{3-66d}$$

In (3-66b)

$$u_o, \, \bar{u}_o = \text{complex coordinates of a reference point } P_o \,, \tag{3-67a}$$

for which

$$P_o \in C_o \quad \text{after (2-138a,b)} \tag{3-67b}$$

is a presupposition, as before for solving transformation problems (2-139) and (2-140).

In contrast to (3-55a), (3-65) and (3-66a) do satisfy a priori the Cauchy-Riemannian differential equations (3-63b) and thus the differential equations (3-64a), which is the main practical advantage compared with the solution described for (3-64). In this way, only the two coefficients of (3-66c) remain for each series term, for the determination of which the two respective equations of (3-64b) are adequate. It would then of course be better to represent these equations in a form which is directly related to the complex coefficients and can be derived analogous to (3-56b) and following. Again, we have the conditions given by (3-54) so that together with (3-54b) the derivatives of the complex coordinates u and $\bar{u}$ with respect to t at $P_o \in C_o$ are known as well, i.e.,

$$\begin{aligned} u^{(m)} &:= (d^m u/dt^m)_o = u^{1(m)}_o + i\,u^{2(m)}_o \,, \\ \bar{u}^{(m)} &:= (d^m \bar{u}/dt^m)_o = \bar{u}^{1(m)}_o + i\,\bar{u}^{2(m)}_o \,. \end{aligned} \tag{3-68a}$$

Observing (3-66a) or (2-165c) we obtain the complex equations at $P_o \in C_o$ equivalent to (3-58b) and (3-64b), i.e.,

$$\bar{u}^{(n)} = n!\,\bar{a}_n\,(u^{(1)})^n + (\ldots) \,, \tag{3-68b}$$

in which $(\ldots)$ stands for the further summands with the coefficients $\bar{a}_m$ with $m < n$. These equations are to be solved stepwise beginning with $n = 1$.

All numerical calculations are generally to be done in real numbers. In addition, the equations of (3-68b) are to be split into real and imaginary parts, and the power series of (3-66a) are to be used according to (2-166). Directly solving for both (3-68a) and (3-66a), which is of considerable formal advantage, is only possible using a computer with complex arithmetic.

We will now treat the example of

$$C_o = u^1\text{-line} = \bar{u}^1\text{-line}\ , \qquad t =: u^1\ , \tag{3-69a}$$

for which the initial conditions of (3-54a) essentially have the form of

$$\Delta\bar{u} = \Delta\bar{u}^1 = f(\Delta u^1)\ , \qquad \text{for } P \in C_o\ , \tag{3-69b}$$

in which $f(\Delta u^1)$ is a given real function. This is then the desired transformation function as well, i.e.,

$$\Delta\bar{u} = f(\Delta u)\ . \tag{3-69c}$$

If (3-69b) has the form of a power series in Δu^1, i.e.,

$$\Delta\bar{u}^1 = \sum_{n=1}^{n'} \bar{a}^1_n\ (\Delta u^1)^n \tag{3-69d}$$

with real coefficients $\bar{a}^1_n$, the result of (3-69c) reads

$$\Delta\bar{u} = \sum_{n=1}^{n'} \bar{a}^1_n\ \Delta u^n\ , \qquad \Delta\bar{u},\ \Delta u \ \text{ after (3-66b)}\ . \tag{3-69e}$$

In contrast to the general solution of (3-66a), only the real parts of the coefficients of (3-66c) appear here, and the imaginary parts vanish, i.e.,

$$\bar{a}^2_n = 0\ . \tag{3-69f}$$

We naturally arrive at this conclusion with the generally applicable method described for (3-68), too. At points $P \in C_o$

$$u^{(n)} = \begin{cases} 1 & \text{for } n = 1 \\ 0 & \phantom{\text{for }} n > 1 \end{cases}\ , \qquad \bar{u}^{(n)} = n!\ \bar{a}^1_n\ , \tag{3-69g}$$

whereby the equations of (3-68b) become

$$\bar{a}_n = \bar{a}^1_n \qquad \rightarrow \qquad \bar{a}^2_n = 0\ . \tag{3-69h}$$

The <u>calculation of the metric tensor</u> in accordance with (3-61b) leads to

$$\bar{G} = G/((\bar{u}^1_{,1})^2 + (\bar{u}^1_{,2})^2)\ , \tag{3-70}$$

whereby the values of $\bar{u}^\alpha_{,\beta}$ can be substituted by means of (3-63b).

4. SURFACE COORDINATES ON ELLIPSOIDS OF REVOLUTION

4.1 PRELIMINARY REMARKS

The most important application of the general fundamentals of surface coordinates presented in Chapters 2 and 3 is the use of these coordinates on ellipsoids of revolution as reference surfaces corresponding to the explanations given for (1-2). Particular problems will be treated in the following subchapters as examples of the methods developed in Chapters 2 and 3. Above all, principles will be emphasized here that in part differ considerably from those which dominate in other relevant geodetic literature. Yet, the results obtained here will naturally agree with those obtained in other works.

4.2 ELLIPSOIDS OF REVOLUTION AND THEIR REPRESENTATION USING GEOGRAPHIC COORDINATES

Ellipsoids of revolution are generated by rotating plane ellipses around one of their axes. Thus, an ellipsoid of revolution is spanned using a one-parameter family of ellipses, the so-called meridian ellipses with

$$a \text{ and } b = \text{semimajor and semiminor axis} . \qquad (4\text{-}1a)$$

Only oblate ellipsoids of revolution whose meridian ellipses rotate around the semiminor axis b are suitable for use as geodetic reference surfaces (1-1b) and (1-2) [see Fig. 1.1]. Mathematically defined, the flattening of an ellipsoid of revolution is given by

$$\alpha = (a-b)/a , \qquad (4\text{-}1b)$$

whereby in place of (4-1a) both parameters

$$a, \alpha \qquad (4\text{-}1c)$$

can be used for determining the meridian ellipses and thus the ellipsoids of revolution. The ellipsoid parameters (4-1c) are given for different geodetic reference ellipsoids in (4-2).

Common ellipsoid parameters (4-2)

Designation	a in m	α
Bessel	6 377 397	1/299.15
Hayford	6 378 388	1/297.00
System 1967	6 378 160	1/298.25
System 1980	6 378 137	1/298.26

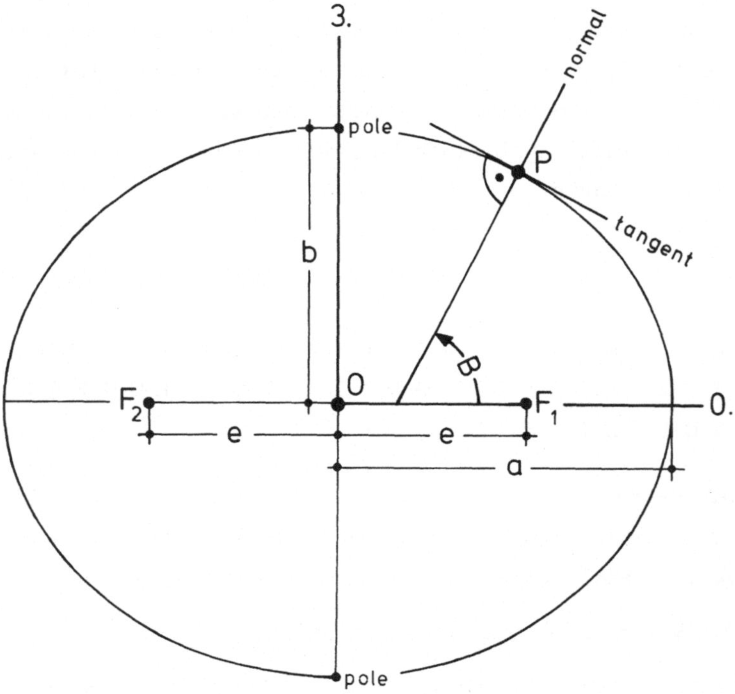

Fig. 4.1. Meridian ellipses

Generally speaking, ellipsoids of revolution are surfaces of the second degree whose center formula in Cartesian coordinates according to Fig. 1.1 is

$$(x_{1.}^2 + x_{2.}^2)/a^2 + x_{3.}^2/b^2 = 1 \ . \tag{4-3a}$$

Together with

$$x_{o.} := (x_{1.}^2 + x_{2.}^2)^{1/2} \tag{4-3b}$$

we then obtain the equation for meridian ellipses, i.e.,

$$x_{o.}^2/a^2 + x_{3.}^2/b^2 = 1 \ . \tag{4-3c}$$

Meridian ellipses are conic sections with the following eccentricities:

$$e = (a^2-b^2)^{1/2} = \text{linear eccentricity} > 0 \,, \tag{4-4a}$$

$$e' = e/a = \text{first eccentricity} \,, \tag{4-4b}$$

$$e'' = e/b = \text{second eccentricity} \,. \tag{4-4c}$$

The focal points F_1 and F_2 of the ellipses on the axis 0. are determined by e [see Fig. 4.1]. For points P of the ellipses

$$\overline{F_1-P} + \overline{F_2-P} = 2a \,. \tag{4-4d}$$

Besides the aforementioned <u>ellipsoid parameters</u> (4-1), i.e.,

$$a,\ b,\ \alpha \tag{4-5a}$$

the following parameters are also used:

$$c = a^2/b,\quad d = b^2/a,\quad e',\quad e'' \,. \tag{4-5b}$$

The following relationships exist between these parameters:

$$e'^2 = e''^2(1+e''^2)^{-1} = \alpha(2-\alpha) \,, \tag{4-5c}$$

$$e''^2 = e'^2(1-e'^2)^{-1} = \alpha(2-\alpha)/(1-\alpha(2-\alpha)) \,, \tag{4-5d}$$

$$\alpha = 1-(1-e'^2)^{1/2} = 1-(1+e''^2)^{-1/2} \,, \tag{4-5e}$$

$$b/a = (1-e'^2)^{1/2} = (1+e''^2)^{-1/2} = e'/e'' = 1-\alpha \,, \tag{4-5f}$$

$$\begin{aligned} c &= a/(1-e'^2)^{1/2} = a(1+e''^2)^{1/2} = a/(1-\alpha) \\ &= b/(1-e'^2) = b(1+e''^2) \\ &= d/(1-e'^2)^{3/2} = d(1+e''^2)^{3/2} \,. \end{aligned} \tag{4-5g}$$

Examples of the independent pairs of parameters are

$$a,\ \alpha\ ;\quad a,\ e';\quad c,\ e'' \,. \tag{4-5h}$$

Based on (4-3c) the <u>slope of the tangent</u> on the ellipse at point $P(x_{o.}, x_{3.})$ becomes

$$dx_{o.}/dx_{3.} = -\tan B = -(a^2/b^2)x_{3.}/x_{o.} \,. \tag{4-6a}$$

The angle of elevation B of the ellipsoid normal at P introduced here [see Fig. 4.1] is designated by

$$B = \text{ellipsoidal latitude} \,. \tag{4-6b}$$

This is also used as a curve coordinate for meridian ellipses. Observing (4-5) and introducing the latitude-dependent functions

$$V = (1+e''^2\cos^2 B)^{1/2} \,,\qquad W = (1-e'^2\sin^2 B)^{1/2} \tag{4-6c}$$

the solutions to (4-3c) and (4-6a) with respect to $x_{o.}$ and $x_{3.}$ are

$$x_{o.} = (c/V)\cos B = (a/W)\cos B \ , \tag{4-6d}$$
$$x_{3.} = (b/V)\sin B = (d/W)\sin B \ . \tag{4-6e}$$

The first V-dependent expressions of (4-6d,e) are preferred for our purposes. For reducing them back to the independent ellipsoid parameters c and e''

$$b/V = (c/V)/(1+e''^2)$$

is inserted in (4-6e), giving

$$x_{3.} = (c/(V(1+e''^2)))\sin B \ . \tag{4-6f}$$

If we introduce

<u>Geographic coordinates</u> (4-7)

$$u^1 =: L = \text{ellipsoidal longitude},$$
$$u^2 =: B = \text{ellipsoidal latitude}$$

as surface coordinates on the ellipsoid of revolution corresponding to (1-9) and Figs. 1.1 and 4.1, then

$$x_{1.} = x_{o.}\cos L \ , \qquad x_{2.} = x_{o.}\sin L \ .$$

If (4-6d) is substituted for $x_{o.}$, (4-6d,e) yields the following <u>transformation equations for L and B into $x_{i.}$</u> :

$$x_{i.} = ((c/V)\cos B \cos L,\ (c/V)\cos B \sin L,\ (b/V)\sin B) \ . \tag{4-8}$$

This is the primary way of representing the position vectors of an ellipsoid of revolution in the sense of (2-1) and (3-1c), most used in geodetic applications. c/V and b/V are also replaced by other expressions, such as those in (4-6d-f) for instance.

The names for the <u>coordinate lines</u> of the geographic coordinates of (4-7) are

$$\begin{aligned} &\text{L-lines} = \text{latitude or parallel circles} \ , \\ &\text{B-lines} = \text{meridians} \ . \end{aligned} \tag{4-9}$$

The latitude circle with $B = 0$ is the <u>equator</u> of an ellipsoid of revolution, and the two points of intersection of the meridians with $B = \pm\pi/2$ are the <u>poles</u>. Due to $g_{12} = 0$ it follows from (4-12) that the coordinate lines of (4-9) form <u>orthogonal families of curves</u>.

The geographic coordinates of (4-7) allow an especially simple representation of the normal vectors of (2-4) and (2-25) on the ellipsoid of revolution, i.e.,

$$n_{i.} = (\cos B \cos L,\ \cos B \sin L,\ \sin B)\ , \tag{4-10a}$$

and the partial derivatives with respect to L and B are

$$n_{i,1} = (\ -\sin L,\ \cos L,\ 0\ )\cos B\ , \tag{4-10b}$$

$$n_{i,2} = (-\sin B \cos L,\ -\sin B \sin L,\ \cos B)\ . \tag{4-10c}$$

$n_{i.}$, $n_{i,1}$, and $n_{i,2}$ are pairwise perpendicular to one another, i.e.,

$$n_{i.}n_{i,\alpha} = 0_\alpha\ , \qquad n_{i,1}n_{i,2} = 0\ , \tag{4-10d}$$

and their absolute values are

$$|n_{i.}| = 1\ , \quad |n_{i,1}| = \cos B\ , \quad |n_{i,2}| = 1\ . \tag{4-10e}$$

With these normal vectors the covariant and contravariant basis vectors of (2-32a,b) can be represented in the following way:

$$b_{i.\alpha} = n_{i,\alpha}R_{(\alpha)}\ , \tag{4-11a}$$

$$b_{i.}^{\alpha} = n_{i,(\alpha)}((R_1\cos^2B)^{-1},\ R_2^{-1})^{\alpha} \tag{4-11b}$$

with the principal radii of curvature

$$R_1 = c/V\ , \qquad R_2 = c/V^3 \tag{4-11c}$$

in accordance with (4-18d).

Using (4-10) and (4-11) the components of both fundamental tensors can be calculated according to (2-32d) and (2-47b). The components for the first fundamental tensor or metric tensor are

$$\begin{aligned} g_{11} &= 1/g^{11} = R_1^2\cos^2B\ , \quad & g_{22} &= 1/g^{22} = R_2^2\ , \\ g_{12} &= g^{12} = 0\ , & g &= |g_{\alpha\beta}| = (R_1R_2\cos B)^2\ , \end{aligned} \tag{4-12}$$

and those for the second fundamental tensor are

$$\begin{aligned} L_{11} &= -R_1\cos^2B\ , \quad L_{12} = L_{21} = 0\ , \quad L_{22} = -R_2\ , \\ L_1^1 &= -R_1^{-1}\ , \quad L_1^2 = L_2^1 = 0\ , \quad L_2^2 = -R_2^{-1}\ . \end{aligned} \tag{4-13}$$

The first partial derivatives of the metric tensor of (4-12) with respect to L and B are

$$g_{11,1} = 0 \,, \qquad g_{11,2} = -(R_1/V)^2 \sin 2B \,, \qquad\qquad (4\text{-}14)$$
$$g_{22,1} = 0 \,, \qquad g_{22,2} = 3\, R_2^2 (e''/V)^2 \sin 2B \,.$$

For the Christoffel's symbols of the first and second kind we obtain the the following in accordance with (2-48b,d) and (2-94), respectively:

$$\Gamma_{11|1} = \Gamma_{12|2} = \Gamma_{22|1} = 0 \,,$$
$$\Gamma_{11|2} = -\Gamma_{12|1} = (1/2)(R_1/V)^2 \sin 2B \,, \qquad\qquad (4\text{-}15a)$$
$$\Gamma_{22|2} = (3/2) R_2^2 (e''/V)^2 \sin 2B \,;$$

$$\Gamma^1_{11} = \Gamma^2_{12} = \Gamma^1_{22} = 0 \,, \qquad \Gamma^2_{11} = (V^2/2)\sin 2B \,, \qquad\qquad (4\text{-}15b)$$
$$\Gamma^1_{12} = -(1/V^2)\tan B \,, \qquad \Gamma^2_{22} = (3/2)(e''/V)^2 \sin 2B \,.$$

For calculating the geodesic curvatures of the coordinate lines of (4-9) we refer back to (2-68). These surface curves can be represented by geographic coordinates using the equations

$$\text{L-lines:} \quad u^\alpha(L) = (L\,, B_o)^\alpha \,, \qquad\qquad (4\text{-}16a)$$

$$\text{B-lines:} \quad u^\alpha(B) = (L_o, B\,)^\alpha \,, \qquad\qquad (4\text{-}16b)$$

whereby L_o and B_o are constant values. Consequently,

$$\text{L-lines:} \quad \dot{u}^\alpha = \delta^\alpha_1 \,, \qquad \ddot{u}^\alpha = 0 \,, \qquad\qquad (4\text{-}16c)$$

$$\text{B-lines:} \quad \dot{u}^\alpha = \delta^\alpha_2 \,, \qquad \ddot{u}^\alpha = 0 \,. \qquad\qquad (4\text{-}16d)$$

Based on (2-68) together with (4-12)-(4-15) we thus obtain

$$\text{L-lines:} \quad \kappa_{g1} = (1/R_1)\tan B \,, \qquad\qquad (4\text{-}16e)$$

$$\text{B-lines:} \quad \kappa_{g2} = 0 \,. \qquad\qquad (4\text{-}16f)$$

The L-lines or latitude circles, except the equator, all have a geodesic curvature not equal to zero. The B-lines or meridians, on the contrary, are always geodesics so that the geographic coordinates represent geodesic surface coordinates (2-110). Due to the poles at the axis of revolution of the ellipsoid they can be interpreted as both polar coordinates (2-112) and parallel coordinates (2-116).

As a result of the choice of coordinates made in (4-7) $n_{i.}$ is the outer ellipsoid normal according to (4-10). In line with the definitions given in Chapters 2.1.6 and 2.1.7 this leads to negative normal curvatures. To avoid this

$$\kappa_n := -\kappa_n \quad \text{after (2-63), (2-72)} \tag{4-17a}$$

is employed so that the quadratic condition equation for the principal curvatures of (2-77b) together with (4-12) and (4-13), whereby R_1 and R_2 are first replaced by the expressions of (4-11c), becomes

$$\kappa_n^2 - \kappa_n V(1+V^2)/c + (V^2/c)^2 = 0 \ . \tag{4-17b}$$

The solutions of (4-17b) are the principal curvatures, i.e.,

$$\kappa_{n1} = 1/R_1 = V/c \ , \qquad \kappa_{n2} = 1/R_2 = V^3/c \ . \tag{4-18a}$$

Based on (2-77a) we herewith obtain the accompanying principal directions of (2-74a), i.e.,

$$\nu_1 = (dB/dL)_1 = 0 \ , \qquad \nu_2 = (dB/dL)_2 = \infty \ , \tag{4-18b}$$

which thus coincide with the directions of the latitude circles and meridians. Accordingly,

$$\text{coordinate lines (4-9)} \ = \ \text{curvature lines (2-80)} \ , \tag{4-18c}$$

and κ_{n1} and κ_{n2} are the normal curvatures of the latitude circles and meridians, respectively. In this way we also obtain the designations for the principal curvature radii, i.e.,

$$\begin{aligned} R_1 &= c/V \ \ = \text{radius of curvature of the prime vertical} \ , \\ R_2 &= c/V^3 = \text{radius of curvature of the meridian} \ , \end{aligned} \tag{4-18d}$$

which were already used in (4-11)ff.

Based on (2-78) or (2-81a) we obtain

$$K = (1/R_1)(1/R_2) \quad = (V^2/c)^2 \ , \tag{4-19a}$$

$$H = (1/R_1 + 1/R_2)/2 = (V/2)(1+V^2)/c \tag{4-19b}$$

for Gaussian and mean curvatures. The direction angle θ introduced in (2-82) is designated here by

$$A = \text{ellipsoidal azimuth} \ . \tag{4-20}$$

Thereby, the theorem of Euler of (2-83) reads

$$1/R = (\sin^2 A)/R_1 + (\cos^2 A)/R_2 = (\sin^2 A + V^2\cos^2 A)V/c \ . \tag{4-21}$$

R is the normal radius of curvature of the ellipsoid of revolution for the normal section with the azimuth A relative to the meridian.

4.3 TRANSFORMATIONS BETWEEN GEODESIC POLAR COORDINATES AND GEOGRAPHIC COORDINATES

4.3.1 TRANSFORMING THE COORDINATES

In the system of geographic coordinates of (4-7), i.e.,

$$u^\alpha = (L,B)^\alpha \ , \tag{4-22}$$

geodesic polar coordinates are defined as follows corresponding to (2-112) and (2-113):

$$(u^\alpha)_o = (L_o, B_o)^\alpha = \text{geographic coordinates of the pole } P_o, \tag{4-23a}$$

$$\nu^\alpha_o = \nu^\alpha_{o2} = (1/R_2)_o \delta^\alpha_2 = \text{northern meridian direction at } P_o, \tag{4-23b}$$

$$\bar{u}^\alpha = (A,r)^\alpha = \text{geodesic polar coordinates} \tag{4-23c}$$

with

A = ellipsoidal azimuth of the geodesic line $P_o - P$ relative to (4-23b) , (4-23d)

r = length of the geodesic line $P_o - P$ (4-23e)

[see Fig. 4.2 with $A_o =: A$]. Combined with (4-12), (2-113g) yields the tangent vector of the geodesic $P_o - P$ in P_o, i.e.,

$$(u^{\alpha'})_o = ((R_1 \cos B)^{-1}_o \sin A, \ R^{-1}_{2o} \cos A)^\alpha , \tag{4-24a}$$

so that

$$\bar{v}^\alpha = ((R_1 \cos B)^{-1}_o \sin A, \ R^{-1}_{2o} \cos A)^\alpha r = (\cos^{-1} B_o \sin A, \ V^2_o \cos A)^\alpha r / R_{1o} \tag{4-24b}$$

are the normal coordinates of (3-15a) of the geodesic $P_o - P$ with reference to P_o.

(4-24) and

$$\Delta L = L - L_o \ , \qquad \Delta B = B - B_o \tag{4-25a}$$

yield the solution to the transformation problem of (2-114) in the form of the Legendrean series of (3-15d), i.e.,

$$\Delta u^\alpha = (\Delta L, \Delta B)^\alpha = \bar{v}^\alpha + \sum_{n=2}^{n'} b^\alpha_{n_1 n_2} (\bar{v}^1)^{n_1} (\bar{v}^2)^{n_2} \ , \quad n_1 + n_2 = n \ . \tag{4-25b}$$

The $b^\alpha_{n_1 n_2}$ coefficients can be determined in the manner given for (3-13) and (3-14) while observing (3-7d). They are a function of the Christoffel's symbols of the second kind, which are all given here by (4-15b) as functions of B. The abbreviations

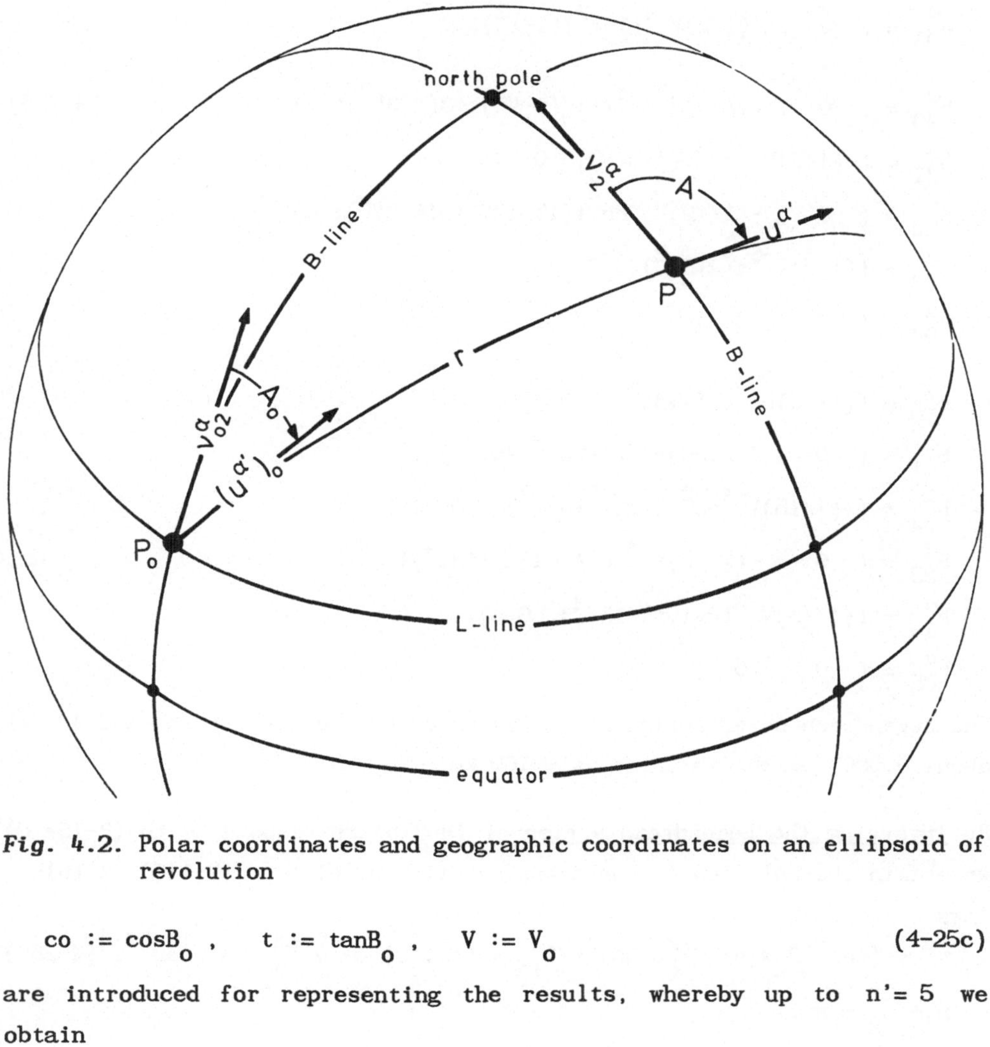

Fig. 4.2. Polar coordinates and geographic coordinates on an ellipsoid of revolution

$$co := \cos B_o \ , \quad t := \tan B_o \ , \quad V := V_o \tag{4-25c}$$

are introduced for representing the results, whereby up to n'= 5 we obtain

$$b^\alpha_{2o} = (\ 0 \ , -(1/2)V^2 co^2 t) \ ,$$
$$b^\alpha_{11} = (V^{-2} t, \ 0 \ ) \ ,$$
$$b^\alpha_{02} = (\ 0 \ , (3/2)V^{-2} t(1-V^2)) \ ,$$

$$b^\alpha_{30} = (-(1/3)co^2 t^2, \ 0 \ ) \ , \tag{4-25d}$$

$$b^{\alpha}_{21} = (\ 0\ ,\ -(1/6)co^2(V^2+3(4-3V^2)t^2))\ ,$$
$$b^{\alpha}_{12} = ((1/3)V^{-4}(V^2+3t^2),\ 0\ )\ ,$$
$$b^{\alpha}_{03} = (\ 0\ ,\ (1/2)V^{-4}(1-V^2)(1-t^2))\ ,$$

$$b^{\alpha}_{40} = (\ 0\ ,\ (1/24)V^2co^4t(1-(1-V^2)t+3(4-3V^2)t^2))\ , \qquad (4\text{-}25d)$$
$$b^{\alpha}_{31} = (-(1/3)V^{-2}co^2t(1+3t^2),\ 0\ )\ ,$$
$$b^{\alpha}_{22} = (\ 0\ ,\ -(1/12)V^{-2}co^2t(17-13V^2+3(5-3V^2)t^2))\ ,$$
$$b^{\alpha}_{13} = ((1/3)V^{-6}t(2+3t^2),\ 0\ )\ ,$$
$$b^{\alpha}_{04} = (\ 0\ ,\ 0\ )\ ,$$

$$b^{\alpha}_{50} = ((1/15)co^4t^2(1+3t^2),\ 0\ )\ ,$$
$$b^{\alpha}_{41} = (\ 0\ ,\ (1/120)co^4(1+30t^2+45t^4))\ ,$$
$$b^{\alpha}_{32} = (-(1/15)V^{-4}co^2(1+20t^2+30t^4),\ 0\ )\ ,$$
$$b^{\alpha}_{23} = (\ 0\ ,\ -(1/30)V^{-4}co^2(2+15t^2+15t^4))\ ,$$
$$b^{\alpha}_{14} = ((1/15)V^{-8}(2+15t^2+15t^4)\ \ 0\ )\ ,$$
$$b^{\alpha}_{05} = (\ 0\ ,\ 0\ )\ .$$

The Legendrean power series of (4-25) have an accuracy for ΔL and ΔB of about 0.0003" with distances $s = 100$ km.

By <u>inverting the Legendrean series</u> of (4-25b) corresponding to (3-15e-g) we obtain the solution to the transformation problem of (2-115) as follows

$$\bar{v}^{\alpha} = (\cos^{-1}B_o\sin A,\ V_o^2\cos A)^{\alpha} r/R_{1o} = (\Delta L,\Delta B)^{\alpha} + \sum_{n=2}^{n'} \bar{b}^{\alpha}_{n_1n_2}\Delta L^{n_1}\Delta B^{n_2}, \qquad (4\text{-}26a)$$
$$n_1 + n_2 = n\ .$$

According to (3-7) the coefficients are calculated with (4-25c,d). Up to $n' = 4$ they become

$$\bar{b}^{\alpha}_{20} = (\ 0\ ,\ (1/2)V^2co^2t)\ ,$$
$$\bar{b}^{\alpha}_{11} = (-t(3-3V^2+V^4),\ 0\ )\ ,$$
$$\bar{b}^{\alpha}_{02} = (\ 0\ ,\ -(3/2)V^2t(3-5V^2+2V^4))\ ,$$

$$\bar{b}^{\alpha}_{30} = (-(1/6)co^2t^2,\ 0\ )\ , \qquad (4\text{-}26b)$$

$$\bar{b}^{\alpha}_{21} = (\ 0\ ,\ (1/6)V^2co^2(1-3(3-3V^2+V^4)t^2))\ ,$$
$$\bar{b}^{\alpha}_{12} = (-(1/6)(6-6V^2+2V^4-9(3-5V^2+2V^4)t^2),\ 0\ )\ , \qquad (4\text{-}26b)$$
$$\bar{b}^{\alpha}_{03} = (\ 0\ ,\ -(1/2)V^2(3-5V^2+2V^4-(8-15V^2+7V^4)t^2))\ ,$$

$$\bar{b}^{\alpha}_{40} = (\ 0\ ,\ (1/24)V^2co^4t(V^2-t^2))\ ,$$
$$\bar{b}^{\alpha}_{31} = (-(1/6)co^2t(1-(2-V^2)t^2),\ 0\ )\ ,$$
$$\bar{b}^{\alpha}_{22} = (\ 0\ ,\ -(1/12)V^2co^2t(9-5V^2-9(1-V^2)t^2))\ ,$$
$$\bar{b}^{\alpha}_{13} = ((1/6)t(1-V^2)(7-3t^2),\ 0\ )\ ,$$
$$\bar{b}^{\alpha}_{04} = (\ 0\ ,\ (1/2)V^2t(1-V^2))\ .$$

Based on the normal coordinates $\bar{v}^{\alpha}$ calculated using (4-26a,b) the polar coordinates of (4-23c) with the abbreviations of (4-25c) are obtained as follows:

$$A = \arctan(V^2co\ \bar{v}^1/\bar{v}^2)\ , \qquad r = (c/V)((\bar{v}^1co)^2 + (\bar{v}^2/V^2)^2)^{1/2}\ . \qquad (4\text{-}26c)$$

For solving the transformation problem of (2-115) the essentially better converging Gaussian mid-coordinate formulas of (3-26) are usually preferred over inverted Legendrean series (4-26). Here, their coefficients are solely functions of the mean latitude, which is why they are named <u>mid-latitude formulas</u>. In the following, their derivation is explained with reference to (3-19)-(3-27) and restricted to $n' = 4$. The coordinate differences for the points designated according to (3-19), i.e.,

$$P_a := P_o\ , \qquad P_b := P\ , \qquad (4\text{-}27a)$$

in the u^{α}-system are

$$\Delta u^{\alpha} = (\Delta L,\ \Delta B)^{\alpha}\ , \qquad \Delta L = L_b - L_a\ , \qquad \Delta B = B_b - B_a\ , \qquad (4\text{-}27b)$$

and for the mid-coordinates of (3-21a) we obtain

$$(u^{\alpha})_m = (L_m,\ B_m)^{\alpha}, \qquad L_m = (L_a + L_b)/2\ , \quad B_m = (B_a + B_b)/2\ . \qquad (4\text{-}27c)$$

In accordance with (4-24) the normal coordinates of (3-22a) become

$$(\bar{v}^{\alpha})_p = (\cos^{-1}B_p\sin A_p,\ V_p^2\cos A_p)^{\alpha}r/R_{1p}\ , \qquad p \in \{a,b\}\ . \qquad (4\text{-}27d)$$

Due to $n' = 4$ in (3-23) and (3-24), (3-25) is limited to $q' = 2$. Therefore and as a result of the sole B dependency of the $\bar{b}_{n_1n_2}$ coefficients, (3-25) is employed here in the form of

$$\left.\begin{matrix}\bar{b}^{\alpha}_{n_1n_2a}\\ \bar{b}^{\alpha}_{n_1n_2b}\end{matrix}\right\} = \bar{b}^{\alpha}_{n_1n_2m}\left\{\begin{matrix}-\\+\end{matrix}\right\} c^{\alpha}_{n_1n_201}\Delta B/2 + c^{\alpha}_{n_1n_202}(\Delta B/2)^2 \tag{4-27e}$$

with

$$c^{\alpha}_{n_1n_20q} = (1/q!)(d^q\bar{b}^{\alpha}_{n_1n_2}/dB^q)_m\ , \quad n_1+n_2 = 2\ , \quad q \in \{1,2\}\ . \tag{4-27f}$$

The coefficients of (4-27f) are calculated proceeding from (4-26b), whereby beginning with the fourth power e" is neglected in those coefficients which are only relevant for the fourth order in $\Delta L\ \Delta B$ so that

$$V^4 \approx -1 + 2V^2\ , \qquad V^6 \approx -2 + 3V^2\ . \tag{4-27g}$$

With the abbreviations analogous to those in (4-25c), i.e.,

$$co_m := \cos B_m\ , \quad t_m := \tan B_m\ , \quad V^2_m := 1 + e''^2\cos^2 B_m\ , \tag{4-27h}$$

we obtain the following results for the relevant coefficients of (4-27f):

$$\begin{aligned}
c^{\alpha}_{2001} &= (\ 0\ ,\ (1/2)co^2_m(V^2_m+(2-3V^2_m)t^2_m))\ ,\\
c^{\alpha}_{1101} &= (-3+3V^2_m-V^4_m+(3-7V^2_m+3V^4_m)t^2_m,\ 0\ )\ ,\\
c^{\alpha}_{0201} &= (\ 0\ ,\ -(3/2)(3V^2_m-5V^4_m+2V^6_m+(6-23V^2_m+27V^4_m-10V^6_m)t^2_m)\\
\\
c^{\alpha}_{2002} &= (\ 0\ ,\ (1/2)co^2_m t_m(3+5V^2_m-3(1-V^2_m)t^2_m)\ ,\\
c^{\alpha}_{1102} &= (t_m(1-2V^2_m-t^2_m),\ 0\ )\ ,\\
c^{\alpha}_{0202} &= (\ 0\ ,\ 3\ t_m(1-V^2_m))\ ,\\
\\
c^{\alpha}_{3001} &= (-(1/3)co^2_m\ t_m,\ 0\ )\ ,\\
c^{\alpha}_{2101} &= (\ 0\ ,\ -(2/3)co^2_m t_m(1+V^2_m)\ ,\\
c^{\alpha}_{1201} &= ((11/3)t_m(1-V^2_m),\ 0\ )\ ,\\
c^{\alpha}_{0301} &= (\ 0\ ,\ 2t_m(1-V^2_m))\ .
\end{aligned} \tag{4-27i}$$

Together with (4-27e) the equations of (3-24) can be written as

$$\begin{aligned}(\bar{v}^{\alpha})_m = (\Delta L,\ \Delta B)^{\alpha} &- (1/2)\sum_{n_1+n_2=2} c^{\alpha}_{n_1n_201}\Delta L^{n_1}\Delta B^{n_2+1}\\
&+ \sum_{n_1+n_2=3} \bar{b}^{\alpha}_{n_1n_2m}\ \Delta L^{n_1}\Delta B^{n_2}\ ,\end{aligned}$$

$$\delta\bar{v}^{\alpha}_{ab} = -\sum_{n_1+n_2=2} \bar{b}^{\alpha}_{n_1n_2m}\, \Delta L^{n_1} \Delta B^{n_2} - (1/4) \sum_{n_1+n_2=2} \bar{c}^{\alpha}_{n_1n_202} \Delta L^{n_1} \Delta B^{n_2+2}$$

$$+ (1/2) \sum_{n_1+n_2=3} c^{\alpha}_{n_1n_201} \Delta L^{n_1} \Delta B^{n_2+1} - \sum_{n_1+n_2=4} \bar{b}^{\alpha}_{n_1n_2m}\, \Delta L^{n_1} \Delta B^{n_2} .$$

With these equations as well as (4-26b) and (4-27i) we obtain the <u>mid-latitude formulas</u> in accordance with (3-26a,b), i.e.,

$$(\bar{v}^{\alpha})_m = ((\bar{v}^{\alpha})_a + (\bar{v}^{\alpha})_b)/2 \tag{4-28a}$$

$$= (\Delta L(1+A^1_{30m}\Delta L^2+A^1_{12m}\Delta B^2),\ \Delta B(1+A^2_{21m}\Delta L^2+A^2_{03m}\Delta B^2))^{\alpha},$$

$$\delta\bar{v}^{\alpha}_{ab} = ((\bar{v}^{\alpha})_b - (\bar{v}^{\alpha})_a)/2 = (\Delta L\ \Delta B(B^1_{11m}+B^1_{31m}\Delta L^2+B^1_{13m}\Delta B^2)\ , \tag{4-28b}$$

$$\Delta L^2(B^2_{20m}+B^2_{40m}\Delta L^2+B^2_{22m}\Delta B^2) + \Delta B^2(B^2_{02m}+B^2_{04m}\Delta B^2))^{\alpha} .$$

The nonvanishing coefficients of $A^{\alpha}_{n_1n_2}$ and $B^{\alpha}_{n_1n_2}$ are

$$A^1_{30m} = \bar{b}^1_{30m} = -(1/6)co^2_m\, t^2_m ,$$

$$A^1_{12m} = \bar{b}^1_{12m} - (1/2)c^1_{1101} = (1/6)(3-3V^2_m+V^4_m+3(6-8V^2_m+3V^4_m)t^2_m) ,$$

$$A^2_{21m} = \bar{b}^2_{21m} - (1/2)c^2_{2001} = -(1/12)co^2_m(V^2_m+3(2+3V^2_m-6V^4_m+2V^6_m)t^2_m) , \tag{4-28c}$$

$$A^2_{03m} = \bar{b}^2_{03m} - (1/2)c^2_{0201} = (1/4)(V^2_m(3-5V^2_m+2V^4_m)+(18-53V^2_m+51V^4_m-16V^6_m)t^2_m).$$

$$B^1_{11m} = -\bar{b}^1_{11m} = t_m(3-3V^2_m+V^4_m) ,$$

$$B^1_{31m} = -\bar{b}^1_{31m} + (1/2)c^1_{3001} = -(1/6)co^2_m t^3_m(2-V^2_m) ,$$

$$B^1_{13m} = -\bar{b}^1_{13m} - (1/4)c^1_{1102} + (1/2)c^1_{1201} = (1/12)t_m(5-2V^2_m+3(3-2V^2_m)t^2_m) ,$$

$$B^2_{20m} = -\bar{b}^2_{20m} = -(1/2)V^2_m co^2_m t_m ,$$

$$B^2_{02m} = -\bar{b}^2_{02m} = (3/2)V^2_m t_m(3-5V^2_m+2V^4_m) , \tag{4-28d}$$

$$B^2_{40m} = -\bar{b}^2_{40m} = (1/24)co^4_m t_m(1-2V^2_m+V^2_m t^2_m) ,$$

$$B^2_{22m} = -\bar{b}^2_{22m} - (1/4)c^2_{2002} + (1/2)c^2_{2101} = -(1/24)co^2_m t_m(7-5V^2_m+15(1-V^2_m)t^2_m),$$

$$B^2_{04m} = -\bar{b}^2_{04m} - (1/4)c^2_{0202} + (1/2)c^2_{0301} = -(1/4)t_m(1-V^2_m) .$$

If $(\bar{v}^{\alpha})_m$ and $\delta\bar{v}^{\alpha}_{ab}$ are calculated according to (4-28) proceeding from the geographic coordinates of (4-22) of points P_a and P_b or their differences and mean values of (4-27b,c), we subsequently continue in accordance with (3-27c,d,f). First, the normal coordinates at P_a and P_b are determined, i.e.,

$$(\bar{v}^{\alpha})_a = (\bar{v}^{\alpha})_m - \delta\bar{v}^{\alpha}_{ab} \ , \qquad (\bar{v}^{\alpha})_b = (\bar{v}^{\alpha})_m + \delta\bar{v}^{\alpha}_{ab} \ , \tag{4-29a}$$

whereby together with the metric tensor of (4-12) the polar coordinates of (4-23c) are obtained according to (3-27f,d) in the following way:

$$\begin{aligned} A_p &= \arctan(V_p^2 \cos B_p (\bar{v}^1)_p/(\bar{v}^2)_p) \ , \\ r &= (c/V_p)(((\bar{v}^1)_p \cos B_p)^2 + ((\bar{v}^2)_p/V_p^2)^2)^{1/2} \ , \quad p \in \{a,b\} \ . \end{aligned} \tag{4-29b}$$

A_a and A_b are the ellipsoidal azimuths of the geodesic line $P_a - P_b$ at P_a and P_b, respectively.

As previously mentioned, the mid-latitude formulas converge much better than the inverted Legendrean series of (4-26). With the equations of (4-28), which are only valid up to the fourth order, the error in r attains a maximum value of 1 cm up to $r \approx 300$ km with average latitudes of 50°; up to $r \approx 100$ km the terms of the fourth order in (4-28b) can be neglected. If the polar coordinates for several points P_b having a common pole P_a are calculated according to the mid-latitude formulas, the coefficients of (4-28d) for every $P_a - P_b$ geodesic must then be calculated anew. Herein lies the disadvantage as compared with (4-26).

4.3.2 TRANSFORMING THE METRIC TENSOR

The metric tensor of polar coordinates is generally calculated in the way given for (3-28) to (3-31). With the abbreviations introduced to (4-25c) and based on (4-12) and (4-24b) we obtain

$$(g_{22}/g_{11})_o^{1/2} = 1/(V^2 co) \ , \quad (g_{11}/g_{22})^{1/2} = V^2 co \ , \tag{4-30a}$$

$$\bar{v}^{\alpha} = ((V/co)\sin A, \ V^3 \cos A) r/c \ , \tag{4-30b}$$

and furthermore

$$r^2 = (g_{11})_o(\bar{v}^1)^2 + (g_{22})_o(\bar{v}^2)^2 = (c/V)^2 co^2(\bar{v}^1)^2 + (c/V^3)^2(\bar{v}^2)^2 . \qquad (4\text{-}30c)$$

These special results are to be regarded in the power series expansion of (3-30a) for $u^{\alpha}_{,\bar{1}}$, which is then inserted into (3-31b). Furthermore, g_{11} and g_{22} are to be represented in (3-30b) as power series in $\bar{v}^{\alpha}$. For this, power series in B will first be set up, i.e.,

$$g_{(\alpha\alpha)} = g_{(\alpha\alpha)o} + \sum_{m=1}^{m'} (1/m!)(d^m g_{(\alpha\alpha)}/dB^m)_o \Delta B^m , \qquad (4\text{-}31a)$$

whose coefficients are calculated based on (4-12) and (4-14). By inserting (4-25b), i.e.,

$$\Delta B = \bar{v}^2 + \sum_{n=2}^{n'} b^2_{n_1 n_2}(\bar{v}^1)^{n_1}(\bar{v}^2)^{n_2} , \qquad (4\text{-}31b)$$

(4-31a) is converted into the desired power series in $\bar{v}^{\alpha}$. Thus, we now have met all conditions for representing

$$h^2 = (u^1_{,\bar{1}})^2 g_{11} + (u^2_{,\bar{1}})^2 g_{22} \qquad (4\text{-}31c)$$

by a power series in $\bar{v}^{\alpha}$. When, in place of the normal coordinates $\bar{v}^{\alpha}$ [see (4-30b)], the polar coordinates of (4-23c), i.e.,

$$\bar{u}^{\alpha} = (A_o, r)^{\alpha} , \qquad A_o := A \text{ after (4-23d)} , \qquad (4\text{-}31d)$$

are directly inserted, the result reads

$$h^2 = r^2(1-(1/3)(V^2/c)^2 r^2 - (2/3)(V/c)^3(1-V^2)t \cos A_o r^3 + (2/45)(V/c)^4 r^4 + \dots) . \qquad (4\text{-}32a)$$

The root of this, the so-called <u>reduced length of the geodesic lines</u>, becomes

$$h = r\,(1-(1/6)(V^2/c)^2 r^2 - (1/3)(V/c)^3(1-V^2)t \cos A_o r^3 + (1/120)(V/c)^4 r^4 + \dots) . \qquad (4\text{-}32b)$$

Observing (4-19a) it can clearly be seen that both terms of (4-32b) are in agreement with (2-125d) and (3-37a).

The first fundamental form of (2-122c) with (4-31d) reads

$$ds^2 = h^2 dA_o^2 + dr^2 , \qquad h^2 \text{ after (4-32a)} . \qquad (4\text{-}33a)$$

For geodesic circles around P_o, i.e.,

$$r = \text{constant} \quad \rightarrow \quad dr = 0 \qquad (4\text{-}33b)$$

the following is given:

$$ds =: ds_A = h\, dA_o \, , \qquad (4\text{-}33c)$$

$$h = ds_A/dA_o \quad \text{after (4-32b)} \, . \qquad (4\text{-}33d)$$

4.3.3 TANGENT VECTORS AND AZIMUTHS OF GEODESIC R-LINES

When using geodesic polar coordinates

$$u^{\alpha'} = du^{\alpha}/dr \qquad (4\text{-}34a)$$
$$= \text{tangent vector on the r-line through } P, \quad u^{\alpha} \text{ after (4-7)}$$

and/or

$$A = \text{azimuth of the r-line through } P \qquad (4\text{-}34b)$$

most frequently be calculated for arbitrary points P.

In keeping with (4-31d)

$$A_o = \text{azimuth of the r-line at the pole } P_o \qquad (4\text{-}35a)$$

is also used here for distinguishing it from (4-34b). Calculating the quantities of (4-34) is done either as a function of the polar coordinates of (4-31d) or normal coordinates of (4-26b), i.e.,

$$\bar{v}^{\alpha} = (V_o/c)(\cos^{-1}B_o \sin A_o,\ V_o^2 \cos A_o)^{\alpha}\, r \, , \qquad (4\text{-}35b)$$

or as a function of the geographic coordinate differences of (4-25a,b), i.e.,

$$\Delta u^{\alpha} = (\Delta L,\ \Delta B)^{\alpha}. \qquad (4\text{-}35c)$$

For the <u>tangent vector of (4-34a) as a function of $\bar{v}^{\alpha}$</u> (3-16a) is generally true, i.e.,

$$u^{\alpha'} = (u^{\alpha'})_o + (1/r) \sum_{n=2}^{n'} n\, b^{\alpha}_{n_1 n_2} (\bar{v}^1)^{n_1} (\bar{v}^2)^{n_2} \, , \qquad (4\text{-}36a)$$

whereby in this particular case

$$u^{\alpha'} = (V/c)(\cos^{-1}B \sin A\, ,\ V^2 \cos A\,) \, , \qquad (4\text{-}36b)$$

$$(u^{\alpha'})_o = (V_o/c)(\cos^{-1}B_o \sin A_o,\ V_o^2 \cos A_o) \, , \qquad (4\text{-}36c)$$

$$b^{\alpha}_{n_1 n_2} \quad \text{after (4-25d)} \, . \qquad (4\text{-}36d)$$

(4-36b,c) follow from (2-113g) with (4-12), and (4-36c) is identical with (4-24a).

If the tangent vector is given by (4-36), then the azimuth of (4-34b) can be calculated by solving (4-36b) corresponding to (4-16d) as follows:

$$A = \arctan((g_{11}/g_{22})^{1/2} u^{1'}/u^{2'}) = \arctan(V^2 \cos B\, u^{1'}/u^{2'}) . \qquad (4\text{-}37)$$

Herein, g_{11} and g_{22} as well as V and $\cos B$ are primarily functions of B. For these terms we must proceed according to (4-31a,b) if A is to be represented as a power series in $\bar{v}^{\alpha}$. This will not be pursued further here. A direct series expansion as in (3-17) and (3-18) is more practical for representing the azimuth of (4-34b) as a function of $\bar{v}^{\alpha}$. The result corresponding to (3-18b) up to $n' = 4$ reads

$$\Delta A = A - A_o = \sum_{n=1}^{4} c_{n_1 n_2} (\bar{v}^1)^{n_1} (\bar{v}^2)^{n_2} , \quad n_1 + n_2 = n . \qquad (4\text{-}38a)$$

The abbreviations of (4-25c) are changed here into

$$co_o := \cos B_o , \quad t_o := \tan B_o , \quad V_o := (1 + e''^2 \cos^2 B_o)^{1/2} , \qquad (4\text{-}38b)$$

whereby we obtain the following results for the coefficients in (4-38a) not equaling zero:

$$\begin{aligned}
c_{10} &= co_o t_o , & c_{11} &= (1/2) co_o (1+2(2-V_o^2) t_o^2) , \\
c_{30} &= -(1/6) co_o^3 t_o (1+2t_o^2) , & c_{12} &= (1/6) V_o^{-4} co_o t_o (5+6t_o^2) , \\
c_{31} &= -(1/24) V_o^{-2} co_o^3 (1+20t_o^2+24t_o^4) , \\
c_{13} &= (1/24) V_o^{-6} co_o (5+28t_o^2+24t_o^4) .
\end{aligned} \qquad (4\text{-}38c)$$

The tangent vector of (4-34a) as a function of Δu^{α} is obtained using the mid-latitude formulas (4-27) and (4-28). Moreover, the designations of (4-27a) are canceled for purposes of consistency with those used here, i.e.,

$$P_o := P_a , \qquad P := P_b , \qquad (4\text{-}39a)$$

so that

$$B_m = (B_o + B)/2 \qquad (4\text{-}39b)$$

is employed in place of (4-27c). The second equation of (4-28e) combined with (4-39a) becomes

$$\bar{v}^{\alpha} = (\bar{v}^{\alpha})_m + \delta \bar{v}_o^{\alpha} \qquad (4\text{-}39c)$$

and directly produces a tangent vector in P as a function of Δu^α [see (4-35c)]. In contrast to (4-34a), however, $\bar{v}^\alpha$ is not a unit vector. By definition

$$u^{\alpha'} = \bar{v}^\alpha / r \; . \tag{4-39d}$$

r is calculated using the second equation of (4-28f) which, like (4-39b), is solely a function of geographic coordinates u^α.

The previously described method of (4-39) can also be used for calculating the azimuth of (4-34b) as a function of Δu^α, whereby the first equation of (4-29b) directly yields

$$A = \arctan(V^2 \cos B \; \bar{v}^1/\bar{v}^2), \quad \bar{v}^\alpha \text{ after (4-39c)} \; . \tag{4-40a}$$

A power series expansion for A in Δu^α is obtained by inserting the inverse Legendrean series of (4-26a) into (4-38a). Up to n' = 3 and by sorting the powers of ΔL and ΔB the result reads

$$\Delta A = A - A_o = \sum_{n=1}^{3} d_{n_1 n_2} \Delta L^{n_1} \Delta B^{n_2} \; , \quad n_1 + n_2 = n \tag{4-40b}$$

with the following nonvanishing coefficients:

$$d_{10} = c o_o t_o \; , \qquad d_{11} = (1/2) c o_o \; ,$$
$$d_{30} = (1/12) c o_o^3 t_o V_o^2 \; , \qquad d_{21} = -(1/12) c o_o t_o (1 - V_o^2) \; . \tag{4-40c}$$

Another possibility of calculating the azimuth is given by Clairaut's equation in (2-108b), which has the form of

$$R_1 \cos B \sin A = (c/V) \cos B \sin A = \text{constant} \tag{4-40d}$$

in geographic coordinates on the ellipsoid of revolution. Accordingly,

$$(c/V) \cos B \sin A = (c/V_o) \cos B_o \sin A_o$$

with which we obtain

$$A = \arcsin((V \cos B_o / (V_o \cos B)) \sin A_o) \; . \tag{4-40e}$$

Besides A, (4-40e) also contains the solution $\pi - A$ so that choosing the correct solution for $A \approx \pi/2$ may be difficult.

4.3.4 TRANSFORMATIONS BETWEEN THE ARC LENGTH OF A MERIDIAN AND THE ELLIPSOIDAL LATITUDE

Here, two points on one and the same meridian ellipse will be considered [see (4-1)]. These points are designated by

$$P_o,\ P = \text{points on a meridian ellipse} \ . \qquad \text{(4-42a)}$$

Furthermore, the terms

$$B_o,\ B = \text{ellipsoidal latitudes of } \ P_o,\ P \ , \qquad \text{(4-42b)}$$

$$\bar{r} = \text{meridian arc length } \ P_o - P \qquad \text{(4-42c)}$$

are introduced. The difference in the ellipsoidal longitudes of P_o and P is equal to zero by definition, i.e.,

$$\Delta L = L - L_o = 0 \ , \qquad \text{(4-42d)}$$

and we let

$$\Delta B = B - B_o \neq 0 \ . \qquad \text{(4-42e)}$$

The transformations

$$\Delta B = f(\bar{r}) \ , \qquad \bar{r} = \bar{f}(\Delta B) \qquad \text{(4-43)}$$

involve special cases of the transformations between polar coordinates and geographic coordinates treated in Chapter 4.3.1 with the assumptions corresponding to (4-42b,c) of

$$\Delta L = 0 \ , \quad A = 0 \ , \quad r =: \bar{r} \ . \qquad \text{(4-44a)}$$

Thereby, (4-24b) yields

$$\bar{v}^{\alpha} = R_{2o}^{-1}\ \bar{r}\ \delta_2^{\alpha} = (V_o^3/c)\bar{r}\ \delta_2^{\alpha} \ , \qquad \text{(4-44b)}$$

whereby (4-25b) produces the following series for the first transformation of (4-43):

$$\Delta B = (V^3/c)\bar{r} + \sum_{n=2}^{n'} a_{0\ n}^{2}((V^3/c)\bar{r})^n \ , \qquad \text{(4-45)}$$

$$V := V_o \ \text{ after (4-25c)}, \quad n := n_2, \quad a_{0\ n}^{2} \ \text{ after (4-25d)} \ .$$

We can use (4-26a) in an analogous way for the inverse transformation, i.e., the second transformation of (4-43). With (4-44) we obtain

$$\bar{r} = (c/V^3)(\Delta B + \sum_{n_2=2}^{n'} \bar{a}^{-2}_{0\,n} \Delta B^n) \;, \quad \bar{a}^{-2}_{0\,n} \text{ after (4-26b)} \;. \tag{4-46}$$

It is of course better to use the mid-latitude formulas of (4-27) and (4-28) instead, whereby the expressions given in (4-39) are employed. Only series (4-28a) is required here, which due to

$$(\bar{v}^{\alpha})_m = ((V_o^3+V^3)/(2c))\bar{r}\;\delta_2^{\alpha} \;, \qquad \Delta L = 0$$

yields

$$\bar{r} = (2c/(V_o^3+V^3))(\Delta B + A^2_{03}\Delta B^3) \;, \quad A^2_{03} \text{ after (4-28c)} \;. \tag{4-47}$$

The previously given power series in the coordinates have limited operational ranges, i.e., using (4-45) and (4-46) up to about 100 km, and (4-47) up to about 300 km. Greater ranges can be reached by adding further terms to the power series, leading however to extraordinarily voluminous expansions. Series expansion with respect to e' or e" [see (4-4b,c)] are more appropriate for great P_o - P distances. For the power series in e" we begin with

$$B' = dB/d\bar{r} = V^3/c$$

according to (4-24a) with (4-44a). Separating $\bar{r}$ and B produces

$$d\bar{r} = c\,V^{-3}dB = c(1+e''^2\cos^2B)^{-3/2}dB \tag{4-48a}$$

so that the meridian arc $\bar{r}$ from the equatorial point

$$P_o:\; B_o = 0 \tag{4-48b}$$

to a point P with the latitude B is determined using the integral

$$\bar{r} = c\int_0^B (1+e''^2\cos^2B)^{-3/2}dB \;. \tag{4-48c}$$

This elliptic integral is solved using the series expansion of the integrand with respect to the powers of e''^2 around the point $e''^2 = 0$, which corresponds to a spherical solution with the radius c. Applying the binomial series yields

$$V^{-3} = (1+e''^2\cos^2B)^{-3/2} = 1 + \sum_{n=1}^{n'} (-(3/2);\, n)(e''^2\cos^2B)^n, \tag{4-49a}$$

$$(-(3/2);\, n) := (-(3/2)(-(3/2)-1)\ldots(-(3/2)-n+1)/n! \;. \tag{4-49b}$$

Due to

$$\int_0^B \cos^{2n}B\, dB = 2^{-2n}[(2n;n)B + \sum_{\nu=0}^{n-1}((2n;\nu)/(n-\nu))\sin(2(n-\nu)B)] \,. \qquad (4\text{-}49c)$$

$$(m;k) := m!/(k!(m-k)!) = (m(m-1)\ldots(m-(k-1)))/k! \qquad (4\text{-}49d)$$

and based on (4-48c) with (4-49a,b) we obtain the <u>trigonometric series for $\bar{r}$ as a function of B</u>, i.e.,

$$\bar{r} = A_o B + \sum_{p=1}^{p'} A_{2p}\sin(2p\, B) \qquad (4\text{-}50a)$$

with the following coefficients:

$$A_o = c(\, 1 + \sum_{p=1}^{n'} 2^{-2n}(-(3/2);\, n)(2n;\; n\,)e''^{2n}) \,, \qquad (4\text{-}50b)$$

$$A_{2p} = c((1/p) \sum_{n=p}^{p'} 2^{-2n}(-(3/2);\, n)(2n;\, n-p)e''^{2n}) \,, \quad (2n;\, 0) =: 1 \,. \qquad (4\text{-}50c)$$

Up to $n' = 3$ (4-50b,c) yield

$$\begin{aligned} A_o &= c(1-(3/4)e''^2 + (45/64)e''^4 - (175/256)e''^6 + \ldots) \,, \\ A_2 &= c(\, -(3/8)e''^2 + (15/32)e''^4 -(525/1024)e''^6 + \ldots) \,, \\ A_4 &= c((15/256)e''^4 -(105/1024)e''^6 + \ldots) \,, \\ A_6 &= c(-\, (35/3072)e''^6 + \ldots) \,. \end{aligned} \qquad (4\text{-}50d)$$

$B = \pi/2$ for <u>meridian quadrants</u> so that based on (4-50a) its length becomes

$$\bar{r}_Q = A_o \pi/2 \,. \qquad (4\text{-}51)$$

When inverting (4-50a) $\bar{r}$ is replaced by the variable

$$\bar{B} = \bar{r}/A_o \,, \qquad (4\text{-}52a)$$

which represents a first approximation of B according to (4-50a). We then obtain a <u>trigonometric series for B as a function of $\bar{B}$</u> in the form of

$$B = \bar{B} + \sum_{p=1}^{p'} \bar{A}_{2p}\sin(2p\, \bar{B}) \qquad (4\text{-}52b)$$

with the following coefficients up to $p' = 3$:

$$\bar{A}_2 = (3/8)e''^2 - (3/16)e''^4 + (213/2048)e''^6 + \dots ,$$
$$\bar{A}_4 = (21/256)e''^4 - (21/256)e''^6 + \dots , \qquad (4\text{-}52c)$$
$$\bar{A}_6 = (151/6144)e''^6 + \dots .$$

For comparison see Helmert (1880) and especially König & Weise (1951), in which the results of (4-52c) are given as a function of

$$n = (a-b)/(a+b) .$$

If the series

$$n = (1/4)e''^2 - (1/8)e''^4 + (5/64)e''^6 - \dots$$

is substituted for n, we then obtain (4-52c).

In addition to the direct application of the trigonometric series of (4-50) and (4-52) we can also use the series of (4-50) which can be employed up to every arbitrary degree of p' without difficulty for directly approximating $\bar{r}(B)$ and $B(\bar{r})$ with polynomials in B and $\bar{r}$, respectively. In this case (4-50) is only used for calculating the paired values

$$B_\nu , \quad \bar{r}_\nu , \qquad \nu = 1,2,3\dots\nu' \qquad (4\text{-}53)$$

at suitably selected points P_ν, which serve for the polynomial approximation [see Mittermayer 1963].

4.4 TRANSFORMATIONS BETWEEN SOLDNER'S PARALLEL COORDINATES AND GEOGRAPHIC COORDINATES

4.4.1 TRANSFORMING THE COORDINATES

In the system of geographic coordinates of (4-7), i.e.,

$$u^\alpha = (L, B)^\alpha , \qquad (4\text{-}54)$$

Soldner's parallel coordinates are defined corresponding to (2-116) as follows:

$$(u^\alpha)_o = (L_o, B_o)^\alpha \qquad (4\text{-}55a)$$

= geographic coordinates of the reference point P_o ,

C_o = abscissa line (2-116b) = meridian through P_o , (4-55b)

$$\nu^\alpha_{Co} = (g_{22})_o^{-1/2}\delta^\alpha_2 = (1/R_{2o})\delta^\alpha_2 \,, \tag{4-55c}$$

$$\bar{u}^\alpha = (x,\, y)^\alpha = \text{Soldner's parallel coordinates} \,, \tag{4-55d}$$

$$x = \text{abscissa of a point } P \text{ after (2-116d)} \,, \tag{4-55e}$$

$$y = \text{ordinate of a point } P \text{ after (2-116e)} \tag{4-55f}$$

[see Fig. 4.3]. The x-lines are geodesic parallels to the abscissa line C_o, which is identical with the B-line for $L = L_o$, i.e., the meridian through P_o. The x-lines are closed surface curves, but the y-lines are generally nonclosed surface curves and usually have no poles as well.

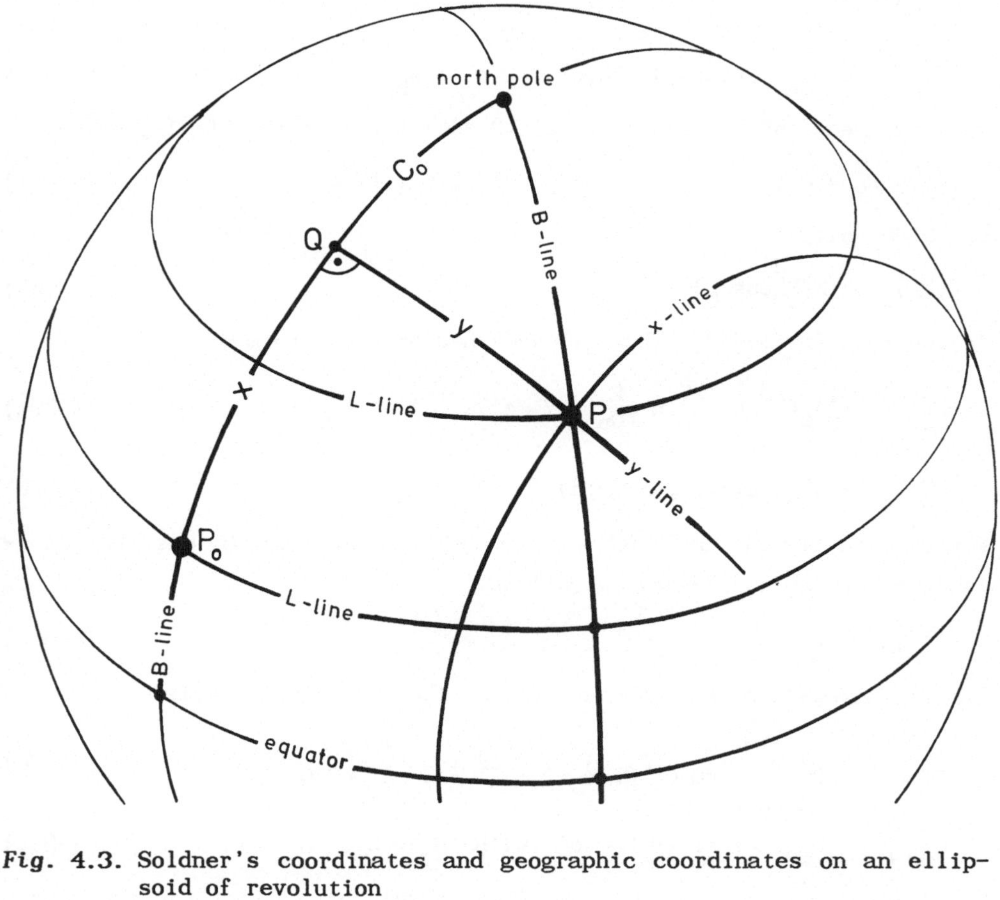

Fig. **4.3.** Soldner's coordinates and geographic coordinates on an ellipsoid of revolution

Various methods for determining the <u>transformation equations</u> between u^α and $\bar{u}^\alpha$ in the form of power series in coordinate differences with respect to P_o have been examined in Chapters 3.4.1 and 3.4.2. Here, we will only apply the method of (3-46) and (3-47) for transforming $\bar{u}^\alpha \to u^\alpha$, which proceeds from (3-44). The inverse transformation of (3-48) is acquired by series inversion.

In the following the abbreviations introduced in (4-25c), i.e.,

$$co := \cos B_o \,, \qquad t := \tan B_o \,, \qquad V := V_o \,, \tag{4-56}$$

will be made use of. (3-44b) combined with (4-55c,e) and (4-56) becomes

$$(\bar{v}^\alpha)_o = (x/R_{2o})\delta_2^\alpha = (V^3/c)x\ \delta_2^\alpha \,, \tag{4-57a}$$

whereby based on (3-44c) up to $n' = 4$ we obtain

$$\Delta L_{oQ} = 0 \,, \qquad \Delta B_{oQ} = (V^3/c)x + \sum_{n=2}^{4} a^2_{0n\ o}((V^3/c)x)^n \,, \tag{4-57b}$$

$$a^2_{0n\ o} := b^2_{0n} \text{ after (4-25c,d)} \,, \qquad n := n_2 \,. \tag{4-57c}$$

The components of the ϵ-tensors of (2-44c) together with (4-12) become

$$\epsilon_1^1 = \epsilon_2^2 = 0 \,, \qquad \epsilon_2^1 = -1/\epsilon_1^2 = 1/(V^2\cos B) \,, \tag{4-58a}$$

of which only

$$(\epsilon_2^1)_Q = 1/(V_Q^2\cos B_Q) \tag{4-58b}$$

is required in (3-44d). Thus, (3-44d) has here the form of

$$L'_{QP} = (\epsilon_2^1)_Q(V^3/c + \sum_{n=2}^{4} n\, a^2_{0n\ o}(V^3/c)^n x^{n-1}) \,, \quad B'_{QP} = 0 \,, \tag{4-59a}$$

$$a^2_{0n\ o} := b^2_{0n} \text{ after (4-25c,d)} \,, \qquad n := n_2 \,, \tag{4-59b}$$

whereby the normal coordinates of P with respect to Q must be calculated according to (3-44e) as follows:

$$(\bar{v}^\alpha)_Q = L'_{QP}y\ \delta_1^\alpha \,. \tag{4-59c}$$

The Legendrean series of (3-44f) yields the following results:

$$\Delta L_{QP} = L'_{QP}y + \sum_{n=2}^{4} a^1_{n0\ Q}(L'_{QP}y)^n \,, \qquad \Delta B_{QP} = \sum_{n=2}^{4} a^2_{n0\ Q}(L'_{QP}y)^n \,, \tag{4-59d}$$

$$a^\alpha_{n0\ Q} := b^\alpha_{n0} \text{ after (4-25c,d) with } B_o := B_Q,\quad n := n_1 \,. \tag{4-59e}$$

We ultimately obtain the desired final result of (3-44g) for the <u>transformation equations of $u^\alpha(\bar{u}^\beta)$</u> by adding (4-57b) and (4-59d), i.e.,

$$L = L_o + \Delta L_{QP} \,, \qquad B = B_o + \Delta B_{oQ} + \Delta B_{QP} \,. \tag{4-60}$$

The previously described method based indirectly on power series expansions can be transformed into a <u>direct power series expansion in x and y</u>

in the form of (3-46), i.e.,

$$(\Delta L,\ \Delta B)^{\alpha} = (L-L_o,\ B-B_o)^{\alpha} = \sum_{n=1}^{4} b^{\alpha}_{n_1 n_2} x^{n_1} y^{n_2}, \qquad n_1 + n_2 = n \ , \tag{4-61a}$$

corresponding to the commentary given for (3-47). Series (3-47a) and (3-47c) have the following special forms resulting from (4-58b) and (4-59d) with (4-57a):

$$a^{\alpha}_{n\ 0Q} = a^{\alpha}_{n\ 0o} + \sum_{m=1}^{4-n} a^{\alpha}_{n\ 0\ 0\ m}((V^3/c)x)^m \ , \tag{4-61b}$$

$$(\epsilon^1_2)_Q = (\epsilon^1_2)_o + \sum_{m=1}^{3} e^1_{20\ m}((V^3/c)x)^m \ . \tag{4-61c}$$

We will not go into the calculation of the series coefficients according to (3-47b,c) here. Using (4-61b,c) the coefficients of (4-61a) can be calculated in the manner described for (3-47d). We obtain

$$\begin{aligned}
b^{\alpha}_{10} &= (\ 0\ ,\ V^3/c)\ ,\\
b^{\alpha}_{01} &= ((V/c)^2 co(1+t^2),\ 0\ )\ ,\\
\\
b^{\alpha}_{20} &= (\ 0\ ,\ (3/2)(V^2/c)^2 t(1-V^2))\ ,\\
b^{\alpha}_{11} &= ((V/c)^2 co\ t(1+t^2),\ 0\ )\ ,\\
b^{\alpha}_{02} &= (\ 0\ ,\ -(1/2)(V^2/c)^2 t)\ ,\\
\\
b^{\alpha}_{30} &= (\ 0\ ,\ (1/2)(V/c)^3(1-V^2)(1-t^2))\ ,\\
b^{\alpha}_{21} &= ((1/2)(V/c)^3 co(1+t^2)(V^2+2t^2),\ 0\ )\ ,\\
b^{\alpha}_{12} &= (\ 0\ ,\ (1/2)(V/c)^3(1-2V^2-(3-2V^2)t^2))\ ,\\
b^{\alpha}_{03} &= (-(1/3)(V/c)^3 co\ t^2(1+t^2),\ 0\ )\ ,\\
\\
b^{\alpha}_{40} &= (\ 0\ ,\ 0\ )\ ,\\
b^{\alpha}_{31} &= ((1/6)(V/c)^4 co\ t(1+t^2)(5+6t^2),\ 0\ )\ ,\\
b^{\alpha}_{22} &= (\ 0\ ,\ -(1/2)(V/c)^4 t(1+t^2))\ ,\\
b^{\alpha}_{13} &= (-(1/3)(V/c)^4 co\ t(1+t^2)(2+3t^2),\ 0\ )\ ,\\
b^{\alpha}_{04} &= (\ 0\ ,\ (1/24)(V/c)^4 t(1+3t^2))\ .
\end{aligned} \tag{4-61d}$$

By inverting series (4-61a,d) in accordance with (3-7) and (3-48) we obtain a direct power series expansion in ΔL and ΔB for the transformation equations of $\bar{u}^{\alpha}(u^{\beta})$ as follows:

$$(x,\ y)^{\alpha} = \sum_{n=1}^{4} \bar{b}^{\alpha}_{n_1 n_2} \Delta L^{n_1} \Delta B^{n_2} \ , \qquad n_1 + n_2 = n \tag{4-62a}$$

with the coefficients

$$\begin{aligned}
\bar{b}^{\alpha}_{10} &= (\ 0\ ,\ (c/V)co)\ ,\\
\bar{b}^{\alpha}_{01} &= (c/V^3,\ 0\ )\ ,\\[1ex]
\bar{b}^{\alpha}_{20} &= ((1/2)(c/V)co^2 t,\ 0\ )\ ,\\
\bar{b}^{\alpha}_{11} &= (\ 0\ ,\ -(c/V^3)co\ t)\ ,\\
\bar{b}^{\alpha}_{02} &= (-(3/2)(c/V^5)t(1-V^2),\ 0\ )\ ,\\[1ex]
\bar{b}^{\alpha}_{30} &= (\ 0\ ,\ -(1/6)(c/V)co^3 t^2)\ ,\\
\bar{b}^{\alpha}_{21} &= ((1/2)(c/V)co^2(1-(2-V^2)t^2),\ 0\ )\ ,\\
\bar{b}^{\alpha}_{12} &= (\ 0\ ,\ -(1/2)(c/V)co(2-V^2-3(1-V^2)t^2))\ ,\\
\bar{b}^{\alpha}_{03} &= (-(1/2)(c/V)(1-V^2)(1-t^2),\ 0\ )\ ,\\[1ex]
\bar{b}^{\alpha}_{40} &= ((1/24)(c/V)co^4 t(5-t^2),\ 0\ )\ ,\\
\bar{b}^{\alpha}_{31} &= (\ 0\ ,\ -(1/6)(c/V)co^3 t(2-t^2))\ ,\\
\bar{b}^{\alpha}_{22} &= (-(c/V)co^2 t,\ 0\ )\ ,\\
\bar{b}^{\alpha}_{13} &= (\ 0\ ,\ (1/6)(c/V)co\ t)\ ,\\
\bar{b}^{\alpha}_{04} &= (\ 0\ ,\ 0\ )\ .
\end{aligned} \tag{4-62b}$$

4.4.2 TRANSFORMING THE METRIC TENSOR

For calculating the metric tensor of Soldner's surface coordinates we will proceed in line with (3-51). (3-51d) becomes

$$n^2 := \bar{g}_{11} = (\Delta L_{,\bar{1}})^2 g_{11} + (\Delta B_{,\bar{1}})^2 g_{22}\ , \tag{4-63}$$

in which n is the so-called reduction factor of the abscissa. The transformation matrices together with the coefficients of (4-61d) become according to (3-51a)

$$(\Delta L_{,\bar{1}},\ \Delta B_{,\bar{1}})^{\alpha} = \sum_{n=1}^{4} n_1 b^{\alpha}_{n_1 n_2} x^{n_1-1} y^{n_2}\ , \tag{4-64a}$$

and a power series in ΔB will first be set up for the components of the metric tensor constructed at the test point as follows:

$$g_{(\alpha\alpha)} = g_{(\alpha\alpha)o} + \sum_{m=1}^{4} (1/m!)(d^m g_{(\alpha\alpha)}/dB^m)_o \Delta B^m\ . \tag{4-64b}$$

The coefficients of this power series are calculated based on (4-12) and (4-14). If the series resulting from (4-61a), i.e.,

$$\Delta B = \sum_{n=1}^{4} b^2_{n_1 n_2} x^{n_1} y^{n_2}\ , \tag{4-64c}$$

is inserted in (4-64b), we then obtain the desired representation of $g_{(\alpha\alpha)}$ as power series in x and y. Herewith and with (4-64a), all conditions for representing (4-63) by a power series in x and y are fulfilled. The solution reads

$$n^2 = 1 - (V/c)^2 y^2 (V^2 + 4(V/c)t(1-V^2)x - (1/3)(V/c)^2 y^2)\ . \tag{4-65a}$$

The <u>abscissa-reduction factor</u> as the root of (4-65a) becomes

$$n = 1 - (1/2)(V/c)^2 y^2 (V^2 + 4(V/c)t(1-V^2)x - (1/12)(V/c)^2 y^2)\ . \tag{4-65b}$$

In (4-65a,b), (4-56) is valid for t and V. The first fundamental form of (2-110g) together with (4-63) reads

$$ds^2 = n^2 dx^2 + dy^2\ , \qquad n^2 \text{ after (4-65a)}\ . \tag{4-66a}$$

For the geodesic parallels to the abscissa line C_o, i.e.,

$$y = \text{constant} \qquad \rightarrow \qquad dy = 0\ , \tag{4-66b}$$

it follows that

$$ds =: ds_x = n\ dx\ , \tag{4-66c}$$

$$n = ds_x/dx \quad \text{after (4-65b)}\ . \tag{4-66d}$$

4.4.3 MERIDIAN CONVERGENCE

In the system of Soldner's coordinates $\bar{u}^\alpha$ [see (4-55d)] the <u>direction of the meridian or B-line at a point P</u> can be given by the contravariant unit vector, i.e.,

$$\bar{\upsilon}^\alpha_B = \text{tangent unit vector in the direction of the B-line at P} . \quad (4\text{-}67a)$$

Equivalently, the direction of the meridian can also be obtained with the angle

$$\begin{aligned} \gamma &= \text{meridian convergence} \quad (4\text{-}67b) \\ &= \text{direction angle of the B-line relative to the x-line at P} \end{aligned}$$

[see Fig. 4.4]. Due to the orthogonality of both the u^α- and the $\bar{u}^\alpha$-systems γ is also the angle between the L-line and the y-line at P.

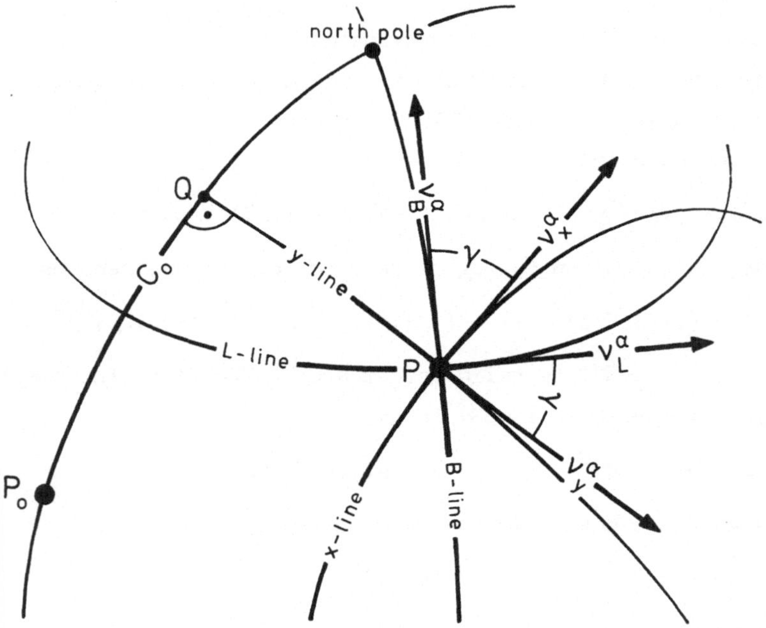

Fig. 4.4. Tangent vectors of the coordinate lines and meridian convergence

For further considerations the contravariant unit vectors are introduced, i.e.,

$$\begin{aligned} &\upsilon^\alpha_p , \quad p \in \{L,B,x,y\} \quad (4\text{-}68a) \\ &= \text{tangent unit vectors on the p-lines in the } u^\alpha\text{-system} , \end{aligned}$$

$$\bar{\upsilon}^{\alpha}_{p} = \text{components of the } \upsilon^{\alpha}_{p} \text{ in the } \bar{u}^{\alpha}\text{-system} . \tag{4-68b}$$

With (4-12) and (4-63) we obtain

$$\upsilon^{\alpha}_{L} = g_{11}^{-1/2}\delta^{\alpha}_{1} = (V/(c\cos B))\delta^{\alpha}_{1} , \qquad \upsilon^{\alpha}_{B} = g_{22}^{-1/2}\delta^{\alpha}_{2} = (V^3/c)\delta^{\alpha}_{2} , \tag{4-68c}$$

$$\bar{\upsilon}^{\alpha}_{x} = \bar{g}_{11}^{-1/2}\delta^{\alpha}_{1} = (1/n)\delta^{\alpha}_{1} , \qquad \bar{\upsilon}^{\alpha}_{y} = \bar{g}_{22}^{-1/2}\delta^{\alpha}_{2} = \delta^{\alpha}_{2} . \tag{4-68d}$$

Thereby, the solution for (4-67a) can be given through a transformation of (2-14) as follows:

$$\bar{\upsilon}^{\alpha}_{B} = \bar{u}^{\alpha}_{,\beta}\upsilon^{\beta}_{B} = (V^3/c)\bar{u}^{\alpha}_{,2} , \tag{4-69a}$$

whereby

$$\bar{u}^{\alpha}_{,2} = (x_{,2}, y_{,2})^{\alpha} = \sum_{n=1}^{4} n_2 \bar{b}^{\alpha}_{n_1 n_2} \Delta L^{n_1} \Delta B^{n_2 - 1} \tag{4-69b}$$

is calculated based on (4-62). If the reference point P_o [see (4-55a)] is selected as the intersection of L-line and C_o, then

$$B_o = B \quad \rightarrow \quad \Delta B = 0 , \tag{4-69c}$$

and (4-69b) can be simplified as follows:

$$\bar{u}^{\alpha}_{,2} = (x_{,2}, y_{,2})^{\alpha} = \sum_{n=0}^{4} \bar{b}^{\alpha}_{n\,1} \Delta L^{n} . \tag{4-69d}$$

γ [see (4-67)] can be calculated as a scalar product (2-30d) of the tangent vectors of (4-68) in different ways. The meridian convergence is $\gamma < 50'$ for $B \approx 50°$ restricted to the convergence domain of series (4-61) and (4-62), i.e., for $y \lesssim 100$ km. Hence, the calculation of $\sin\gamma$ is recommended over $\cos\gamma$, i.e.,

$$\sin\gamma = -g_{\alpha\beta}\upsilon^{\alpha}_{B}\upsilon^{\beta}_{y} = g_{\alpha\beta}\upsilon^{\alpha}_{L}\upsilon^{\beta}_{x} = -\bar{g}_{\alpha\beta}\bar{\upsilon}^{\alpha}_{B}\bar{\upsilon}^{\beta}_{y} = \bar{g}_{\alpha\beta}\bar{\upsilon}^{\alpha}_{L}\bar{\upsilon}^{\beta}_{x} , \tag{4-70}$$

$g_{\alpha\beta}$, $\bar{g}_{\alpha\beta}$ after (4-12), (4-63) , υ^{α}_{p} , $\bar{\upsilon}^{\alpha}_{p}$ after (4-68) .

Observing the transformation laws of (2-14) and proceeding from the first form of (4-70) we obtain

$$\sin\gamma = -g_{\alpha\beta}\upsilon^{\alpha}_{B}u^{\beta}_{,\bar{\gamma}}\bar{\upsilon}^{\gamma}_{y} = -(c/V^3)\Delta B_{,\bar{2}} \tag{4-71a}$$

and based on (4-61a,b)

$$\Delta B_{,\bar{2}} = \sum_{n=1}^{4} n_2 b^{2}_{n_1 n_2} x^{n_1} y^{n_2 - 1} . \tag{4-71b}$$

Using the third form of (4-70) we have

$$\sin\gamma = -\bar{g}_{\alpha\beta}\bar{u}^{\alpha}_{,\gamma}\upsilon^{\gamma}_{B}\ \bar{\upsilon}^{\beta}_{y} = -(V^3/c)y_{,2} \tag{4-72a}$$

with

$$y_{,2} = \sum_{n=1}^{4} n_2\bar{b}^2_{n_1 n_2}\ \Delta L^{n_1}\Delta B^{n_2-1} \tag{4-72b}$$

based on (4-62). In (4-71) and (4-72) V is a function of the latitude of the test point, i.e.,

$$V = (1+e''^2\cos^2 B)^{1/2}\ . \tag{4-73}$$

Accordingly, in (4-71) $\sin\gamma$ is represented as a function of B, x, and y and in (4-72) as a function of L and B.

4.5 DEFINING ISOTHERMAL SURFACE COORDINATES IN THE GEOGRAPHIC COORDINATE SYSTEM

In the following a review of the special definitions of geodesically relevant isothermal surface coordinates on reference ellipsoids in the system of geographic coordinates of (4-7) is given, i.e.,

$$u^{\alpha} = (L,\ B)^{\alpha}\ , \tag{4-74a}$$

for which the metric tensor $g_{\alpha\beta}$ as a position function according to (4-12) is known. As an alternative to (4-74a),

$$u^{\alpha} = (B,\ L)^{\alpha} \tag{4-74b}$$

will also be used, for which g_{11} and g_{22} in (4-12) are to be substituted for one another. For defining

$$\bar{u}^{\alpha} = \text{isothermal surface coordinates} \tag{4-74c}$$

the metric tensor is determined by

$$\bar{G} := \bar{g}_{11} = \bar{g}_{22}\ , \qquad \bar{g}_{12} = 0 \tag{4-74d}$$

corresponding to (2-132).

Only the <u>initial value problems of the differential equations of (2-134) or (3-53)</u>, as described for (2-138) and (3-54)ff, respectively, will be considered in defining isothermal coordinates. The surface curve C_o [see (2-138a)] is selected as an L-line or B-line of the geographic coordinates through a reference point, i.e.,

$$P_o \in C_o \,, \qquad (u^\alpha)_o = (L_o, B_o) \,, \qquad (4\text{-}75a)$$

corresponding to (3-59a). As in (3-59b) the curve coordinate of C_o is t so that the equation for C_o in the u^α-system of (4-74a) or (4-74b) is either

<u>C_o = parallel circle through P_o</u> (4-75b)

$$u_o^\alpha(L) = (L\ ,\ B_o) \,, \qquad t =: L$$

or

<u>C_o = meridian through P_o</u> (4-75c)

$$u_o^\alpha(B) = (B\ ,\ L_o) \,, \qquad t =: B$$

in the sense of (3-54a,b). The isothermal coordinates $\bar{u}^\alpha$ are defined in accordance with (2-138d), i.e.,

<u>$C_o = \bar{u}^1$-line = abscissa line</u> (4-75d)

$$\bar{u}_o^\alpha = (f(t),\ 0)$$

with the freely available function

$$f(t), \quad t \in \{L,B\} \text{ after (4-75b,c)} \,. \qquad (4\text{-}75e)$$

The assumptions made for (4-75) hold also for all systems of isothermal surface coordinates treated in Chapter 4.6 and following. These systems differ from one another only in the function selected for (4-75e), with which the metric tensor component $\bar{G}_o$ along C_o [see (4-75b,c)] is determined in accordance with (2-138c). (2-138c) with (2-75) becomes

$$\bar{G}_o(t) = (g_{11})_o/(df/du^1)^2, \quad du^1 = dt \in \{dL,\ dB\} \qquad (4\text{-}76a)$$

when coordinates (4-74a) and (4-74b) are used in (4-75b) and (4-75c), respectively. For geodetic applications we generally strive to attain

$$\bar{G} \approx 1 \qquad (4\text{-}76b)$$

for the effective region of an isothermal coordinate system. For fulfilling this requirement the function of (4-75e) or, equivalently, $\bar{G}_o$ [see (4-76a)] must be given in a suitable form.

The <u>differential equations of isothermal surface coordinates</u> are to be solved proceeding from (4-75e) as initial values along the surface curve C_o according to (4-75b) or (4-75c). The forms given in (3-53b) are decisive for geographic coordinates. Applying (4-74a) together with (4-12) these forms become

$$\bar{u}^1_{,1} = V^2\cos B\, \bar{u}^2_{,2}\,, \qquad \bar{u}^2_{,1} = -V^2\cos B\, \bar{u}^1_{,2}\,, \tag{4-77a}$$

$$u^2_{,\bar{2}} = V^2\cos B\, u^1_{,\bar{1}}\,, \qquad u^2_{,\bar{1}} = -V^2\cos B\, u^1_{,\bar{2}}\,. \tag{4-77b}$$

Contrarily, applying (4-74b) for the geographic coordinates these differential equations read

$$\bar{u}^2_{,2} = V^2\cos B\, \bar{u}^1_{,1}\,, \qquad \bar{u}^1_{,2} = -V^2\cos B\, \bar{u}^2_{,1}\,, \tag{4-77c}$$

$$u^1_{,\bar{1}} = V^2\cos B\, u^2_{,\bar{2}}\,, \qquad u^1_{,\bar{2}} = -V^2\cos B\, u^2_{,\bar{1}}\,. \tag{4-77d}$$

In Chapter 4.6 and following, examples of initial value problems in the forms given in (4-75) for the differential equations of (4-77) are treated. For this, the solutions of (4-77) will be represented by analytic functions in the form of the power series in the coordinates or coordinate differences. The general foundation necessary for this is found in Chapter 3.5.

4.6 TRANSFORMATIONS BETWEEN ISOTHERMAL GEOGRAPHIC COORDINATES AND GEOGRAPHIC COORDINATES

4.6.1 PRELIMINARY REMARKS

Isothermal geographic coordinates $\bar{u}^\alpha$ in the system of geographic coordinates u^α [see (4-74a,b)] are defined in accordance with (3-60). This involves particularly simple definitions for the primary construction of isothermal coordinates, which only have intermediate character in geodesy. They are used as initial isothermal coordinates for constructing geodetically relevant, isothermal coordinates using analytic functions of complex variables according to (3-65)ff.

4.6.2 ISOTHERMAL LATITUDE

It was proven in (2-148) that the meridians and latitude circles of surfaces of revolution can be coordinate lines of isothermal surface coordinates. The accompanying isothermal coordinates for ellipsoids of revolution are given as follows:

$$\bar{u}^\alpha = (L, q) , \tag{4-78a}$$

$$L = \text{geographic longitude (4-7)} , \quad q = \text{isothermal latitude} \tag{4-78b}$$

with

$$\bar{G} = g_{11} = (c/V)^2 \cos^2 B , \tag{4-78c}$$

$$dq = (g_{22}/g_{11})^{1/2} dB = (1/(V^2 \cos B)) dB . \tag{4-78d}$$

The integral of (4-78d) from $B = 0$ to B is

$$q = \ln \tan(\pi/4+B/2) + (e'/2)\ln((1-e'\sin B)/(1+e'\sin B)) . \tag{4-78e}$$

Geographic coordinates L and B are transformed into isothermal geographic coordinates (4-78a,b) using (4-78e), and the metric tensor is transformed according to (4-78c).

$\bar{u}^\alpha$ [see (4-78a,b)] in the u^α-system [see (4-74a)] will now be defined as an initial value problem (4-75) for the differential equations of (4-77a) as follows:

$$C_o = \text{latitude circle through } P_o \text{ after (4-75c)} , \tag{4-79a}$$
$$\text{geographic latitude } B_o , \quad \text{isothermal latitude } q_o = q(B_o);$$

the transformation equations for C_o [see (4-75e)] are

$$\bar{u}^1_o = f(L) = L , \qquad \bar{u}^2_o = 0 . \tag{4-79b}$$

A power series (3-4) about the point P_o with

$$\begin{aligned} \Delta u^1 &= \Delta L = L - L_o, & \Delta u^2 &= \Delta B = B - B_o , \\ \Delta\bar{u}^1 &= \bar{u}^1 - L_o , & \Delta\bar{u}^2 &= \Delta q = q - q_o \end{aligned} \tag{4-80a}$$

is chosen for solving (4-77a), i.e.,

$$\Delta\bar{u}^\alpha = (\bar{u}^1 - L_o, \Delta q)^\alpha = \sum_{n=1}^{n'} \bar{d}^\alpha_{n_1 n_2} \Delta L^{n_1} \Delta B^{n_2} , \tag{4-80b}$$

$$\bar{d}^\alpha_{n_1 n_2} = (1/(n_1! n_2!))(\partial^n \bar{u}^\alpha/(\partial L^{n_1} \partial B^{n_2}))_{P_o} . \tag{4-80c}$$

Corresponding to (3-60c,d), (4-79b) directly yields the partial derivatives

$$\bar{u}^{\alpha\prime} = \delta^\alpha_1 , \quad \bar{u}^{\alpha(n)} = 0^\alpha , \quad n > 1 , \quad \text{in } P_o \in C_o . \tag{4-80d}$$

Herewith and observing (4-77a) we obtain the following general results at $P_o \in C_o$ using the method described for (3-56) to (3-58):

$$\bar{u}^2_{,2} = 1/(V^2\cos B)\ ,\qquad \bar{u}^2_{,22\ldots2} \neq 0\ ,$$
$$(\bar{u}^2_{,22\ldots2})_{,1} = 0\ ,\qquad \bar{u}^1_{,22\ldots2} = 0\ ,\qquad (4\text{-}80e)$$

with which (4-80b) assumes the following form:

$$\Delta\bar{u}^1 = \bar{u}^{1\bullet} - L_o = \Delta L \ \rightarrow\ \bar{u}^1 = L\ ,\qquad (4\text{-}81a)$$

$$\Delta\bar{u}^2 = q - q_o = \Delta q = \sum_{n=1}^{n'} \bar{d}_n \Delta B^n\ ,\quad \bar{d}_n := \bar{d}^2_{0\ n} = (1/n!)\bar{u}^2_{,nx2}\ .\qquad (4\text{-}81b)$$

Under the especially simple conditions of (4-80e) we obtain $\bar{u}^2_{,22\ldots2}$ by continuously differentiating $\bar{u}^2_{,2}$ with respect to B. With the abbreviations

$$co := \cos B_o\ ,\quad t := \tan B_o\ ,\quad V := V_o = (1+e''^2\cos^2 B_o)^{1/2}\qquad (4\text{-}81c)$$

the coefficients $\bar{d}_n$ become

$$\bar{d}_1 = V^{-2}co(1+t^2)\ ,$$
$$\bar{d}_2 = -(1/2)V^{-4}co\ t(1+t^2)(2-3V^2)\ .$$
$$\bar{d}_3 = -(1/6)V^{-6}co(1+t^2)(2V^2-3V^4-2(4-9V^2+6V^4)t^2)\ ,\qquad (4\text{-}81d)$$
$$\bar{d}_4 = -(1/24)V^{-8}co\ t(1+t^2)(14-19V^2+6(3-4V^2)t^2)\ ,$$
$$\bar{d}_5 = (1/120)co(1+t^2)(5+28t^2+24t^4)\ .$$

(4-81b,d) is equivalent to a power series expansion of (4-78e) in ΔB around the point with the latitudes B_o and q_o.

The inverse transformation for (4-80b) can be solved analogously based on the differential equations of (4-77b) together with (4-79). However, a simpler procedure is the inversion of series (4-81a,b), whereby only (4-81b) must be inverted in accordance with (2-169) to (2-171). The result reads

$$L = \bar{u}^1\ ,\qquad \Delta B = B - B_o = \sum_{n=1}^{n'} d_n \Delta q^n\qquad (4\text{-}82a)$$

with the coefficients

$$d_1 = V^2 co\ ,$$
$$d_2 = (1/2)V^2 co^2 t(2-3V^2)\ ,\qquad (4\text{-}82b)$$

$$d_3 = -(1/6)co^3(3-9V^2+7V^4-(15-41V^2+27V^4)t^2) , \qquad (4\text{-}82b)$$

$$d_4 = -(1/24)co^4t(51-56V^2-(39-40V^2)t^2) ,$$

$$d_5 = (1/120)co^5(5-18t^2+t^4) ,$$

for which (4-81c) is valid. (4-82a,b) is equivalent to a power series expansion of the inverse function for (4-78e) in Δq around the point with the latitudes B_o and q_o.

4.6.3 ISOTHERMAL LONGITUDE

The isothermal surface coordinates

$$\bar{u}^\alpha =: p^\alpha = \text{isothermal longitude} \qquad (4\text{-}83a)$$

in the u^α-system (4-74b) are defined as an initial value problem (4-75) for the differential equations of (4-77a) in the following manner:

$$C_o = \text{meridian through } P_o \text{ after (4-75b),} \quad \text{geographic longitude } L_o ; \qquad (4\text{-}83b)$$

the transformation equations for C_o [see (4-75e)] are

$$p_o^1 = f(B) = B , \qquad p_o^2 = 0 . \qquad (4\text{-}83c)$$

This definition differs fundamentally from that of (4-79) by the fact that a meridian in accordance with (4-75e) is selected for C_o instead of a latitude circle according to (4-75c). The transformation equation for C_o is defined in the simplest way in both cases [see (4-79b) and (4-83c)].

Power series for the transformation equations between p^α and the geographic coordinates of (4-74b) are obtained analogous to (4-80)ff. For the transformation between the isothermal longitude p^α and the isothermal geographic coordinates $\bar{u}^\alpha$ [see (4-78a,b)] the transformation function can also be represented as an analytic function of a complex variable according to (3-65)ff. This procedure will be described in the following.

The coordinates are defined as the complex variables

$$p = p^1 + i\,p^2, \qquad \bar{u} = q + i\,L . \tag{4-84a}$$

Their differences with respect to

$$P_o \in C_o \quad \text{with the coordinates } p_o^\alpha,\ q_o,\ L_o,\ B_o,$$
$$p_o = p_o^1 = B_o , \qquad \bar{u}_o = q_o + i\,L_o \tag{4-84b}$$

are

$$\Delta p = p - p_o , \qquad \Delta\bar{u} = \bar{u} - \bar{u}_o . \tag{4-84c}$$

We let the accompanying analytic transformation functions be

$$\Delta\bar{u} = f(\Delta p) , \qquad \Delta p = g(\Delta\bar{u}) . \tag{4-85a}$$

For C_o [see (4-83b,c)]

$$\Delta p = \Delta p^1 = \Delta B = B - B_o , \quad \Delta\bar{u} = \Delta\bar{u}^1 = \Delta q = q - q_o , \tag{4-85b}$$

whereby based on (4-84d)

$$\Delta q = f(\Delta B) , \qquad \Delta B = g(\Delta q) \tag{4-85c}$$

or for $B_o = q_o = 0$

$$q = f(B) \qquad B = g(q) . \tag{4-85d}$$

Accordingly, the transformation function f is identical with the function of (4-78e), and g is the inverse function of (4-78e).

The power series (4-81b-d) and (4-82a,b) for (4-85c) are already at hand. According to (3-69) these series are also formally true for (4-85a), i.e.,

$$\Delta\bar{u} = \Delta q + i\,\Delta L = \sum_{n=1}^{n'} \bar{d}_n \Delta p^n \tag{4-86a}$$

$$\bar{d}_n = \text{real coefficients (4-81d)} , \tag{4-86b}$$

$$\Delta p = \Delta p^1 + i\,p^2 = \sum_{n=1}^{n'} d_n \Delta\bar{u}^n , \quad \Delta p^1 = p^1 - B_o , \tag{4-87a}$$

$$d_n = \text{real coefficients (4-82b)} . \tag{4-87b}$$

In (4-86) and (4-87) the coefficients $\bar{d}_n$ and d_n are formed at an arbitrary point B_o. In particular

$$B_o = B \rightarrow \Delta B = 0 , \quad \Delta q = 0 , \quad \Delta p^1 = p^1 - B , \tag{4-88a}$$

are selected, and series (4-87a,b) thus becomes

$$\Delta p = \Delta p^1 + i\, p^2 = \sum_{n=1}^{n'} d_n(B)(i\,\Delta L)^n \,. \tag{4-88b}$$

Splitting (4-88b) into real and imaginary parts yields

$$\Delta p = p^1 - B = \sum_{n=1}^{n'} (-1)^n d_{(2n)} \Delta L^{2n} \,, \tag{4-88c}$$

$$p^2 = \sum_{n=0}^{n'} (-1)^n d_{(2n+1)} \Delta L^{2n+1} \,. \tag{4-88d}$$

p can also be represented as a line integral corresponding to (2-164):

$$p = p_o + \int_{P_o}^{P} dp = B_o + \int_{P_o}^{P} (dg/d\bar{u})d\bar{u} \,, \tag{4-89a}$$

which is independent of the integration path because the isothermal coordinates p and u satisfy the Cauchy-Riemannian differential equations. Observing (4-78d)

$$dg/d\bar{u} = (dB/dq)_{\text{for } B=:p} = V(p)^2 \cos p \tag{4-89b}$$

follows from the results of (4-85) so that (4-89a) can be written as

$$p = B_o + \int_{P_o}^{P} V(p)^2 \cos p (dq + i\, dL) \,. \tag{4-89c}$$

With the particular choice of

$$B_o = B \qquad \text{and} \qquad P_o - P = \text{parallel circle} \tag{4-89d}$$

(4-89c) becomes

$$p = B + i \int_{P_o}^{P} V(p)^2 \cos p \, dL \,. \tag{4-89e}$$

In (4-89b-e)

$$V(p)^2 = 1 + e''^2 \cos^2 p \,, \qquad p = p^1 + i\, p^2 \,. \tag{4-89f}$$

(4-89e) is very appropriate especially for constructing the isothermal longitude p by numerical integration. For more on this subject and the general significance of isothermal longitude see Glasmacher (1987).

4.7 TRANSFORMATIONS BETWEEN GAUSSIAN ISOTHERMAL COORDINATES AND GEOGRAPHIC COORDINATES

4.7.1 DIRECTLY TRANSFORMING THE COORDINATES

The system of geographic coordinates is chosen in the form of (4-74b), i.e.,

$$u^\alpha = (B,L)^\alpha \, , \tag{4-90}$$

so that the expressions for g_{11} and g_{22} in (4-12) must be exchanged. In this system the Gaussian isothermal coordinates, also called Gauss-Krüger coordinates, are defined corresponding to (4-75) as follows:

$(u^\alpha)_o = (B_o, L_o)^\alpha$ (4-91a)
$\quad$ = geographic coordinates of the reference point P_o ,

C_o = abscissa line through P_o (4-91b)
$\quad$ = meridian through P_o after (4-75c) ,

$\bar{u}^\alpha$ = $(x, y)^\alpha$ = Gaussian coordinates with reference to P_o, (4-91c)

x = abscissa of a point P , (4-91d)

y = ordinate of a point P ; (4-91e)

the transformation equations for C_o [see (4-75e)] are

$$\bar{u}_o^1 = f(B) = m\, \Delta s(B) \, , \qquad \bar{u}_o^2 = y_o = 0 \, , \tag{4-91f}$$

$$\Delta s(B) = s(B) - s_o \, , \qquad s_o := s(B_o) \, ,$$

$s(B)$ = meridian arc between the equatorial point $B = 0$ and a point of latitude B , (4-91g)

m = constant scale factor . (4-91h)

This definition is consistent with (2-138d-f) as well as (3-59) and (3-60a). The choice of the abscissa line C_o corresponds to that of Soldner's coordinates [see (4-55b)], and the transformation equations for C_o are also identical, except the factor m in (4-91f). Seen qualitatively, Fig. 4.3 is a depiction of Gaussian coordinates, whereby the y-lines, of course, are not generally geodesics. From (4-91f) we obtain

$$df/dB = m\, ds/dB \tag{4-92a}$$

for points of C_o, and consequently

$$\bar{G}_o = 1/m^2 = \text{constant} \tag{4-92b}$$

according to (4-76a). Thus, the requirement of (4-76b), i.e., $\bar{G} \approx 1$, can be best achieved by using Gaussian coordinates for strip-shaped areas in the direction of the meridian C_o. The following values for m are common:

$$m = 1 \quad \rightarrow \quad \bar{G}_o = 1 \quad : \text{Gauss-Krüger}, \tag{4-92c}$$

$$m = 0.9996 \quad \rightarrow \quad \bar{G}_o = 1.00080048 : \text{UTM}\ . \tag{4-92d}$$

The Gauss-Krüger coordinates of (4-92c) used in the Federal Republic of Germany are separated into strips differing in longitude by 3°. The strips of the UTM coordinates (Universal Transverse Mercator coordinates) used in the United States differ in longitude by 6°. For better observing (4-76b) the UTM coordinates with the scale factor m according to (4-92d) are chosen for these wider strips.

For the <u>transformation function $\bar{u}^\alpha(u^\beta)$</u> we will begin with a power series setup like (3-4) which together with

$$\Delta u^\alpha = (B-B_o, L-L_o)^\alpha = (\Delta B, \Delta L)^\alpha \tag{4-93a}$$

and in accordance with (4-90), (4-91a), and (4-91c) can be written in the following way:

$$\bar{u}^\alpha = (x,y)^\alpha = \sum_{n=1}^{n'} \bar{a}^\alpha_{n_1 n_2} \Delta B^{n_1} \Delta L^{n_2}\ , \quad n_1 + n_2 = n\ , \tag{4-93b}$$

$$\begin{aligned} \bar{a}^\alpha_{n_1 n_2} &= (1/(n_1!n_2!))(\partial^n \bar{u}^\alpha/(\partial B^{n_1} \partial L^{n_2}))_{P_o} \\ &= (1/(n_1!n_2!))(\bar{u}^\alpha_{,n_1 x1\ n_2 x2})_{P_o}\ . \end{aligned} \tag{4-93c}$$

The coefficients of (4-93c) are determined in the way described for (3-56) to (3-58) whereby the simplifications of (3-59) ensue owing to the specific initial condition of (4-91). Consequently,

$$u^{\alpha'} = \delta^\alpha_1\ , \quad u^{\alpha(n)} = 0^\alpha\ , \quad n > 1\ , \qquad \text{for } P_o \in C_o \tag{4-93d}$$

in accordance with (3-59c,d) and observing (4-91f-h) and (4-92a) we obtain the following for (3-59e):

$$\bar{u}^{\alpha(n)} = m(d^n s/dB^n)_{P_o} \delta^\alpha_1 = m\ c(d^{n-1}(V^{-3})/dB^{n-1})_{P_o} \delta^\alpha_1\ . \tag{4-93e}$$

The equations of (4-93d,e) contain the prerequisites required for setting up the equations of (3-58b). Due to u^α as prescribed by (4-90) the equations of (3-58a) are to be formed proceeding from (4-77c). With the abbreviations

$$co := \cos B_o, \quad t := \tan B_o, \quad V := V_o = (1+e''^2\cos^2 B_o)^{1/2} \tag{4-93f}$$

the coefficients for $m = 1$ up to $n' = 5$ read

$$\bar{a}^\alpha_{10} = (c/V^3,\ 0\ ),$$
$$\bar{a}^\alpha_{01} = (\ 0\ ,\ (c/V)co),$$

$$\bar{a}^\alpha_{20} = (-(3/2)(c/V^5)t(1-V^2),\ 0\ ),$$
$$\bar{a}^\alpha_{11} = (\ 0\ ,\ -(c/V^3)co\ t),$$
$$\bar{a}^\alpha_{02} = ((1/2)(c/V)co^2t,\ 0\ ),$$

$$\bar{a}^\alpha_{30} = (-(1/2)(c/V)(1-V^2)(3-2V^2)-(8-7V^2)t^2),\ 0\ ),$$
$$\bar{a}^\alpha_{21} = (\ 0\ ,\ -(1/2)(c/V)co(3-3V^2+V^4-3(3-5V^2+2V^4)t^2)),$$
$$\bar{a}^\alpha_{12} = ((1/2)(c/V)co^2(1-(3-3V^2+V^4)t^2),\ 0\ ),$$
$$\bar{a}^\alpha_{03} = (\ 0\ ,\ (1/6)(c/V)co^3(V^2-t^2)),$$

(4-93g)

$$\bar{a}^\alpha_{40} = ((1/2)(c/V)t(1-V^2),\ 0\ ),$$
$$\bar{a}^\alpha_{31} = (\ 0\ ,\ (1/6)(c/V)co\ t(11-10V^2-3(1-V^2)t^2)),$$
$$\bar{a}^\alpha_{22} = (-(1/4)(c/V)co^2t(7-3V^2-3(1-V^2)t^2),\ 0\ ),$$
$$\bar{a}^\alpha_{13} = (\ 0\ ,\ -(1/6)(c/V)co^3t(1+4V^2-(2-V^2)t^2)),$$
$$\bar{a}^\alpha_{04} = (-(1/24)(c/V)co^4t(4-9V^2+t^2),\ 0\ ),$$

$$\bar{a}^\alpha_{50} = (\ 0\ ,\ 0\ ),$$
$$\bar{a}^\alpha_{41} = (\ 0\ ,\ (1/24)(c/V)co),$$
$$\bar{a}^\alpha_{32} = (-(1/3)(c/V)co^2(1-t^2),\ 0\ ),$$
$$\bar{a}^\alpha_{23} = (\ 0\ ,\ -(1/12)(c/V)co^3(5-13t^2)),$$
$$\bar{a}^\alpha_{14} = ((1/24)(c/V)co^4(5-18t^2+t^4),\ 0\ ),$$
$$\bar{a}^\alpha_{05} = (\ 0\ ,\ (1/120)(c/V)co^5(5-18t^2+t^4)).$$

For $m \neq 1$

$$\bar{a}^{\alpha}_{n_1 n_2} := m\, \bar{a}^{\alpha}_{n_1 n_2} \tag{4-93h}$$

is true. If we select

$$B_o = B = \text{latitude of the test point}, \tag{4-94a}$$

(4-93b) then becomes

$$\bar{u}^{\alpha} = (x,y)^{\alpha} = \sum_{n=1}^{n'} \bar{a}^{\alpha}_{0\,n} \Delta L^{n}, \quad n_2 =: n\ , \tag{4-94b}$$

$\bar{a}^{\alpha}_{0\,n}$ after (4-93g) for $B_o = B$.

The power series setup for the transformation function $u^{\alpha}(\bar{u}^{\beta})$ reads

$$\Delta u^{\alpha} = (\Delta B,\ \Delta L)^{\alpha} = \sum_{n=1}^{n'} a^{\alpha}_{n_1 n_2} x^{n_1} y^{n_2}\ , \tag{4-95a}$$

$$\begin{aligned} a^{\alpha}_{n_1 n_2} &= (1/(n_1!n_2!))(\partial^n u^{\alpha}/(\partial x^{n_1} \partial y^{n_2}))_{P_o} \\ &= (1/(n_1!n_2!))(u^{\alpha}_{,n_1 x\bar{1}\ n_2 x\bar{2}})_{P_o}\ . \end{aligned} \tag{4-95b}$$

This involves the inverse transformation for (4-93) so that the coefficients of (4-95b) can be determined by series inversion according to (3-7). On the other hand, however, they can also be determined by solving the initial value problem (4-91) for the differential equations of (4-77d) with the method described for (3-56) to (3-58), i.e., analogous to the explanation for (4-93d,e). These equations are also true here when, again, $t =: B$ is the curve coordinate for C_o. The equations of (3-58a) are to be substituted by the appropriate equations resulting from (4-77d), and in (3-58b) u^{α} und $\bar{u}^{\alpha}$ are to be substituted for one another. With the abbreviations of (4-93f) the coefficients for $m = 1$ up to $n' = 5$ read

$$\begin{aligned} a^{\alpha}_{10} &= (V^3/c,\ \ 0\ )\ , \\ a^{\alpha}_{01} &= (\ 0\ \ ,\ (V/c)co(1+t^2))\ , \\[1ex] a^{\alpha}_{20} &= ((3/2)(V^2/c)^2 t(1-V^2),\ \ 0\ )\ , \\ a^{\alpha}_{11} &= (\ 0\ \ ,\ (V/c)^2 co\ t(1+t^2))\ , \\ a^{\alpha}_{02} &= (-(1/2)(V^2/c)^2 t,\ \ 0\ )\ , \end{aligned} \tag{4-95c}$$

$$
\begin{aligned}
a^{\alpha}_{30} &= (-(1/2)(V/c)^3)(1-V^2)(1-2V^2-(5-6V^2)t^2)\ ,\ 0\ )\ ,\\
a^{\alpha}_{21} &= (\ 0\ ,\ (1/2)(V/c)^3 co(1+t^2)(V^2+2t^2))\ ,\\
a^{\alpha}_{12} &= (-(1/2)(V/c)^3 V^2(V^2+(4-3V^2)t^2),\ 0\ )\ ,\\
a^{\alpha}_{03} &= (\ 0\ ,\ -(1/3)a^2_{21})\ ,\\
\\
a^{\alpha}_{40} &= (-(1/2)(V/c)^4 t(1-V^2),\ 0\ )\ ,\\
a^{\alpha}_{31} &= (\ 0\ ,\ (1/6)(V/c)^4 co\ t(1+t^2)(4+V^2+6t^2))\ ,\\
a^{\alpha}_{22} &= (-(1/4)(V/c)^4 t(11-9V^2+(3-V^2)t^2),\ 0\ )\ ,\\
a^{\alpha}_{13} &= (\ 0\ ,\ -a^2_{31})\ ,\\
a^{\alpha}_{04} &= (-(1/24)(V/c)^4 t(1-6V^2-3(3-2V^2)t^2),\ 0\ )\ ,\\
\\
a^{\alpha}_{50} &= (\ 0\ ,\ 0\ )\ ,\\
a^{\alpha}_{41} &= (\ 0\ ,\ (1/24)(V/c)^5 co(1+t^2)(5+28t^2+24t^4))\ ,\\
a^{\alpha}_{32} &= (-(1/6)(V/c)^5(1+4t^2+3t^4),\ 0\ )\ ,\\
a^{\alpha}_{23} &= (\ 0\ ,\ -2a^2_{41})\ ,\\
a^{\alpha}_{14} &= ((1/24)(V/c)^5(5+14t^2+9t^4),\ 0\ )\ ,\\
a^{\alpha}_{05} &= (\ 0\ ,\ (1/5)a^2_{41})\ .
\end{aligned}
\tag{4-95c}
$$

For $m \neq 1$

$$a^{\alpha}_{n_1 n_2} := a^{\alpha}_{n_1 n_2} / m^{n_1+n_2} \tag{4-95d}$$

is true. By selecting

$$B_o = B_f = \text{latitude of the ordinate foot point } Q : y_Q = 0 \tag{4-96a}$$

$x = 0$ corresponding to the definition of (4-91c) so that (4-95a) becomes

$$u^{\alpha} = (B-B_f,\ \Delta L)^{\alpha} = \sum_{n=1}^{n'} a^{\alpha}_{0n}\, y^n\ ,\quad n_2 := n\ , \tag{4-96b}$$

a^{α}_{0n} after (4-95c) for $B_o = B_f$.

4.7.2 INDIRECTLY TRANSFORMING THE COORDINATES

In place of the procedure for directly determining the transformation equations of (4-93) and (4-95) described in the previous chapter it is naturally possible as well to derive these equations by the intermediate use of isothermal geographic coordinates. The fundamentals of this method, which is only treated in the geodetic literature, are explained in the following.

Regarding isothermal geographic coordinates we will select the isothermal latitude and the geographic longitude of (4-78), which are combined into the complex variable

$$v = q + i\,L \,. \qquad (4\text{-}97a)$$

For the isothermal Gaussian coordinates of (4-91c-e) the complex variable

$$\bar{u} = x + i\,y \qquad (4\text{-}97b)$$

is introduced. Thereby, the intermediate transformation equations between the Gaussian and the isothermal geographic coordinates can be represented as analytic functions of complex variables corresponding to (3-65). With the coordinates

$$v_o = q_o + i\,L_o \qquad (4\text{-}97c)$$

of the reference point P_o [see (4-91a)] and their respective differences

$$\Delta v = v - v_o = \Delta q + i\,\Delta L \,, \quad \Delta q = q - q_o \,, \quad \Delta L = L - L_o \qquad (4\text{-}97d)$$

we can begin with transformation functions in the form of

$$\bar{u} = f(\Delta v) \,, \qquad \Delta v = g(\bar{u}) \,. \qquad (4\text{-}98a)$$

Due to the initial conditions of (4-91f-h)

$$x = m\,\Delta s = f(\Delta q) \,, \qquad \Delta q = g(x) = g(m\,\Delta s) \qquad (4\text{-}98b)$$

must be true for points of C_o. Excluding the constant factor m these initial value problems are equivalent to the example of (3-69) and can be solved in the same manner using power series for the transformation functions of (4-98a) as follows:

$$\bar{u} = \sum_{n=1}^{n'} \bar{b}_n \Delta v^n \,, \quad \bar{b}_n = \bar{b}^1_n = (1/n!)(d^n x/dq^n)_o \,, \quad \bar{b}^2_n = 0 \,, \qquad (4\text{-}98c)$$

and

$$\Delta v = \sum_{n=1}^{n'} b_n\,\bar{u}^n \,, \quad b_n = b^1_n = (1/n!)(d^n q/dx^n)_o \,, \quad b^2_n = 0 \,. \qquad (4\text{-}98d)$$

The coefficients $\bar{b}_n$ and b_n are real in this case. Proceeding from (4-98b) the first derivatives required for (4-98c,d) are to be formed as follows:

$$ds/dq = m(ds/dB)_o(dB/dq)_o \ , \qquad dq/dx = (dq/dB)_o(dB/ds)_o/m \ . \qquad (4\text{-}98e)$$

Herein,

$$ds/dB = c/V^3 \ , \qquad dq/dB = 1/(V^2\cos B) \qquad (4\text{-}98f)$$

in accordance with (4-48a) and (4-78d). Proceeding from (4-98e,f) we obtain the higher derivatives of x with respect to q or of q with respect to x necessary for calculating the coefficients of (4-98c,d) by continued differentiation.

Using (4-98c,d) only the intermediate transformations between the coordinates $\bar{u}$ and v are determined. To acquire the transformation equations of (4-93) and (4-94) between the Gaussian coordinates $\bar{u}$ and the geographic coordinate differences (4-93a), i.e.,

$$\Delta u^\alpha = (\Delta B, \ \Delta L)^\alpha \ , \qquad (4\text{-}99)$$

(4-93c,d) must first be split into real and imaginary parts according to (2-166). Then, series (4-81) is substituted for Δq in the right side of (4-98c). In (4-98d), on the contrary, the power series obtained for Δq is to be inserted into series (4-82).

4.7.3 TRANSFORMING THE METRIC TENSOR

A method similiar to that of (4-63)ff is used for calculating the metric tensor of Gaussian coordinates. $\bar{G}$ should be represented as a function of x and y so that as for (4-63) we will begin with the first equation of (3-2b) offering two alternatives here:

$$\begin{aligned} \bar{G} := \bar{g}_{11} &= (\Delta B_{,\bar{1}})^2 g_{11} + (\Delta L_{,\bar{1}})^2 g_{22} \\ = \bar{g}_{22} &= (\Delta B_{,\bar{2}})^2 g_{11} + (\Delta L_{,\bar{2}})^2 g_{22} \ . \end{aligned} \qquad (4\text{-}100a)$$

The second equation of (3-2b) used in (3-61) is only appropriate when $\bar{G}$ is to be determined as a function of ΔB and ΔL. Based on (4-95) the transformation matrices for the first equation of (4-100a) become

$$(\Delta B_{,\bar{1}},\ \Delta L_{,\bar{1}}^{\alpha}) = \sum_{n=1}^{5} n_1 a^{\alpha}_{n_1 n_2} x^{n_1 - 1} y^{n_2} , \qquad (4\text{-}100b)$$

and the components $g_{(\alpha\alpha)}$ of the metric tensor of the geographic coordinates to be formed at the test point are to be expanded into a power series in x and y corresponding to (4-64b,c). We now have all the presuppositions for representing (4-100a) by a power series in x and y. Assuming m = 1 for (4-91h) the result reads

$$\bar{G} = 1 - (V/c)^2 y^2 (V^2 + 4(V/c)t(1-V^2)x + (1/6)(V/c)^2(5-9V^2)y^2) , \qquad (4\text{-}101a)$$

and from this

$$\bar{G}^{1/2} = 1 - (1/2)(V/c)^2 y^2 (V^2 + 4(V/c)t(1-V^2)x + (1/12)(V/c)^2(7-12V^2)y^2) , \qquad (4\text{-}101b)$$

$$\bar{G}^{-1/2} = 1 + (1/2)(V/c)^2 y^2 (V^2 + 4(V/c)t(1-V^2)x + (1/12)(V/c)^2 y^2) \qquad (4\text{-}101c)$$

ensue.

4.7.4 M E R I D I A N C O N V E R G E N C E

The comments here are identical with those made in Chapter 4.4.3 [see (4-67)ff] with the exception of obtaining the transformation matrices. Figure 4.4 is particularly illustrative here as well. Analogous to (4-67a), in the system of Gaussian coordinates $\bar{u}^{\alpha}$ [see (4-91a)] the direction of the meridian or of the B-line in a point P can be given by the contravariant unit vector

$$\bar{\nu}_B^{\alpha} = \text{tangent unit vector in the direction of the B-line in P} \qquad (4\text{-}102a)$$

and correponding to (4-67b) is defined as

$$\begin{aligned}\gamma &= \text{meridian convergence} \qquad (4\text{-}102b)\\ &= \text{direction angle of the B-line relative to the x-line at P.}\end{aligned}$$

Due to the orthogonality of both the u^{α}-system and the $\bar{u}^{\alpha}$-system γ is also the angle between the L-line and y-line at P.

The arrangement of the geographic coordinates of (4-74b) chosen here requires an exchange of L and B in (4-68a,c), and the metric tensor of the Gaussian coordinates is introduced into (4-68d). Thereby, we obtain for the contravariant unit vectors, i.e.,

$$\nu_p \,, \quad p \in \{B,L,x,y\} \tag{4-103a}$$

= tangent unit vectors on the p-lines in the u^α-system ,

$$\bar{\nu}^{\alpha}_{p} = \text{components of the } \nu^{\alpha}_{p} \text{ in the } \bar{u}^{\alpha}\text{-system} \,, \tag{4-103b}$$

the expressions

$$\nu^{\alpha}_{B} = g_{11}^{-1/2}\delta^{\alpha}_{1} = (V^3/c)\delta^{\alpha}_{1} \,, \quad \nu^{\alpha}_{L} = g_{22}^{-1/2}\delta^{\alpha}_{2} = (V/(c\cos B))\delta^{\alpha}_{2} \,, \tag{4-103c}$$

$$\bar{\nu}^{\alpha}_{x} = \bar{g}_{11}^{-1/2}\delta^{\alpha}_{1} = \bar{G}^{-1/2}\delta^{\alpha}_{1} \,, \quad \bar{\nu}^{\alpha}_{y} = \bar{g}_{22}^{-1/2}\delta^{\alpha}_{2} = \bar{G}^{-1/2}\delta^{\alpha}_{2} \,. \tag{4-103d}$$

By transformation using (2-14) the condition equation for (4-102a) thus takes on the following form:

$$\bar{\nu}^{\alpha}_{B} = \bar{u}^{\alpha}_{,\beta}\nu^{\beta}_{B} = (V^3/c)\bar{u}^{\alpha}_{,1} \,, \tag{4-104a}$$

in which

$$\bar{u}^{\alpha}_{,1} = (x_{,1}, y_{,1})^{\alpha} = \sum_{n=1}^{5} n_1 \bar{a}^{\alpha}_{1 n_1 n_2} \Delta B^{n_1-1} \Delta L^{n_2} \tag{4-104b}$$

is to be calculated based on (4-93). If reference point P_o [see (4-91a)] is chosen as the point of intersection between the L-line and C_o, then

$$B_o = B \qquad \rightarrow \qquad \Delta B = 0 \,, \tag{4-104c}$$

and (4-104b) is simplified as follows:

$$\bar{u}^{\alpha}_{,1} = (x_{,1}, y_{,1})^{\alpha} = \sum_{n=1}^{5} \bar{a}^{\alpha}_{1\,n} \Delta L^{n} \,. \tag{4-104d}$$

The meridian convergence γ [see (4-102b)] is also calculated in accordance with (4-70) to (4-72). If, for example, the first form of (4-70) is selected, we then obtain

$$\sin\gamma = -g_{\alpha\beta}\nu^{\alpha}_{B}\, u^{\beta}_{,\bar{\gamma}}\, \bar{\nu}^{\gamma}_{y} = -(c/V^3)\bar{G}^{-1/2}\Delta B_{,\bar{2}} \,. \tag{4-105a}$$

Due to (4-95)

$$\Delta B_{,\bar{2}} = \sum_{n=1}^{5} n_2 a^2_{2 n_1 n_2} x^{n_1} y^{n_2-1} \,, \tag{4-105b}$$

and $\bar{G}^{-1/2}$ is to be inserted into (4-105a) according to (4-101c). V is calculated as a function of the latitude B of the test point.

4.8 TRANSFORMATIONS BETWEEN GAUSSIAN ISOTHERMAL COORDINATES AND GEODESIC POLAR COORDINATES

4.8.1 DIRECTLY TRANSFORMING THE COORDINATES

In the system of Gaussian isothermal coordinates (4-91c)

$$u^\alpha = (x,y)^\alpha \tag{4-106a}$$

with the metric tensor

$$G(u^\alpha) := \bar{G} \quad \text{after (4-101)} \tag{4-106b}$$

the geodesic polar coordinates are defined corresponding to (2-112) and (2-113) whereby the direction angle from the u^1-lines is measured instead of (2-113b) [see Fig. 4.5]. The following special symbols and definitions will be used here:

$$(u^\alpha)_a = (x_a,\ y_a)^\alpha = \text{Gaussian coordinates of the pole } P_a\ , \tag{4-107a}$$

$$\begin{aligned} \nu_a^\alpha &= \nu_{a1}^\alpha = G^{-1/2}\delta_1^\alpha \\ &= \text{northern direction of the x-line at } P_a\ , \end{aligned} \tag{4-107b}$$

$$\begin{aligned} \bar{u}^\alpha &= (\beta_a,\ r_{ab})^\alpha \\ &= \text{geodesic polar coordinates of } P_b \text{ with respect to } P_a \end{aligned} \tag{4-107c}$$

with

$$\begin{aligned} \beta_a &= \text{direction angle of the geodesic line } P_a - P_b \\ &\quad\ \text{relative to (4-107b)}, \end{aligned} \tag{4-107d}$$

$$r_{ab} = \text{length of the geodesic line } P_a - P_b\ . \tag{4-107e}$$

Corresponding to (2-113g) and observing (4-107d) and (4-106b) we obtain the direction of the geodesic line $P_a - P_b$ in P_a, i.e.,

$$(u^{\alpha'})_a = (x'_a,\ y'_a)^\alpha = G_a^{-1/2}(\cos\beta_a,\ \sin\beta_a)^\alpha\ , \tag{4-108a}$$

so that

$$(\bar{v}^\alpha)_a = G_a^{-1/2}(\cos\beta_a,\ \sin\beta_a)^\alpha r_{ab} \tag{4-108b}$$

are the normal coordinates (3-15a) of the geodesic $P_a - P_b$ with respect to P_a.

The following way of determining the transformation equations in the form of power series between u^α and $\bar{u}^\alpha$ is formally the same as that given in Chapter 4.3.1. Together with (4-108) and

$$\Delta x = x_b - x_a \ , \qquad \Delta y = y_b - y_a \tag{4-109a}$$

the solution to the transformation problem of (2-114) becomes a <u>Legendrean series</u> (3-15c) analogous to (4-25b), i.e.,

$$\Delta u^\alpha = (\Delta x,\ \Delta y)^\alpha = (\bar{v}^\alpha)_a + \sum_{n=2}^{n'} b^\alpha_{n_1 n_2} (\bar{v}^1)_a^{\,n_1} (\bar{v}^2)_a^{\,n_2} \ , \quad n_1 + n_2 = n \ . \tag{4-109b}$$

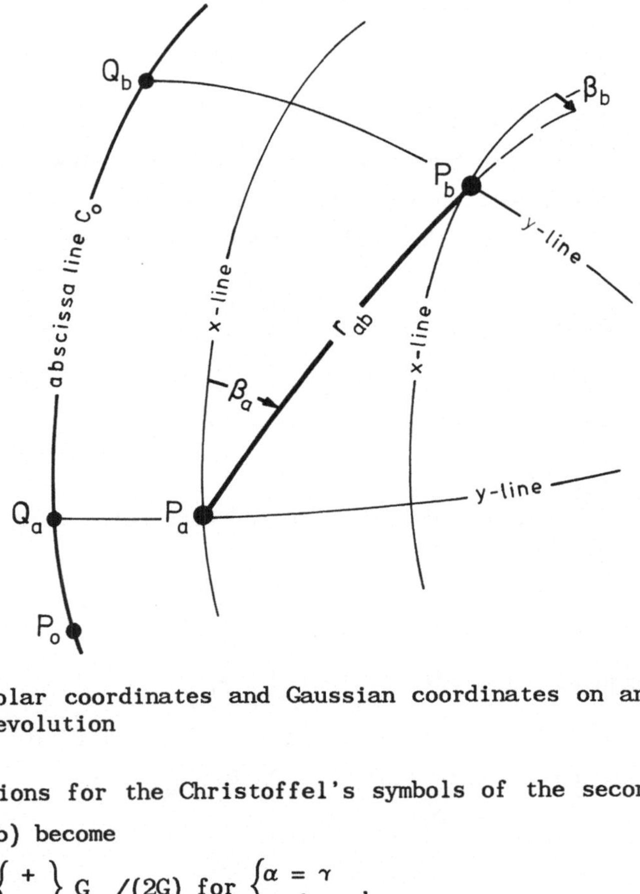

Fig. 4.5. Polar coordinates and Gaussian coordinates on an ellipsoid of revolution

The expressions for the Christoffel's symbols of the second kind (2-94) with (4-106b) become

$$\Gamma^\gamma_{(\alpha\alpha)} = \left\{ \begin{matrix} + \\ - \end{matrix} \right\} G_{,\gamma}/(2G) \ \text{for} \ \begin{cases} \alpha = \gamma \\ \alpha \neq \gamma \end{cases} \ , \tag{4-109c}$$

$$\Gamma^1_{12} = G_{,2}/(2G) \ , \qquad \Gamma^2_{12} = G_{,1}/(2G) \ ,$$

whereby we obtain the following for the coefficients of (4-109b) up to $n' = 3$ based on (3-14):

$$b^\alpha_{20} = (-G_{,1},\ G_{,2})^\alpha_a/(4G_a) \ , \quad b^\alpha_{11} = (-G_{,2},\ -G_{,1})^\alpha_a/(2G_a) \ ,$$

$$b^\alpha_{02} = (\ G_{,1},\ -G_{,2})^\alpha_a/(4G_a) \ , \tag{4-109d}$$

$$b^\alpha_{30} = (2G^2_{,1} - G^2_{,2} - G\,G_{,11},\ -3G_{,1}G_{,2} + G\,G_{,12})^\alpha_a/(12G^2_a)\ , \qquad (4\text{-}109d)$$
$$b^\alpha_{21} = (9G_{,1}G_{,2} - 3G\,G_{,12},\ 5G^2_{,1} - 4G^2_{,2} - 2G\,G_{,11} + G\,G_{,22})^\alpha_a/(12G^2_a)\ ,$$
$$b^\alpha_{12} = (5G^2_{,2} - 4G^2_{,1} - 2G\,G_{,22} + G\,G_{,11},\ 9G_{,1}G_{,2} - 3G\,G_{,12})^\alpha_a/(12G^2_a)\ ,$$
$$b^\alpha_{03} = (-3G_{,1}G_{,2} + G\,G_{,12},\ 2G^2_{,2} - G^2_{,1} - G\,G_{,22})^\alpha_a/(12G^2_a)\ .$$

Due to (4-101a) the fundamental quantity G and its partial derivatives are calculated as follows:

$$G_a = 1-(V/c)^2 y^2_a(V^2+4(V/c)t(1-V^2)x_a+(1/6)(V/c)^2(5-9V^2)y^2_a), \qquad (4\text{-}110a)$$

$$(G_{,1})_a = -4(V/c)^3 t\ (1-V^2)y^2_a,$$
$$(G_{,2})_a = -2(V/c)^2 y_a(V^2+4(V/c)t(1-V^2)x_a+(1/3)(V/c)^2(5-9V^2)y^2_a), \qquad (4\text{-}110b)$$

$$(G_{,11})_a = 0\ , \qquad (G_{,12})_a = -8(V/c)^3 t(1-V^2)y_a\ , \qquad (4\text{-}110c)$$
$$(G_{,22})_a = -2(V/c)^2\ (V^2+4(V/c)t(1-V^2)x_a+(V/c)^2(5-9V^2)y^2_a)\ ;$$

t and V are calculated at the reference point P_o of the Gaussian coordinates of (4-91a), i.e.,

$$t := \tan B_o\ , \qquad V := (1+e''^2\cos^2 B_o)^{1/2}\ . \qquad (4\text{-}110d)$$

When applying Gaussian coordinate systems to geodetic surveying, we can frequently assume

$$y < 150 \text{ km}\ , \qquad r < 20 \text{ km}\ . \qquad (4\text{-}111a)$$

If

$$x_a = 0 \qquad \rightarrow \qquad B_o = B_{Qa} \approx B_a \qquad (4\text{-}111b)$$

is used as well [B_Q after (4-96a)], then by inserting Gaussian curvature (4-19a), i.e.,

$$K_a = (V^2_a/c)^2\ , \qquad (4\text{-}111c)$$

the expressions of (4-110) can be approximated accurately enough with

$$G_a = 1 - K_a y^2_a\ , \qquad (4\text{-}111d)$$

$$(G_{,2})_a = -2\ K_a y_a,\quad (G_{,22})_a = -2\ K_a\ , \qquad (4\text{-}111e)$$
$$(G_{,1})_a = (G_{,11})_a = (G_{,12})_a = 0\ .$$

If the coefficients of (4-109d) are calculated in the same way,

$$K^2_a \rightarrow 0 \qquad (4\text{-}111f)$$

can also be assumed, with which the following results are ultimately obtained:

$$b^{\alpha}_{20} = (\ 0\ ,\ -(1/2)K_a)y_a\ ,\quad b^{\alpha}_{11} = (\ K_a\ ,\ 0\ )y_a\ ,$$
$$b^{\alpha}_{02} = (\ 0\ ,\ \ (1/2)K_a)y_a\ ,$$
$$b^{\alpha}_{30} = (\ 0\ ,\ 0\ )\ ,\qquad b^{\alpha}_{21} = (\ 0\ ,\ -(1/6)K_a)\ ,\qquad (4\text{-}111g)$$
$$b^{\alpha}_{12} = ((1/3)K_a,\ 0\ )\ ,\qquad b^{\alpha}_{03} = (\ 0\ ,\ \ (1/6)K_a)\ .$$

Due to (4-111f) the normal coordinates of (4-108b) can be calculated accurately enough with $G_a = 1$ in (4-109b) for $n > 1$, i.e.,

$$(\bar{v}^{\alpha})_a \approx (\cos\beta_a,\ \sin\beta_a)^{\alpha} r_{ab}\ . \qquad (4\text{-}112a)$$

Yet, for $n = 1$ (4-111d) must be considered anyway, i.e.,

$$(\bar{v}^{\alpha})_a = (1+(1/2)K_a y_a^2)(\cos\beta_a,\ \sin\beta_a)^{\alpha} r_{ab}\ . \qquad (4\text{-}112b)$$

Only roughly approximated values are required for calculating the Gaussian curvature of (4-111c) so that a mean value of K_a is adequate for areas only extending up to 100 km in a north-south direction.

By inverting the Legendrean series of (4-109b) corresponding to (3-15e-g) we obtain the following solution to the transformation problem of (2-115):

$$(\bar{v}^{\alpha})_a = (1+(1/2)K_a y_a^2)(\cos\beta_a,\ \sin\beta_a)^{\alpha} r_{ab}$$
$$= (\Delta x,\ \Delta y)^{\alpha} + \sum_{n=2}^{n'} \bar{b}^{\alpha}_{n_1 n_2} \Delta x^{n_1} \Delta y^{n_2},\quad n_1 + n_2 = n\ . \qquad (4\text{-}113a)$$

The coefficients are calculated in line with (3-7). With the assumptions made for and results obtained from (4-111) the following is true up to $n' = 3$:

$$\bar{b}^{\alpha}_{n_1 n_2} = -b^{\alpha}_{n_1 n_2} \quad \text{after (4-111g).} \qquad (4\text{-}113b)$$

Based on the normal coordinates $(\bar{v}^{\alpha})_a$ calculated by means of (4-113a,b) the polar coordinates of (4-107c-e) become

$$\beta_a = \arctan((\bar{v}^2)_a/(\bar{v}^1)_a)\ ,$$
$$r_{ab} = (1-(1/2)K_a y_a^2)((\bar{v}^1)_a^2 + (\bar{v}^2)_a^2)^{1/2}\ . \qquad (4\text{-}113c)$$

For solving the transformation problem of (2-115) the essentially better converging <u>Gaussian mid-coordinate formulas</u> of (3-26) are usually preferred over the inverse Legendrean series of (4-113). In the following the mid-coordinate formulas will be derived similar to (3-19) to (3-27) or (4-27) and (4-28) only for $n' = 3$. The coordinate differences between the points P_a and P_b in the u^α-system of (4-106) are

$$\Delta u^\alpha = (\Delta x,\ \Delta y)^\alpha\ ,\quad \Delta x = x_b - x_a\ ,\quad \Delta y = y_b - y_a\ , \tag{4-114a}$$

and for the mid-coordinates of (3-21a) we obtain

$$(u^\alpha)_m = (x_m,\ y_m)^\alpha\ ,\quad x_m = (x_a + x_b)/2\ ,\quad y_m = (y_a + y_b)/2\ . \tag{4-114b}$$

According to (4-112b) the normal coordinates of (3-22a) become

$$(\bar{v}^\alpha)_p = (1+(1/2)K\ y_p^2)(\cos\beta_p,\ \sin\beta_p)^\alpha r_{ab}\ , \tag{4-114c}$$
$$p \in \{a,b\}\ ,\quad K \approx K_a \approx K_b\ .$$

Due to $n' = 3$ in (3-23) and (3-24), (3-25) is limited to $q' = 1$. Thereby and as a result of the sole y-dependency of the coefficients $\bar{b}^\alpha_{n_1n_2}$ (3-25) is employed here in the form of

$$\left.\begin{matrix}\bar{b}^\alpha_{n_1n_2a}\\ \bar{b}^\alpha_{n_1n_2b}\end{matrix}\right\} = \bar{b}^\alpha_{n_1n_2m}\left\{\begin{matrix}-\\+\end{matrix}\right\} c^\alpha_{n_1n_201}\ \Delta y/2 \tag{4-114d}$$

with

$$c^\alpha_{n_1n_201} = (d\bar{a}^\alpha_{n_1n_2}/dy)_m\ ,\quad n_1 + n_2 = 2\ . \tag{4-114e}$$

Based on (4-113b) and (4-111g) these coefficients become

$$c^\alpha_{2001} = (\ 0\ ,\ \ (1/2)K)\ ,\quad c^\alpha_{1101} = (\ -K\ ,\ 0\ )\ ,\quad c^\alpha_{0201} = (\ 0\ ,\ -(1/2)K)\ . \tag{4-114f}$$

The equations of (3-24) together with (4-114d) can be written as

$$\begin{aligned}(\bar{v}^\alpha)_m &= (\Delta x,\ \Delta y)^\alpha - (1/2)\sum_{n_1+n_2=2} c^\alpha_{n_1n_201}\ \Delta x^{n_1}\ \Delta y^{n_2+1}\\ &\quad + \sum_{n_1+n_2=3} \bar{b}^\alpha_{n_1n_2m}\ \Delta x^{n_1}\ \Delta y^{n_2}\\ \delta\bar{v}_{ab} &= \quad - \sum_{n_1+n_2=2} \bar{b}^\alpha_{n_1n_2m}\ \Delta x^{n_1}\ \Delta y^{n_2}\ ,\end{aligned} \tag{4-114g}$$

from which we obtain together with (4-113b) and (4-114f) the <u>mid-coordi-</u>

nate formulas in accordance with (3-26a,b), i.e.,

$$(\bar{v}^{\alpha})_m = ((\bar{v}^{\alpha})_a + (\bar{v}^{\alpha})_b)/2 \tag{4-115a}$$

$$= (\Delta x(1+A^1_{12m}\Delta y^2),\ \Delta y(1+A^2_{21m}\Delta x^2 + A^2_{03m}\Delta y^2))^{\alpha},$$

$$\delta\bar{v}^{\alpha}_{ab} = ((\bar{v}^{\alpha})_b - (\bar{v}^{\alpha})_a)/2 \tag{4-115b}$$

$$= (B^1_{11m}\Delta x\ \Delta y,\ B^2_{20m}\Delta x^2 + B^2_{02m}\Delta y^2)^{\alpha} .$$

The nonvanishing coefficients of $A^{\alpha}_{n_1n_2m}$ and $B^{\alpha}_{n_1n_2m}$ are

$$A^1_{12m} = \bar{a}^1_{12m} - (1/2)c^1_{1101} = (1/6)\,K ,$$
$$A^2_{21m} = \bar{a}^2_{21m} - (1/2)c^2_{2001} = -(1/12)K , \tag{4-115c}$$
$$A^2_{03m} = \bar{a}^2_{03m} - (1/2)c^2_{0201} = (1/12)K .$$

$$B^1_{11m} = -\bar{a}^1_{11m} = K\,y_m ,$$
$$B^2_{20m} = -\bar{a}^2_{20m} = -(1/2)K\,y_m , \tag{4-115d}$$
$$B^2_{02m} = -\bar{a}^2_{02m} = (1/2)K\,y_m .$$

If $(\bar{v}^{\alpha})_m$ and $\delta\bar{v}^{\alpha}_{ab}$ are calculated based on the Gaussian coordinates of (4-106a) of points P_a and P_b or their differences, and mid-values of (4-114a,b) according to (4-115), we then proceed according to (3-27c, d,f). The normal coordinates at P_a and P_b are calculated first, i.e.,

$$(\bar{v}^{\alpha})_a = (\bar{v}^{\alpha})_m - \delta\bar{v}^{\alpha}_{ab} , \qquad (\bar{v}^{\alpha})_b = (\bar{v}^{\alpha})_m + \delta\bar{v}^{\alpha}_{ab} , \tag{4-115e}$$

whereby the polar coordinates of (4-107c,e) are obtained in the following way proceeding from (4-114c):

$$\beta_p = \arctan((\bar{v}^2)_p/(\bar{v}^1_p)) , \quad p \in \{a,b\} ,$$
$$r_{ab} = (1-(1/2)K\,y_p^2)((\bar{v}^1)_p^2 + (\bar{v}^2)_p^2)^{1/2} . \tag{4-115f}$$

β_a and β_b are the direction angles of the geodesic $P_a - P_b$ at P_a and P_b corresponding to Fig. 4.5.

4.8.2 TANGENT VECTORS AND DIRECTION ANGLES OF GEODESIC R-LINES

Analogous to the problems dealt with in Chapter 4.3.3 we will now go into the calculation of

$$u^{\alpha'} = du^{\alpha}/dr \ , \quad u^{\alpha} \text{ after (4-106)} \tag{4-116a}$$

$= \text{tangent vector on the r-line } P_a - P_b \text{ at } P \ ,$

and

$$\beta = \text{direction angle of the r-line at } P \ . \tag{4-116b}$$

P is an arbitrary point on the geodesic line $P_a - P_b$ determined by the polar coordinates

$$\bar{u}^{\alpha} = (\beta_a, r) \ , \qquad r = \text{distance } P_a - P \tag{4-117a}$$

or the normal coordinates

$$(\bar{v}^{\alpha})_a = G_a^{-1/2}(\cos\beta_a, \sin\beta_a)^{\alpha} r \tag{4-117b}$$

with the pole at P_a. We let

$$u^{\alpha} = (x, y)^{\alpha} \tag{4-117c}$$

be the Gaussian coordinates of P, and

$$\Delta u^{\alpha} = u^{\alpha} - (u^{\alpha})_a = (x-x_a, y-y_a) \tag{4-117d}$$

is formed.

For the <u>tangent vector of (4-116a) as a function of $\bar{v}^{\alpha}$</u> the power series (3-16a), which together with the assumptions made here, reads

$$u^{\alpha'} = (u^{\alpha'})_a + (1/r) \sum_{n=2}^{r'} n\, b^{\alpha}_{n_1 n_2} (\bar{v}^1)_a^{n_1} (\bar{v}^2)_a^{n_2} \ , \tag{4-118a}$$

is generally valid. For (4-118a) the following is true:

$$u^{\alpha'} = G^{-1/2}(\cos\beta \ , \sin\beta\ ) \ , \quad G = G(u^{\alpha}) \tag{4-118b}$$

$$(u^{\alpha'})_a = G_a^{-1/2}(\cos\beta_a, \sin\beta_a) \ , \tag{4-118c}$$

$$b^{\alpha}_{n_1 n_2} \quad \text{after (4-109d) and (4-111g)} \ . \tag{4-118d}$$

For the <u>direction angle of (4-116b) as a function of $\bar{v}^{\alpha}$</u> we obtain the following based on (4-118b):

$$\beta = \arctan\,(u^{2'}/u^{1'}) \ , \tag{4-119}$$

in which series (4-118a) must be inserted on the right side. (4-119) is independent of the metric tensor in P so that here we do not have the difficulties mentioned for (4-37). The procedures given for (3-17) and (3-18) are purposeful here for representing β as a direct power series in $\bar{v}^\alpha$. The result corresponding to (3-18b) up to $n' = 2$ reads

$$\Delta\beta = \beta - \beta_a = \sum_{n=1}^{2} c_{n_1 n_2} (\bar{v}^1)_a^{n_1} (\bar{v}^2)_a^{n_2} , \quad n_1 + n_2 = n . \tag{4-120a}$$

On the basis of (4-111a-f) the nonvanishing coefficients become

$$c_{10} = -K_a y_a , \qquad c_{11} = -(1/2)K_a . \tag{4-120b}$$

With this accuracy

$$(v^\alpha)_a \approx (\cos\beta_a, \sin\beta_a) r \approx \Delta u^\alpha \tag{4-120c}$$

is also true whereby the power series

$$\Delta\beta = \beta - \beta_a = \sum_{n=1}^{2} c_{n_1 n_2} (\Delta u^1)^{n_1} (\Delta u^2)^{n_2} \tag{4-120d}$$

for the <u>direction angle of (4-116b) as a function of Δu^α</u> directly follows from (4-120a).

The <u>tangent vector of (4-116a) as a function of Δu^α</u> is obtained via the mid-coordinate formulas of (4-114) and (4-115) with

$$P := P_b , \qquad y_m = (y_a + y)/2 . \tag{4-121a}$$

Thereby, the second equation of (4-115e) assumes the form of

$$\bar{v}^\alpha = u^{\alpha'} r = (\bar{v}^\alpha)_m + \delta\bar{v}^\alpha_a . \tag{4-121b}$$

r is calculated based on the second equation of (4-115f) whereby

$$u^{\alpha'} = \bar{v}^\alpha / r \tag{4-121c}$$

is finally obtained. The <u>direction angle of (4-116b) as a function of Δu^α</u> can also be calculated with the results of (4-121b) for $\bar{v}^\alpha$ according to the first equation of (4-115f) as follows:

$$\beta = \arctan(\bar{v}^2 / \bar{v}^1) . \tag{4-121d}$$

The previous calculations related to an arbitrary point P on the geodesic line $P_a - P_b$ are naturally valid for P_a as well. Then,

$$P := P_b \qquad \rightarrow \qquad r =: r_{ab} \text{ after (4-107e)} \tag{4-122a}$$

is to be used whereby the desired quantities of (4-116a,b) become

$$(u^{\alpha'})_b = \text{tangent vector on the r-line at } P_b ,$$
$$\beta_b = \text{direction angle of the r-line at } P_b \qquad (4\text{-}122b)$$

[see Fig. 4.5].

4.8.3 COORDINATE TRANSFORMATION BY REDUCING DIRECTIONS AND DISTANCES

The transformations discussed in Chapter 4.8.1 will be regarded again in the following from another viewpoint. In addition to the geodesic polar coordinates of (4-107c-e),

$$\bar{w}^{\alpha} = (\bar{\beta}_a, \bar{r}_{ab})^{\alpha} \qquad (4\text{-}123a)$$
$$= \text{reduced polar coordinates of } P_b \text{ relative to } P_a ,$$

$$\bar{\beta}_a = \text{reduced direction angle of the geodesic } P_a - P_b , \qquad (4\text{-}123b)$$

$$\bar{r}_{ab} = \text{reduced length of the geodesic } P_a - P_b \qquad (4\text{-}123c)$$

are introduced whereby the term "reduced length" is not identical with (4-32b). Reduced polar coordinates are defined by the equation

$$\Delta u^{\alpha} = (\Delta x, \Delta y)^{\alpha} = (\cos\bar{\beta}_a, \sin\bar{\beta}_a)^{\alpha}\bar{r}_{ab} . \qquad (4\text{-}123d)$$

The components of the differences between (4-123a) and (4-107c) are referred to as <u>direction and distance reductions</u>, i.e.,

$$\delta\bar{w}^{\alpha} = \bar{u}^{\alpha} - \bar{w}^{\alpha} = (\delta\beta_a, \delta r_{ab})^{\alpha} ,$$
$$\delta\beta_a = \beta_a - \bar{\beta}_a , \qquad \delta r_{ab} = r_{ab} - \bar{r}_{ab} . \qquad (4\text{-}123e)$$

Before we go into the calculation of $\delta\bar{w}^{\alpha}$, its application in the <u>transformation problems of (2-114) and (2-115)</u> should first be explained. In the case of (2-114) the following steps replace (4-109) to (4-112):

- calculating $\delta\bar{w}^{\alpha}$ after (4-127b,c) , (4-124a)
- $\bar{\beta}_a = \beta_a - \delta\beta_a$, $\quad \bar{r}_{ab} = r_{ab} - \delta r_{ab}$, (4-124b)
- calculating Δu^{α} after (4-123d) . (4-124c)

For the inverse transformation (2-115) the following steps replace (4-113) to (4-115):

- calculating $\bar{w}^{\alpha}$ based on (4-123d), i.e., (4-125a)

$$\bar{\beta}_a = \arctan(\Delta u^2/\Delta u^1) , \qquad \bar{r}_{ab} = ((\Delta u^1)^2 + (\Delta u^2)^2)^{1/2} ,$$

- calculating $\delta\bar{w}^{\alpha}$ after (4-127b,c) , (4-125b)

- $\beta_a = \bar{\beta}_a + \delta\beta_a , \qquad r_{ab} = \bar{r}_{ab} + \delta r_{ab} .$ (4-125c)

For <u>calculating the relatively small $\delta\bar{w}^{\alpha}$</u> the right side of (4-123d) is expanded into a power series with respect to $\delta\bar{w}^{\alpha}$ at the point

$$\delta\bar{w}^{\alpha} = 0^{\alpha} \rightarrow \bar{\beta}_a = \beta_a , \qquad \bar{r}_{ab} = r_{ab} . \tag{4-126a}$$

By introducing the Riemannian normal coordinates with respect to P_a, i.e.,

$$v^{\alpha} = (\cos\beta_a, \sin\beta_a)r_{ab} = G_a^{1/2}(\bar{v}^{\alpha})_a , \tag{4-126b}$$

this series can be written as

$$\Delta u^{\alpha} = v^{\alpha} + \sum_{n=1}^{n'} c^{\alpha}_{n_1 n_2}(\delta\beta_a)^{n_1}(\delta r_{ab})^{n_2} , \qquad n_1 + n_2 = n . \tag{4-126c}$$

The coefficients herein are

$$c^{\alpha}_{n_1 n_2} = (-1)^n(1/(n_1!n_2!))\partial^n v^{\alpha}/(\partial\beta_a^{n_1}\partial r_{ab}^{n_2}) , \tag{4-126d}$$

and the partial derivatives of $(v^{\alpha})_a$ can be calculated in a simple way using (4-126b). Up to $n' = 2$

$$c^{\alpha}_{10} = (v^2, -v^1)^{\alpha} , \qquad c^{\alpha}_{01} = -v^{\alpha}/r_{ab} ,$$
$$c^{\alpha}_{20} = (v^1, v^2)^{\alpha}/2 , \qquad c^{\alpha}_{11} = c^{\alpha}_{10}/r_{ab} , \qquad c^{\alpha}_{02} = 0^{\alpha} . \tag{4-126e}$$

The desired result for $\delta\bar{w}^{\alpha}$ is obtained by inverting series (4-126d,e) according to (3-7), i.e.,

$$\delta\bar{w}^{\alpha} = (\delta\beta_a, \delta r_{ab})^{\alpha} = \sum_{n=1}^{n'} \bar{c}^{\alpha}_{n_1 n_2}(\Delta x - v^1)^{n_1}(\Delta y - v^2)^{n_2} , \tag{4-126f}$$

with the following coefficients up to $n' = 2$:

$$\bar{c}^{\alpha}_{10} = (v^2/r, -v^1)^{\alpha}/r , \qquad c^{\alpha}_{01} = (-v^1/r, -v^2)^{\alpha}/r ,$$
$$\bar{c}^{\alpha}_{20} = (-v^1v^2/r, -(v^2)^2/2)^{\alpha}/r^3 ,$$
$$\bar{c}^{\alpha}_{11} = (((v^1)^2 - (v^2)^2)/r, v^1v^2)^{\alpha}/r^3$$
$$\bar{c}^{\alpha}_{02} = (v^1v^2/r, -(v^1)^2/2)^{\alpha}/r^3 . \tag{4-126g}$$

On the right side of (4-126f) either the inverse series (4-113a) is to be substituted for $(\bar{v}^{\alpha})_a$, or the Legendrean series (4-109b) for Δu^{α} in the differences

$$\Delta u^{\alpha} - v^{\alpha} = \Delta u^{\alpha} - G_a^{1/2}(\bar{v}^{\alpha})_a, \quad \Delta u^{\alpha} = (\Delta x, \Delta y)^{\alpha} . \tag{4-126h}$$

Only the first of these possibilities with the accuracy given in (4-111) will be treated further here. For this purpose the series of (4-126f,g) is limited to $n' = 1$, and

$$v^{\alpha} \approx (\Delta x, \Delta y)^{\alpha} , \qquad r \approx (\Delta x^2 + \Delta y^2)^{1/2} \tag{4-127a}$$

can be assumed in the remaining coefficients. Under these conditions we obtain the following results for the direction and distance reductions of (4-123e) by inserting (4-113a,b) into (4-126f,g):

$$\delta\beta_a = \beta_a - \bar{\beta}_a = (1/6)K_a(2y_a + y_b)\Delta x , \tag{4-127b}$$

$$\delta r_{ab} = r_{ab} - \bar{r}_{ab} = -(1/6)K_a(y_a^2 + y_a y_b + y_b^2) r_{ab} . \tag{4-127c}$$

4.9 TRANSFORMATIONS BETWEEN TWO SYSTEMS OF GAUSSIAN ISOTHERMAL COORDINATES

4.9.1 INDIRECTLY TRANSFORMING THE COORDINATES

In the following, two systems of Gaussian coordinates will be considered, which are defined in the system of geographic coordinates of (4-90) corresponding to (4-91) as

$$u^{\alpha} = (B, L)^{\alpha} . \tag{4-128}$$

The following symbols apply to both systems:

$$GS_s = \text{Gaussian coordinate system} \quad s \in \{1,2\} , \tag{4-129a}$$

$$(u^{\alpha})_{os} = (B_{os}, L_{os})^{\alpha} \tag{4-129b}$$

= geographic coordinates of the reference point P_{os} ,

$$C_{os} = \text{abscissa line through } P_{os} \tag{4-129c}$$

= meridian through P_{os} after (4-75c) ,

$$\overset{s}{u}{}^{\alpha} = (\overset{s}{x}, \overset{s}{y})^{\alpha} = \text{Gaussian coordinates in } GS_s . \tag{4-129d}$$

The Gaussian systems GS_s differ essentially in their reference longitudes. The factor m [see (4-91h)] is assumed to have a value of one so that

$$L_{o1} \neq L_{o2} , \qquad m = 1 . \tag{4-129e}$$

In contrast, the reference latitudes B_{os} can have equal or different values.

An indirect method for transformation between $\overset{1}{u}{}^{\alpha}$ and $\overset{2}{u}{}^{\alpha}$ consists of using the transformation functions between Gaussian and geographic coordinates of (4-93) and (4-95). Thereby, the

<u>Transformation problem $\overset{1}{u}{}^{\alpha} \rightarrow \overset{2}{u}{}^{\alpha}$</u> (4-130)

for two Gaussian coordinate systems (4-129), i.e.,

<u>Given</u> are the coordinates $\overset{1}{u}{}^{\alpha}$ of a point P , (4-130a)

<u>Required</u> are the coordinates $\overset{2}{u}{}^{\alpha}$ of the same point P , (4-130b)

can be solved in the following way:

- calculating $(\Delta u^{\alpha})_1 = (B-B_{o1}, L-L_{o1})^{\alpha}$ after (4-95) , (4-131a)
  whereby the coefficients are to be calculated for B_{o1},
- calculating $\overset{2}{u}{}^{\alpha} = (\overset{2}{x}, \overset{2}{y})$ after (4-93) (4-131b)
  with $(\Delta u^{\alpha})_2 = (B-B_{o2}, L-L_{o2})^{\alpha}$,
  whereby the coefficients are to be calculated for B_{o2}.

The inverse transformation of (4-130) is done in the same way.

4.9.2 DIRECTLY TRANSFORMING THE COORDINATES

As power series, equations for the direct transformation between Gaussian coordinates (4-129) generally have the form of (3-4). Corresponding to (3-3) a <u>transformation point P_o</u> is established whose coordinates in the systems of (4-128) and (4-129) involve the following relationships:

$$P_o : (u^{\alpha})_o, \quad (\overset{s}{u}{}^{\alpha})_o = (\overset{s}{x}_o, \overset{s}{y}_o)^{\alpha}, \quad s \in \{1,2\} . \tag{4-132a}$$

For arbitrary test points P the coordinate differences relative to P_o

are given by

$$\Delta u^\alpha = u^\alpha - (u^\alpha)_o = (\Delta B, \Delta L)^\alpha , \quad \Delta \overset{s}{u}^\alpha = \overset{s}{u}^\alpha - (\overset{s}{u}^\alpha)_o = (\Delta \overset{s}{x}, \Delta \overset{s}{y})^\alpha . \qquad (4\text{-}132b)$$

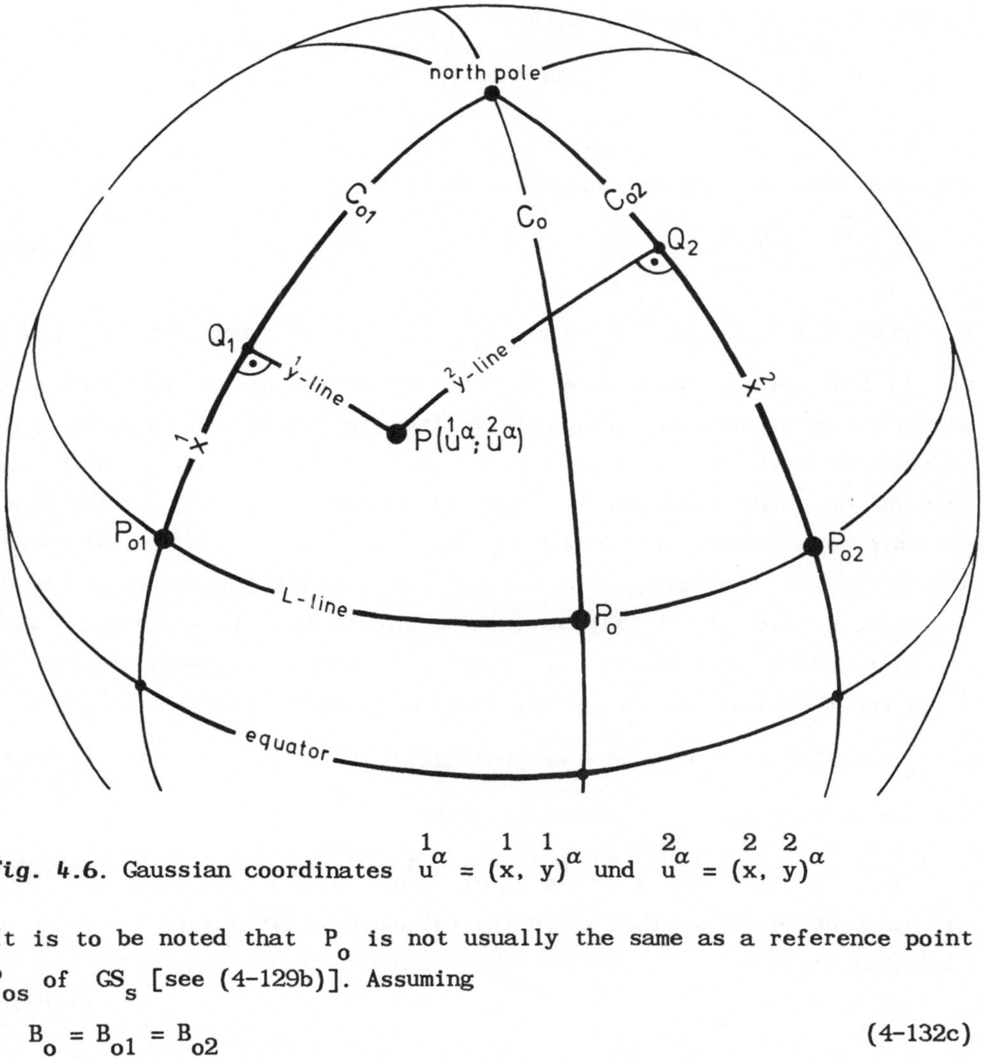

Fig. 4.6. Gaussian coordinates $\overset{1}{u}^\alpha = (\overset{1}{x}, \overset{1}{y})^\alpha$ und $\overset{2}{u}^\alpha = (\overset{2}{x}, \overset{2}{y})^\alpha$

It is to be noted that P_o is not usually the same as a reference point P_{os} of GS_s [see (4-129b)]. Assuming

$$B_o = B_{o1} = B_{o2} \qquad (4\text{-}132c)$$

is generally practical, as a result of which P_o, P_{o1}, and P_{o2} belong to the same latitude circle (L-line) [see Fig. 4.6]. With the assumptions made here the power series belonging to the transformation problem of (4-130) can be written corresponding to (3-4) as follows:

$$ {}^{2}\Delta u = ({}^{2}\Delta x,\ {}^{2}\Delta y) = \sum_{n=1}^{n'} b^{\alpha}_{n_1 n_2}\ {}^{1}\Delta x^{n_1}\ {}^{1}\Delta y^{n_2}\ , \qquad n_1 + n_2 = n\ . \tag{4-133} $$

Both coordinate systems GS_1 and GS_2 are isothermal, and thus (4-133) can also be replaced by a complex power series according to (3-65) and (3-66). This method together with

$$ {}^{s}u = {}^{s}x + i\ {}^{s}y\ , \qquad {}^{s}\Delta u = {}^{s}\Delta x + i\ {}^{s}\Delta y\ , \quad s \in \{1,2\}\ , $$
$$ c_n = c^1_n + i\ c^2_n \tag{4-134a} $$

produces the power series equivalent to (4-133), i.e.,

$$ {}^{2}\Delta u = \sum_{n=1}^{n'} c_n\ {}^{1}\Delta u^n\ . \tag{4-134b} $$

For <u>determining the coefficients $b^{\alpha}_{n_1 n_2}$ or c_n</u> of power series (4-133) and (4-134) we can make more or less extensive use of the fact that Gaussian coordinates are primarily defined in the system of geographic coordinates according to (4-90) and (4-129) and are constructed in the form of the transformation functions of (4-93) and (4-95). Thus, coefficients can be determined based completely on the power series of (4-93) and (4-95) corresponding, for example, to (4-131). However, the power series expansion of (4-133) is more commonly done in accordance with (3-54) to (3-58) whereby here the simpler equations of (3-64) are substituted for the equations of (3-58). It is purposeful to select

$$ C_o = \text{meridian through } P_o \text{ [see (4-132a)]} \tag{4-135a} $$

for the surface curve C_o. Together with

$$ \delta L_s := L_o - L_{os}\ , \qquad B_o = B_{o1} = B_{o2} \tag{4-135b} $$

and based on (4-93) we then yield the following equations in the sense of (3-54a,b) with

$$ t =: \Delta B = B - B_o : \tag{4-135c} $$

$$ {}^{s}u^{\alpha}_{o} = ({}^{s}x,\ {}^{s}y)^{\alpha}_{C_o} = \sum_{n=1}^{n'} \bar{a}^{\alpha}_{n_1 n_2}\ \delta L_s^{n_2}\, \Delta B^{n_1}\ , \qquad n_1 + n_2 = n\ . \tag{4-135d} $$

Thereby, we have not only established the basis for determining the coefficients of (4-133) according to (3-64), but also the basis for determining the coefficients c_n of the complex series (4-134) according to (3-65) to (3-68). Both methods are practically the same in every repect so that it does not matter which one is applied. Assuming

$$L_o = (L_{o1}+ L_{o2})/2 , \quad \delta L := L_{o2}- L_{o1} = 2\delta L_1 = -2\delta L_2 \tag{4-136a}$$

and with the abbreviations of (4-93f), i.e.,

$$co := \cos B_o, \; t := \tan B_o, \; V := (1+e''^2\cos^2 B_o)^{1/2} , \tag{4-136b}$$

the real and imaginary parts c_n^1 and c_n^2 of c_n with $n' = 3$ in (4-134b) become

$$\begin{aligned}
c_1^1 &= 1-(1/2)co^2t^2(1+(1/12)co^2(2-t^2)\delta L^2)\delta L^2 ,\\
c_1^2 &= co\; t(1-(1/12)co^2(2-3V^2+2t^2-(1/20)co^2(2-11t^2+2t^4)\delta L^2)\delta L^2)\delta L ,\\
\\
c_2^1 &= -(1/4)(V/c)co^2t(3V^2+(1/8)co^2(1-13t)\delta L^2)\delta L^2 ,\\
c_2^2 &= (1/2)(V/c)co(V^2-(1/24)co^2(1+31t^2)\delta L^2)\delta L ,\\
\\
c_3^1 &= -(1/12)(V/c)^2co^2(3-4t^2)\delta L^2 ,\\
c_3^2 &= (1/6)\;(V/c)^2co\; t(4-5V^2-(1/12)co^2(37-26t^2)\delta L^2)\delta L .
\end{aligned} \tag{4-136c}$$

The following results are then obtained for the $b^\alpha_{n_1n_2}$ coefficients of (4-133) via (2-166):

$$\begin{aligned}
b^\alpha_{10} &= (\;c_1^1,\; c_1^2)^\alpha , & b^\alpha_{01} &= (-c_1^2,\; c_1^1)^\alpha ,\\
\\
b^\alpha_{20} &= (\;c_2^1,\; c_2^2)^\alpha , & b^\alpha_{11} &= 2(-c_2^2,\; c_2^1)^\alpha ,\\
b^\alpha_{02} &= (-c_2^1,-c_2^2)^\alpha , & &\\
\\
b^\alpha_{30} &= (\;c_3^1,\; c_3^2)^\alpha , & b^\alpha_{21} &= 3(-c_3^2,\; c_3^1)^\alpha ,\\
b^\alpha_{12} &= 3(-c_3^1,-c_3^2)^\alpha , & b^\alpha_{03} &= (\;c_3^2,-c_3^1)^\alpha .
\end{aligned} \tag{4-137}$$

4.10 TRANSFORMATIONS BETWEEN ISOTHERMAL STEREOGRAPHIC COORDINATES AND GEOGRAPHIC COORDINATES

4.10.1 TRANSFORMING THE COORDINATES

The term "stereographic coordinates" is generally used in geodetic applications for isothermal surface coordinates on reference ellipsoids whose metric tensor with respect to an arbitrary eligible point

$$P_o = \text{central point} \tag{4-138a}$$

has the following properties:

$$\bar{G}(P_o) = 1\ , \quad \bar{G}(P) \approx F(\Delta s), \quad \Delta s = \text{distance } P_o - P\ . \tag{4-138b}$$

Accordingly, $\bar{G}$ is nearly constant on concentric geodesic circles around P_o so that (4-76b) is easily attained for circular areas. The definition of stereographic coordinates is thus essentially dependent upon the choice of the function $F(\Delta s)$. There are various setups used for this which, however, only differ slightly in their results.

One possibility for specifying a suitable function $F(\Delta s)$ in the sense of (4-138b) consists of first replacing the ellipsoid of revolution by an osculating sphere at P_o with the Gaussian curvature of (4-19a), i.e.,

$$K_o = 1/R_o^2 = (V_o/c)^2\ . \tag{4-139a}$$

The known laws of conformal stereographic projection are then applied to this sphere. The laws can be fomulated so that the great circles of the sphere through P_o on the plane become straight lines through the projection $\bar{P}_o$ of P_o while maintaining the differences in azimuth, and so that the law of projection

$$\Delta\bar{s} = 2m\ R_o \tan(\Delta s/(2R_o)), \quad m = \text{scale factor} \tag{4-139b}$$

is valid for the great circles [see Hubeny 1953, 1977]. Proceeding on this basis and according to (2-138) we arrive at the isothermal <u>stereographic coordinates $\bar{u}^\alpha$ on the sphere</u> (R_o) when one great circle through P_o is identified with the abscissa line C_o. For this abscissa line

$$\bar{u}_o^1 := \Delta\bar{s}(\Delta s)_{C_o} \quad \text{after (4-139b)} \tag{4-139c}$$

is used in the sense of (2-138b). By differentiating with respect to s we obtain

$$d\bar{u}_o^1 = m \cos^{-2}(\Delta s/(2R_o))\, ds$$

from which

$$\bar{G} = (1/m^2)\cos^4(\Delta s/(2R_o)) = F(\Delta s) \qquad (4\text{-}139d)$$

follows. Accordingly, (4-138b) is completely satisfied for stereographic coordinates on the sphere.

<u>Stereographic coordinates $\bar{u}^\alpha$ on ellipsoids of revolution</u> can thus be defined so that (4-139c,d) can be used for a geodesic abscissa line C_o through P_o. This means of course that only

$$\bar{G} \approx F(\Delta s) \qquad (4\text{-}140a)$$

is true for all other geodesics through P_o with relative deviations in the magnitude of the flattening of the ellipsoid. In the following we will use

$$C_o = \text{meridian through } P_o \text{ with } \bar{u}_o^1 \text{ and } \bar{G}_o \text{ after (4-139c,d).} \qquad (4\text{-}140b)$$

With (4-140b) the necessary requirements for defining a system of stereographic coordinates in the system of geographic coordinates corresponding to the explanations given in Chapter 4.5 are fulfilled. We will proceed from the geographic coordinates as defined by (4-74b), i.e.,

$$u^\alpha = (B,\ L)^\alpha\ , \qquad (4\text{-}141)$$

so that in (4-12) the expressions for g_{11} and g_{22} are to be exchanged for one another. The isothermal <u>stereographic coordinates on the ellipsoid of revolution</u> are defined in (4-141) corresponding to (4-75) as follows:

$$(u^\alpha)_o = (B_o,\ L_o)^\alpha \qquad (4\text{-}142a)$$

= geographic coordinates of the central point P_o,

C_o = abscissa line through P_o (4-142b)

= meridian through P_o after (4-75c),

$\bar{u}^\alpha = (x,\ y)^\alpha$ (4-142c)

= stereographic coordinates with respect to P_o ,

x = abscissa of a point P , (4-142d)

y = ordinate of a point P ; (4-142e)

the transformation equations for C_o [see (4-75e)] are

$$\bar{u}_o^1 = \Delta\bar{s}(\Delta s) \text{ after (4-139c)} =: f(B),\quad \bar{u}_o^2 = y_o = 0\ , \qquad (4\text{-}142f)$$

$$\Delta s(B) = s(B) - s_o\ , \qquad s_o := s(B_o)\ ,$$

s(B) = meridian arc between the equator point $B = 0$ and a point with the latitude B . (4-142g)

This arrangement of abscissas and ordinates corresponds to that of the Gaussian coordinates of (4-91) and the Soldner's coordinates of (4-55) so that Fig. 4.3 is qualitatively true for stereographic coordinates as well.

The <u>transformation function $\bar{u}^{\alpha}(u^{\beta})$</u> is based on the power series setup of (3-4) which together with

$$\Delta u^{\alpha} = (B-B_o,\ L-L_o)^{\alpha} = (\Delta B,\ \Delta L)^{\alpha} \tag{4-143a}$$

according to (4-141), (4-142a) and (4-142c) can be written as follows:

$$\bar{u}^{\alpha} = (x,y)^{\alpha} = \sum_{n=1}^{n'} \bar{a}^{\alpha}_{n_1 n_2}\ \Delta B^{n_1} \Delta L^{n_2} , \quad n_1 + n_2 = n . \tag{4-143b}$$

$$\begin{aligned}\bar{a}^{\alpha}_{n_1 n_2} &= (1/(n_1!n_2!))(\partial^n \bar{u}^{\alpha}/(\partial B^{n_1} \partial L^{n_2}))_{P_o} \\ &= (1/(n_1!n_2!))(\bar{u}^{\alpha}_{,n_1\times 1\ n_2\times 2})_{P_o} .\end{aligned} \tag{4-143c}$$

The coefficients of (4-143c) are determined in the manner described for (3-56) to (3-58).

$$B := t \tag{4-143d}$$

is purposefully used as a curve coordinate for C_o so that in (4-139b) the meridian arc length $\bar{r}$ is substituted for s in accordance with (4-43) and (4-46). In doing so, we now have the special case of (3-59), and according to (3-59c,d)

$$u^{\alpha\prime} = \delta^{\alpha}_1 , \quad u^{\alpha(n)} = 0^{\alpha},\ n > 1 , \quad \text{for } P_o \in C_o . \tag{4-143e}$$

Observing (4-142f,g) we obtain

$$\bar{u}^{\alpha(n)} = (d^n \Delta\bar{s}/dB^n)_{P_o}\, \delta^{\alpha}_1 \tag{4-143f}$$

for (3-59e). The presuppositions necessary for setting up the equations of (3-58b) are contained in (4-143d-f). Due to the arrangement of u^{α} according to (4-141) the equations of (3-58a) are to be calculated proceeding from (4-77c). With the abbreviations

$$co := \cos B_o , \quad t := \tan B_o , \quad V := V_o = (1+e''^2\cos^2 B_o)^{1/2} \tag{4-143g}$$

determining the coefficients for $m = 1$ up to $n' = 5$ results in

$$\bar{a}^{\alpha}_{10} = (c/V^3,\ 0\ ),$$
$$\bar{a}^{\alpha}_{01} = (\ 0\ ,\ (c/V)co),$$

$$\bar{a}^{\alpha}_{20} = (-(3/2)(c/V^5)t(1-V^2),\ 0\ ),$$
$$\bar{a}^{\alpha}_{11} = (\ 0\ ,\ -(c/V^3)co\ t),$$
$$\bar{a}^{\alpha}_{02} = ((1/2)(c/V)co^2 t,\ 0\ ),$$

$$\bar{a}^{\alpha}_{30} = (-(1/12)(c/V)(12-22V^2+9V^4-6(8-15V^2+7V^4)t^2),\ 0\ ),$$
$$\bar{a}^{\alpha}_{21} = (\ 0\ ,\ -(1/4)(c/V)co(3-3V^2+V^4-6(3-5V^2+2V^4)t^2)),$$
$$\bar{a}^{\alpha}_{12} = ((1/4)(c/V)co^2(1-2(3-3V^2+V^4)t^2),\ 0\ ),$$
$$\bar{a}^{\alpha}_{03} = (\ 0\ ,\ (1/12)(c/V)co^3(V^2-2t^2)),$$

$$\bar{a}^{\alpha}_{40} = ((1/8)(c/V)t(1-V^2),\ 0\ ),$$
$$\bar{a}^{\alpha}_{31} = (\ 0\ ,\ (1/12)(c/V)co\ t(4-5V^2-6(1-2V^2)t^2)),$$
$$\bar{a}^{\alpha}_{22} = (-(1/8)(c/V)co^2 t(1+2V^2+6(1-2V^2+V^4)t^2),\ 0\ ),$$
$$\bar{a}^{\alpha}_{13} = (\ 0\ ,\ (1/6)(c/V)co^3 t(2-4V^2+(2-V^2)t^2)), \tag{4-143h}$$
$$\bar{a}^{\alpha}_{04} = ((-1/24)(c/V)co^4 t(4-6V^2+t^2),\ 0\ ),$$

$$\bar{a}^{\alpha}_{50} = ((1/120)(c/V),\ 0\ ),$$
$$\bar{a}^{\alpha}_{41} = (\ 0\ ,\ -(1/24)(c/V)co),$$
$$\bar{a}^{\alpha}_{32} = (-(1/24)(c/V)co^2(1+t^2),\ 0\ ),$$
$$\bar{a}^{\alpha}_{23} = (\ 0\ ,\ -(1/24)(c/V)co^3(2-7t^2)),$$
$$\bar{a}^{\alpha}_{14} = ((1/48)(c/V)co^4(2-11t^2+2t^4),\ 0\ ),$$
$$\bar{a}^{\alpha}_{05} = (\ 0\ ,\ (1/240)(c/V)co^5(2-11t^2+2t^4)).$$

The power series for the <u>transformation function $u^{\alpha}(\bar{u}^{\beta})$</u> reads

$$\Delta u^{\alpha} = (\Delta B,\ \Delta L)^{\alpha} = \sum_{n=1}^{n'} a^{\alpha}_{n_1 n_2}\, x^{n_1} y^{n_2}, \quad n_1 + n_2 = n \tag{4-144a}$$

$$\begin{aligned} a^{\alpha}_{n_1 n_2} &= (1/(n_1!n_2!))(\partial^n u^{\alpha}/(\partial x^{n_1}\partial y^{n_2}))_{P_o} \\ &= (1/(n_1!n_2!))(u^{\alpha}_{,n_1\times 1\ n_2\times 2})_{P_o}. \end{aligned} \tag{4-144b}$$

This involves the inverse transformation of (4-143) so that the coefficients of (4-144b) can be determined by series inversion according to (3-7). On the other hand, they can also be obtained by solving the initial value problem of (4-142) for the differential equations of (4-77d) with the procedure presented for (3-56) to (3-58), i.e., analogous to the explanations given for (4-143d-f), which are also true here. The equations of (3-58a) are to be replaced by the corresponding equations resulting from (4-77d), and in (3-58b) u^α and $\bar{u}^\alpha$ are exchanged for one another. With the abbreviations of (4-143g) determining the coefficients for $m = 1$ up to $n' = 5$ results in

$$\begin{aligned}
a^\alpha_{10} &= (V^3/c,\ 0\ )\,,\\
a^\alpha_{01} &= (\ 0\ ,\ (V/c)co(1+t^2))\,,\\[6pt]
a^\alpha_{20} &= ((3/2)(V^2/c)^2 t(1-V^2)\ ,\ 0\ )\,,\\
a^\alpha_{11} &= (\ 0\ ,\ (V/c)^2 co\ t(1+t^2))\,,\\
a^\alpha_{02} &= (-(1/2)(V^2/c)^2 t,\ 0\ )\,,\\[6pt]
a^\alpha_{30} &= (-(1/12)(V/c)^3(6-18V^2+13V^4-6(5-11V^2+6V^4)t^2,\ 0\ )\,,\\
a^\alpha_{21} &= (\ 0\ ,\ (1/4)(V/c)^3 co(1+t^2)(V^2+4t^2))\,,\\
a^\alpha_{12} &= (-(1/4)(V/c)^3(V^4-2(4-8V^2+3V^4)t^2),\ 0\ )\,,\\
a^\alpha_{03} &= (\ 0\ ,-(1/12)(V/c)^3 co(1+t^2)(V^2+4t^2))\,,\\[6pt]
a^\alpha_{40} &= (-(3/4)(V/c)^4 t(1-V^2),\ 0\ )\,,\\
a^\alpha_{31} &= (\ 0\ ,\ (1/6)(V/c)^4 co(1+t^2)(4-V^2+6t^2))\,,\\
a^\alpha_{22} &= (-(1/4)(V/c)^4 t(9-8V^2+(3-V^2)t^2),\ 0\ )\,,\\
a^\alpha_{13} &= (\ 0\ ,-(1/6)(V/c)^4 co(1+t^2)(4-V^2+6t^2))\,,\\
a^\alpha_{04} &= ((1/24)(V/c)^4 t(1+2V^2+3(3-2V^2)t^2),\ 0\ )\,,\\[6pt]
a^\alpha_{50} &= ((1/80)(V/c)^5,\ 0\ )\,,\\
a^\alpha_{41} &= (\ 0\ ,\ (1/16)(V/c)^5 co(1+t^2)(1+12t^2+16t^4))\,,\\
a^\alpha_{32} &= (-(1/8)(V/c)^5 t^2(3+4t^2),\ 0\ )\,,
\end{aligned}\tag{4-144c}$$

$$a^{\alpha}_{23} = (\ 0\ ,\ -(1/8)(V/c)^5 co(1+t^2)(1+12t^2+16t^4))\ , \qquad (4\text{-}144c)$$
$$a^{\alpha}_{14} = ((1/16)(V/c)^5(1+6t^2+6t^4)\ ,\ 0\ )\ ,$$
$$a^{\alpha}_{05} = (\ 0\ ,\ (1/80)(V/c)^5 co(1+t^2)(1+12t^2+16t^4))\ .$$

4.10.2 TRANSFORMING THE METRIC TENSOR

The metric tensor of stereographic coordinates can be calculated in the same way as for Gaussian coordinates described in Chapter 4.7.3. Especially the equations of (4-100a,b) are formally valid here as well. The result corresponding to (4-101a) reads

$$\bar{G} = 1-(1/2)(V^2/c)^2(x^2+y^2)-4(V/c)^3V^2t(1-V^2)xy^2 \qquad (4\text{-}145a)$$
$$+(1/48)(V/c)^4(9V^4x^4+6(3V^4-8(1-V^2)(1-t^2))x^2y^2+(9V^4+16(1-V^2))y^4).$$

Together with

$$x^2 + y^2 \approx \Delta s^2\ , \qquad (V^2/c)^2 = K$$

in spherical approximation this expression becomes

$$\bar{G} \approx 1-(1/2)K\,\Delta s^2 + (3/16)K^2\Delta s^4\ , \qquad (4\text{-}145b)$$

whereby the properties of stereographic coordinates required by (4-138b) are confirmed.

5. THREE-DIMENSIONAL COORDINATES

5.1 PRELIMINARY REMARKS

Regarding position space geodetic problems are fundamentally of three-dimensional nature so that the accompanying geometric models based on Newtonian mechanics are always constructed in three-dimensional Euclidean space. For this purpose

$$\overset{n}{y}_{i.}, \quad i \in \{1,2,3\} \tag{5-1a}$$

= rectangular, rectilinear coordinates with respect to a system (basis) S_n,

also called Cartesian coordinates or position vectors, play an important role. They have the simplest metric of all thinkable coordinates in Euclidean space. Their metric tensor is the same as the Cartesian δ-tensor, i.e.,

$$\overset{n}{g}_{ij} \equiv \delta_{ij.} = \begin{cases} 1 & \text{for } i=j \\ 0 & \text{for } i \neq j \end{cases}, \tag{5-1b}$$

so that the fundamental form for the distance ds between infinitesimally neighboring points, i.e.,

$$ds^2 = \delta_{ij.} d\overset{n}{y}_{i.} d\overset{n}{y}_{j.} = d\overset{n}{y}_{i.} d\overset{n}{y}_{i.}, \tag{5-1c}$$

is also valid for finite spatial distances s, i.e.,

$$s^2 = \Delta\overset{n}{y}_{i.} \Delta\overset{n}{y}_{i.} . \tag{5-1d}$$

(5-1d) is the fundamental observation equation for determining Cartesian coordinates based on distance measurements. The observation equations for all other geometrically defined quantities can be derived from (5-1d) relatively simply [see, e.g., Heitz 1980-1983, Chapter 11]. Yet, using

$$q^a, \quad a \in \{1,2,3\} \tag{5-2a}$$

= curvilinear coordinates

instead of (5-1a) in definitions suitable for special problems is frequently purposeful.

$$g_{ab} \neq \delta_{ab}. \tag{5-2b}$$

is generally true for their metric tensors so that the accompanying fundamental form of

$$ds^2 = g_{ab} dq^a dq^b \tag{5-2c}$$

can be directly applied only to infinitesimally small distances ds. (5-2c) is the <u>fundamental observation equation of Euclidean geometry</u>, which can be used in different ways for, e.g.,

- determining the coordinates q^a of a field of arbitrarily distant points with the given metric tensor $g_{ab}(q^c)$ based on distance observations (s), (5-3a)

which is a generalization of the determination of Cartesian coordinates given for (5-1c). The method for

- determining the metric tensor $g_{ab}(q^c)$ with the given coordinates q^a of a field of infinitesimally neighboring points based on observations of distance differentials (ds), (5-3b)

which is also possible using (5-2c), is the inverse problem of (5-3a). The complete geometry of a point field must be given for both (5-3a) and (5-3b). In the case of (5-3a) this is done with the metric tensor as an analytic position function, and in the case of (5-3b) with the coordinates of a field of infinitesimally neighboring points. Besides (5-3a,b) many other problems are conceivable, which are always connected with the condition that the geometry of the point field must be given in some form or another.

An especially important example of (5-2) is the representation of point fields of the earth's surface using curvilinear coordinates given for (1-1) and (1-2): the equipotential surfaces and plumb lines of the earth's gravity field can be approximated using the coordinate surfaces and lines of these curvilinear coordinates. In this example the connection to the gravity field is attained through the special <u>selection of the reference surface as a model geoid</u>. When other types of reference surfaces are chosen, the definitions of curvilinear coordinates according to (1-1b-g) can, however, be meaningful to other types of problems. Thus, for problems involving civil or mechanical engineering, for instance, it may be appropriate to identify the <u>reference surfaces of curvilinear coordinates</u> exactly or approximately with <u>surfaces of bodies</u>.

Curvilinear coordinates defined on the basis of reference surfaces corresponding to (1-1) are the main object of the following considerations, whereby only cases involving straight q^3-lines corresponding to (1-1f-h) will be discussed. Coordinates in this form are called surface-normal coordinates. They consist of the surface coordinates u^α of the vertical foot points Q of the reference surface and the height coordinates H so that transformation equations between arbitrary three-dimensional coordinates $\bar{q}^a$ and surface-normal coordinates q^a also contain those between $\bar{q}^a$ and u^α. This involves mapping three-dimensional Euclidean space on reference surfaces by means of surface normals, which play an important role in geodetic surveying and cartography. Transformations of this type will be discussed generally in Chapter 5.3. Chapter 5.4 gives examples of geodetic coordinates (ellipsoidal surface-normal coordinates), and Chapter 5.5 is devoted to transformations between geodetic coordinates. The general fundamentals of differential geometry in three- dimensional Euclidean space necessary for Chapters 5.3 to 5.5 are summarized in Chapter 5.2 in a way similar to Heitz (1980-1983, Chapter 12.2) and Sokolnikoff (1964, Chapter 3).

In the following, coordinates and tensors are given in both three- and two-dimensional spaces, the latter in the form of reference surfaces for surface-normal coordinates. This makes a distinction between the accompanying symbols necessary which is done by designating the indices differently:

$$a,b,c\ldots \in \{1,2,3\} = \text{three-dimensional indices} , \tag{5-4a}$$

$$\alpha,\beta,\gamma\ldots \in \{1,2\} = \text{two-dimensional indices} . \tag{5-4b}$$

Indices of three-dimensional Cartesian coordinates and tensors are preferably designated by $i,j,k\ldots$.

5.2 FUNDAMENTALS OF THREE-DIMENSIONAL EUCLIDEAN GEOMETRY

5.2.1 COORDINATE TRANSFORMATIONS

Two operations are essentially involved in dealing with coordinates, i.e., defining the coordinates and transformations between different coordinates. This was true for the surface coordinates treated in Chapters 1 to 4 and is also the case for three-dimensional coordinates in Euclidean space discussed here. The definition of coordinates q^a should be regarded as given when their metric tensor as an analytic function of the coordinates is known, i.e.,

$$g_{ab} = g_{ab}(q^c) \quad . \tag{5-5a}$$

It is assumed in the following that this condition is met for the two coordinate systems q^a and $\bar{q}^a$, i.e., that in addition to (5-5a) the metric tensor of $\bar{q}^a$ is also defined or determined, i.e.,

$$\bar{g}_{ab} = \bar{g}_{ab}(\bar{q}^c) \quad . \tag{5-5b}$$

The accompanying Riemannian tensors of curvature of (2-50), used here in three dimensions [see (5-4a)], must vanish identically for the metric tensors of Euclidean space. As in (2-50) this results from the symmetry condition for the third partial derivatives of the Cartesian position vectors with respect to the coordinates q^a or $\bar{q}^a$. Consequently, for (5-5a) we have the condition

$$R^d_{abc} = \Gamma^d_{ab,c} - \Gamma^d_{ac,b} + \Gamma^e_{ab}\,\Gamma^d_{ec} - \Gamma^e_{ac}\,\Gamma^d_{eb} \equiv 0 \tag{5-5c}$$

with Christoffel's symbols of the second kind, which are formally defined corresponding to (2-48), i.e.,

$$\Gamma^c_{ab} = g^{cd}(g_{ad,b} + g_{bd,a} - g_{ab,d})/2 \quad . \tag{5-5d}$$

(5-5c,d) are to be used similar to the crossbarred system of (5-5b). Under these conditions the transformation equations of q^a and $\bar{q}^a$, i.e.,

$$q^a = q^a(\bar{q}^b) \;, \qquad \bar{q}^a = \bar{q}^a(q^b) \;, \tag{5-6a}$$

are generally uniquely determined with up to six degrees of freedom for the relative spatial movements of both systems. The basis for this is the law of transformation of the metric tensor, which have forms analogous to (2-27a,b). Here, the forms of (2-27a) are chosen, i.e.,

$$\bar{g}_{ab} = q^c_{,\bar{a}}\, q^d_{,\bar{b}}\, g_{cd} \;, \qquad g_{ab} = \bar{q}^c_{,a}\, \bar{q}^d_{,b}\, \bar{g}_{cd} \;. \tag{5-6b}$$

With covariant metric tensors given according to (5-5a,b) each of these forms involves a system of six linear, partial differential equations of the first order for the transformation functions of (5-6a). Thus, they are each determinable up to the aforementioned six integration constants based on (5-6b). This can be demonstrated by representing the transformation equations by power series as setups for solving (5-6b), which will be discussed in the following.

Let

$$(q^a)_o, \quad (\bar{q}^a)_o \tag{5-7a}$$

be the coordinates of a reference point P_o in the q^a- and $\bar{q}^a$-systems, and

$$\Delta q^a = q^a - (q^a)_o \;, \qquad \Delta\bar{q}^a = \bar{q}^a - (\bar{q}^a)_o \tag{5-7b}$$

the differences in coordinates of arbitrary test points P relative to P_o. As an example, the second transformation equation of (5-6a) is chosen with a <u>power series setup at P_o</u> in the form of

$$\Delta\bar{q}^a = (\bar{q}^a_{,b})_o \Delta q^b + \sum_{n=2}^{n'} (1/n!)(\bar{q}^a_{,b_1b_2\ldots b_n})_o \Delta q^{b_1} \Delta q^{b_2} \ldots \Delta q^{b_n} \;, \tag{5-7c}$$

$$\bar{q}^a_{,b_1b_2\ldots b_n} := \partial^n \bar{q}^a / (\partial q^{b_1} \partial q^{b_2} \ldots \partial q^{b_n}) \;. \tag{5-7d}$$

The <u>six degrees of freedom</u> for the relative spatial movements are represented by one of the coordinate triples of (5-7a), e.g., the coordinates

$$(\bar{q}^a)_o \;, \tag{5-8a}$$

and the transformation matrix at P_o, i.e.,

$$(\bar{q}^a_{,b})_o := (\partial\bar{q}^a / \partial q^b)_o \;, \tag{5-8b}$$

of whose nine components only three are independent due to (5-6b).

When determining the other partial derivatives in (5-7c), i.e.,

$$(\bar{q}^a_{,b_1b_2\ldots b_n})_o \quad \text{for} \quad n > 1 \;, \tag{5-9a}$$

it must be noted that they are symmetrical in all subscripts. Hence, of the collective 3^{n+1} only components independent of

$$N_n = (3/2)(n+1)(n+2) \tag{5-9b}$$

remain, which can be determined at P_o stepwise with the help of the respective (n-1)th partial derivative of the second equation of (5-6b). Thus, in the first step, i.e.,

$$n = 2 \qquad \rightarrow \qquad N_2 = 18\ , \tag{5-9c}$$

we obtain the linear equation system in $(\bar{q}^a_{,b_1 b_2})_o$ through the first partial derivative of the second equation of (5-6b), i.e.,

$$(\bar{q}^c_{,ae}\bar{q}^d_{,b} + \bar{q}^c_{,a}\bar{q}^d_{,be})_o(\bar{g}_{cd})_o = (g_{ab,e})_o - (\bar{q}^c_{,a}\bar{q}^d_{,b}\bar{q}^f_{,e})_o(\bar{g}_{cd,\bar{f}})_o, \tag{5-9d}$$

which can then be solved with respect to the second partial derivatives:

$$(\bar{q}^a_{,bc})_o = (\Gamma^d_{bc}\bar{q}^a_{,d})_o - (\bar{\Gamma}^a_{de}\bar{q}^d_{,c}\bar{q}^e_{,b})_o\ . \tag{5-9e}$$

In this case Γ^d_{bc} and $\bar{\Gamma}^a_{de}$ are Christoffel's symbols of the second kind corresponding to (5-5d). (5-9e) follows from the transformation equations of the Christoffel's symbols of (5-5e), i.e.,

$$\Gamma^a_{bc} = q^a_{,\bar{f}}\bar{q}^d_{,c}\bar{q}^e_{,b}\bar{\Gamma}^f_{de} + q^a_{,\bar{f}}\bar{q}^f_{,bc}\ , \tag{5-9f}$$

after inner multiplication with $\bar{q}^g_{,a}$. The formulas of (5-9e), which are very important for the coordinate transformations, can be traced back to Christoffel [cf. Sokolnikoff 1964, No. 32]. Another derivation of (5-9e) will be disussed in (5-11) to (5-13). The right side of (5-9e) is known from the previous zero step, in this case based on the initial values of (5-8b). The procedure used in (5-7) to (5-9) can also be used in the very same way for the inverse transformation, i.e., the first equation of (5-6a). For this purpose, we will begin with the first differential equation of (5-6b) and the crossbarred and noncrossbarred symbols in (5-7) to (5-9) are exchanged for one another.

Euclidean spaces are distinguished by the existence of the Cartesian coordinates of (5-1a). Assuming

$$\bar{q}^i =: y_{i.} \quad \text{after (5-1a)} \tag{5-10a}$$

the first partial derivatives or transformation matrices become covariant and contravariant basis vectors, i.e.,

$$c_{i.a} := \bar{q}^i_{,a} = \partial y_{i.}/\partial q^a\ , \qquad c^a_{i.} := q^a_{,i} = \partial q^a/\partial y_{i.}\ , \tag{5-10b}$$

and the transformation equations for the metric tensor of (5-6b) become

$$\delta_{ij.} = c^c_{i.} c^d_{j.} g_{cd} , \qquad g_{ab} = c_{i.a} c_{i.b} . \tag{5-10c}$$

The formalities of the power series expansions of (5-7) to (5-9) remain basically unchanged with (5-10a), which will be discussed in more detail in Chapter 5.2.2. Here, only the differential equations of (5-9d) and the accompanying inverse forms are given, i.e.,

$$(c_{i.a,b})_o = (\Gamma^c_{ab})_o (c_{i.c})_o , \quad (c^a_{i,b})_o = -(\Gamma^a_{bc})_o (c^c_{i.})_o . \tag{5-10d}$$

We will now go once again into the transformation between two curvilinear coordinate systems q^a and $\bar{q}^a$ according to (5-6) to (5-9) whereby it is presumed that the transformation equations for both systems are given in the Cartesian coordinate systems of (5-1), i.e.,

$$\begin{aligned} &S: \text{ origin } O, \text{ position vectors } y_{i.} , \\ &\bar{S}: \text{ origin } \bar{O}, \text{ position vectors } \bar{y}_{i.} , \end{aligned} \tag{5-11a}$$

as follows:

$$\begin{aligned} y_{i.} &= y_{i.}(q^a) , \qquad & q^a &= q^a(y_{i.}) , \\ \bar{y}_{i.} &= \bar{y}_{i.}(\bar{q}^a) , \qquad & \bar{q}^a &= \bar{q}^a(\bar{y}_{i.}) . \end{aligned} \tag{5-11b}$$

Let $\bar{S}$ relative to S be determined by

$$\begin{aligned} p_{i.} &= \text{position vector of } \bar{O} \text{ relative to } S , \\ e_{ij.} &= \text{rotation matrix of } \bar{S} \text{ relative to } S , \end{aligned} \tag{5-12a}$$

so that the transformations between S and $\bar{S}$ read

$$\begin{aligned} y_{i.} &= p_{i.} + e_{ij.}\bar{y}_{j.} = \bar{y}_{i.} + p_{i.} + (e_{ij.} - \delta_{ij.})\bar{y}_{j.} , \\ \bar{y}_{i.} &= \bar{p}_{i.} + e_{ji.} y_{j.} = y_{i.} + \bar{p}_{i.} + (e_{ji.} - \delta_{ji.}) y_{j.} \end{aligned} \tag{5-12b}$$

with

$$\bar{p}_{i.} = -e_{ji.} p_{j.} = \text{position vector of } O \text{ relative to } \bar{S} . \tag{5-12c}$$

The result of this is that the transformation matrices between $y_{i.}$ and $\bar{y}_{i.}$ are equal to the rotation matrix, i.e.,

$$y_{i,\bar{j}} = \partial y_{i.}/\partial \bar{y}_{j.} = e_{ij.} , \quad \bar{y}_{i,j} = \partial \bar{y}_{i.}/\partial y_{j.} = e_{ji.} . \tag{5-12d}$$

Together with (5-11) and (5-12) the transformation matrices between q^a and $\bar{q}^a$ can be written as

$$q^a_{,\bar{b}} = c^a_{i.} e_{ij.} \bar{c}_{j.b} , \qquad \bar{q}^a_{,b} = \bar{c}^a_{i.} e_{ji.} c_{j.b} , \tag{5-13a}$$

in which the basis vectors are equal to the partial derivatives of the functions of (5-11b) corresponding to (5-10b). The partial derivatives of the second equation of (5-13a) with respect to q^c are obtained by taking the differential equations of (5-10d) into account, i.e.,

$$\bar{q}^{a}_{,bc} = e_{ji.}(\Gamma^{d}_{bc} c_{j.d} \bar{c}^{a}_{i.} - e_{lk.} \bar{\Gamma}^{a}_{de} \bar{c}^{e}_{i.} \bar{c}^{d}_{k.} c_{j.b} c_{l.c}) \qquad (5\text{-}13b)$$

$$= \Gamma^{d}_{bc} \bar{q}^{a}_{,d} - \bar{\Gamma}^{a}_{de} \bar{q}^{d}_{,c} \bar{q}^{e}_{,b} ,$$

of which the second form independent of the Cartesian coordinates corresponds to (5-9e).

5.2.2 REPRESENTING TRANSFORMATIONS BETWEEN THREE-DIMENSIONAL CURVILINEAR AND CARTESIAN COORDINATES BY POWER SERIES

All of the following considerations are primarily based on the transformation equations

$$y_{i.} = y_{i.}(q^a) . \qquad (5\text{-}14a)$$

For the accompanying covariant and contravariant basis vectors of (5-10b)

$$c_{i.a} c^{a}_{j.} = \delta_{ij.} , \qquad c_{i.a} c^{b}_{i.} = \delta^{b}_{a} \qquad (5\text{-}14b)$$

are true, and according to (5-10c)

$$g_{ab} = c_{i.a} c_{i.b} , \qquad g^{ab} = c^{a}_{i.} c^{b}_{i.} . \qquad (5\text{-}14c)$$

For points P_o and P with the coordinates

$$P_o : (y_{i.})_o, (q^a)_o , \qquad P : y_{i.} , q^a \qquad (5\text{-}15a)$$

the coordinate differences corresponding to (5-7b) are designated by

$$\Delta y_{i.} = y_{i.} - (y_{i.})_o , \qquad \Delta q^a = q^a - (q^a)_o . \qquad (5\text{-}15b)$$

In addition to these representations, three-dimensional polar coordinates relative to P_o are also used, i.e.,

$$r_{i.} := \Delta y_{i.} / |\Delta y_{j.}| = \text{direction vector } P_o \rightarrow P ,$$
$$s := |\Delta y_{i.}| = \text{distance } P_o - P . \qquad (5\text{-}16a)$$

Herewith, the Cartesian coordinate differences of (5-15b), also named <u>Cartesian normal coordinates</u> of P relative to P_o, become

$$\Delta y_{i.} = r_{i.} s \ . \tag{5-16b}$$

Derivatives with respect to s are abbreviated by

$$f^{(n)} := d^n f/ds^n \ . \tag{5-17}$$

Thereby,

$$r_{i.} = \Delta y_{i.}^{(1)} = c_{i.a} q^{a(1)} \tag{5-18a}$$

follows from (5-16b) for all points of the straight line P_o - P, and after inner multiplication by $c^b_{i.}$ we have

$$r^b = c^b_{i.} r_{i.} = q^{b(1)} = dq^b/ds \ . \tag{5-18b}$$

Analogous to (3-16a), <u>contravariant polar coordinates</u> relative to P_o are defined with the contravariant direction vector of (5-18b) in P_o as follows:

$$(r^a)_o = (q^{a(1)})_o, \quad s , \tag{5-18c}$$

and

$$p^a := (r^a)_o s = (q^{a(1)})_o s = (c^a_{i.})_o \Delta y_{i.} \tag{5-18d}$$

are the <u>contravariant normal coordinates</u> of P relative to P_o, which are defined analogous to (3-15a). They are the contravariant components of the Cartesian normal coordinates of (5-16b) at P_o. The polar coordinates can be calculated using p^a as follows:

$$s = ((g_{ab})_o p^a p^b)^{1/2} , \qquad (r^a)_o = p^a/s \ . \tag{5-18e}$$

The direction vector of (5-18a) is constant for a <u>straight line P_o - P</u> so that in the case of Cartesian coordinates the differential equations become

$$r^{(1)}_{i.} = y^{(2)}_{i.} = d^2 y_{i.}/ds^2 = 0_{i.} \ . \tag{5-19a}$$

The derivative of the third form of (5-18a) with respect to s must vanish, i.e.,

$$c_{i.a} q^{a(2)} + c_{i.a,b} q^{b(1)} q^{a(1)} = 0_{i.} \ .$$

After inner multiplication with $c^d_{i.}$ and substitution of the first equation of (5-10d) we obtain the following differential equation of the straight line in curvilinear coordinates after index exchange:

$$q^{c(2)} + \Gamma^c_{ab} q^{a(1)} q^{b(1)} = 0 \ . \tag{5-19b}$$

These equations are the <u>differential equations of geodesic lines</u> in non-Euclidean spaces and hence are also valid for geodesic surface curves [see Chapter 2.1.9]. In Euclidean space, geodesics are straight lines corresponding to the assumption of (5-19).

A power series of (5-7c,d) calculated for the transformation equations of (5-14a) at P_o can be written in the manner for the coordinate differences of (5-15b) as follows:

$$\Delta y_{i.} = \sum_{n=1}^{n'} (1/n!)(y_{i,a_1a_2\ldots a_n})_o \Delta q^{a_1} \Delta q^{a_2} \ldots \Delta q^{a_n}, \tag{5-20a}$$

$$(y_{i,a_1a_2\ldots a_n})_o := (\partial^n y_{i.}/(\partial q^{a_1} \partial q^{a_2} \ldots \partial q^{a_n}))_o . \tag{5-20b}$$

Alternative forms of (5-20a,b) are

$$\Delta y_{i.} = \sum_{n=1}^{n'} \bar{a}_{i.n_1n_2n_3} (\Delta q^1)^{n_1} (\Delta q^2)^{n_2} (\Delta q^3)^{n_3} , \quad n_1 + n_2 + n_3 = n , \tag{5-21a}$$

$$\bar{a}_{i.n_1n_2n_3} = (1/(n_1!n_2!n_3!)) \cdot (\partial^n y_{i.}/((\partial q^1)^{n_1} (\partial q^2)^{n_2} (\partial q^3)^{n_3}))_o . \tag{5-21b}$$

If the metric tensor of q^c [see (5-5a)] is given as a function of q^c, instead of the position vector of (5-14a), the partial derivatives of (5-20b) and (5-21b) can be calculated proceeding from the covariant basis vectors and the accompanying differential equations of (5-10b,d). We yield

$$\begin{aligned}
(y_{i,a_1})_o &= (c_{i.a_1})_o = \delta^d_{a_1} (c_{i.d})_o , \\
(y_{i,a_1a_2})_o &= (\Gamma^d_{a_1a_2})_o (c_{i.d})_o , \\
(y_{i,a_1a_2a_3})_o &= (\Gamma^d_{a_1a_2,a_3} + \Gamma^c_{a_1a_2} \Gamma^d_{ca_3})_o (c_{i.d})_o
\end{aligned} \tag{5-22a}$$

etc.,

whereby the power series of (5-20a) assumes the form of

$$\Delta y_{i.} = \sum_{n=1}^{n'} (\bar{d}^d_{a_1a_2\ldots a_n} \Delta q^{a_1} \Delta q^{a_2} \ldots \Delta q^{a_n})(c_{i.d})_o . \tag{5-22b}$$

$\Delta y_{i.}$ is spanned in this way using the covariant basis at P_o, i.e.,

$$(c_{i.d})_o = (\partial y_{i.}/\partial q^d)_o . \tag{5-22c}$$

The coefficients

$$\bar{d}^{d}_{a_1 a_2 \ldots a_n} \tag{5-22d}$$

are functions of the metric tensor and its partial derivatives with respect to q^c at P_o.

After inner multiplication of (5-22b) with $(c^{a}_{i.})_o$, $\Delta y_{i.}$ becomes the contravariant normal coordinates of (5-18d), for which the power series is

$$p^a = \Delta q^a + \sum_{n=2}^{n'} \bar{d}^{a}_{a_1 a_2 \ldots a_n} \Delta q^{a_1} \Delta q^{a_2} \ldots \Delta q^{a_n} . \tag{5-23a}$$

The form corresponding to (5-21) reads

$$p^a = \Delta q^a + \sum_{n=2}^{n'} \bar{e}^{a}_{n_1 n_2 n_3} (\Delta q^1)^{n_1} (\Delta q^2)^{n_2} (\Delta q^3)^{n_3} . \tag{5-23b}$$

According to (5-18e) the polar coordinates of (5-18c) can be calculated with p^a.

For the inverse transformation of (5-14a), i.e.,

$$q^a = q^a(y_{i.}) , \tag{5-24}$$

we first obtain the following power series setup for the coordinate differences:

$$\Delta q^a = \sum_{n=1}^{n'} (1/n!)(\partial^n q^a/(\partial y_{i_1.} \partial y_{i_2.} \ldots \partial y_{i_n.}))\Delta y_{i_1.} \Delta y_{i_2.} \ldots \Delta y_{i_n.} \tag{5-25a}$$

or

$$\Delta q^a = \sum_{n=1}^{n'} a^{a}_{n_1 n_2 n_3} (\Delta y_{1.})^{n_1} (\Delta y_{2.})^{n_2} (\Delta y_{3.})^{n_3} , \tag{5-25b}$$

$$a^{a}_{n_1 n_2 n_3} := (1/(n_1! n_2! n_3!)) \cdot (\partial^n q^a/((\partial y_{1.})^{n_1} (\partial y_{2.})^{n_2} (\partial y_{3.})^{n_3}))_o , \tag{5-25c}$$

which are the inverse power series of (5-20) and (5-21).

If the metric tensor of q^a is given instead of (5-24), the partial derivatives of q^a with respect to $y_{i.}$ are then calculated proceeding from the contravariant basis vectors and accompanying differential equations of (5-10b,d). We obtain

$$(\partial q^a/\partial y_{i_1 .})_o = \delta^a_{a_1}(c^{a_1}_{i_1 .})_o ,$$

$$(\partial^2 q^a/(\partial y_{i_1 .}\partial y_{i_2 .}))_o = -(\Gamma^a_{a_1a_2})_o (c^{a_1}_{i_1 .})_o (c^{a_2}_{i_2 .})_o , \tag{5-26a}$$

$$(\partial^3 q^a/(\partial y_{i_1 .}\partial y_{i_2 .}\partial y_{i_3 .}))_o = -(\Gamma^a_{a_1a_2,a_3} - 2\Gamma^a_{a_1c}\Gamma^c_{a_2a_3})_o \cdot (c^{a_1}_{i_1 .})_o (c^{a_2}_{i_2 .})_o (c^{a_3}_{i_3 .})_o$$

etc.,
whereby the power series of (5-25a,b) together with

$$(c^{a_m}_{i_m .})_o \Delta y_{i_m .} = p^{a_m} \tag{5-26b}$$

assume the following forms in accordance with (5-18d):

$$\Delta q^a = p^a + \sum_{n=2}^{n'} d^a_{a_1a_2\ldots a_n} p^{a_1} p^{a_2} \ldots p^{a_n} \tag{5-26c}$$

or

$$\Delta q^a = p^a + \sum_{n=2}^{n'} e^a_{n_1n_2n_3} (p^1)^{n_1} (p^2)^{n_2} (p^3)^{n_3} , \tag{5-26d}$$

which are the inverse series of (5-23a,b).

(5-26c,d) could also be derived from (5-23a,b) by series inversion corresponding to (3-7), which is naturally possible in the reverse case as well. We will discuss series of (5-23a,b) and (5-26c,d), which involve the solutions to the direct and inverse geodetic problems in three-dimensional Euclidean space, again in reference to (5-35). Here, it will only be mentioned that the results of (5-22a), (5-23a), and (5-26a,c) for coefficients $\bar{d}$ and d are formally identical with the results of $\bar{a}$ and a in accordance with (3-15e,f) and (3-15b,c). The coefficients of the fourth order given there are also valid here.

5.2.3 SPACE CURVES

In curvilinear coordinates q^a a space curve C as a function of

$$s = \text{arc length of } C \tag{5-27a}$$

is represented as

$$q^a = q^a(s) . \tag{5-27b}$$

C is a generally curved space line so that s does not usually correspond to the definition of (5-16a). If the transformation equations of (5-14a) are given, we then obtain

$$y_{i.}(s) := y_{i.}(q^{a}(s)) \tag{5-27c}$$

for the position vector of C. The definitions given in (5-4) and (5-17) are used here.

In terms of differential geometry space curves are considered based on a Cartesian moving trihedron $e_{ij.L}$ [see Heitz 1980-1983, Chapter 3.1; Laugwitz 1965, § 1.4; Sokolnikoff 1964, No. 49]:

$$\begin{aligned} e_{i1.L} &= y_{i.}^{(1)} && = \text{tangent vector} , \\ e_{i2.L} &= y_{i.}^{(2)}/|y_{j.}^{(2)}| && = \text{principal normal vector} , \\ e_{i3.L} &= \epsilon_{ijk.}e_{j1.L}e_{k2.L} && = \text{binormal vector} . \end{aligned} \tag{5-28a}$$

Coordinate systems of

$$\begin{aligned} &S_L = \text{local systems or bases,} \\ &\quad \text{axes } j.L \text{ in the directions of } e_{ij.L} \end{aligned} \tag{5-28b}$$

are spanned in these orthonormal trihedrons. In these systems the basis vectors of q^a can be written as

$$\overset{L}{c}_{j.a} = e_{ij.L}c_{i.a} , \qquad \overset{L}{c}{}^{a}_{j.} = e_{ij.L}c^{a}_{i.} , \tag{5-29a}$$

which in the following are termed covariant and contravariant moving trihedrons of C, respectively. The inversion of (5-29a) yields

$$c_{i.a} = e_{ij.L}\overset{L}{c}_{j.a} , \qquad c^{a}_{i.} = e_{ij.L}\overset{L}{c}{}^{a}_{j.} . \tag{5-29b}$$

Inner multiplication of (5-29a) by $c^{a}_{k.}$ or $c_{k.a}$ produces

$$e_{kj.L} = c^{a}_{k.}\overset{L}{c}_{j.a} = c_{k.a}\overset{L}{c}{}^{a}_{j.} \tag{5-29c}$$

due to (5-14b). Observing the first equation of (5-28a) and based on the second equation of (5-29a) for $j = 1$ we obtain

$$\overset{L}{c}{}^{a}_{1.} = c^{a}_{i.}y^{(1)}_{i.} = dq^{a}/ds = q^{a(1)} . \tag{5-29d}$$

By introducing the rotation vector of Darboux, i.e.,

$$\overset{L}{d}_{i.} = (\tau,0,\kappa) , \quad \kappa,\ \tau = \text{curvature, torsion of } C , \tag{5-30a}$$

the <u>equations of Frenet</u> can be written as follows:

$$e^{(1)}_{ij.L} = de_{ij.L}/ds = \epsilon_{jkl.}e_{ik.L}\overset{L}{d}_{l.} \quad . \tag{5-30b}$$

Differentiating the second equation of (5-29a) with respect to ds and in view of (5-30b) and (5-29c) yields the <u>contravariant equations of Frenet</u>, i.e.,

$$\overset{L}{c}{}^{a(1)}_{i.} = d\overset{L}{c}{}^{a}_{i.}/ds = c^{a}_{j,c}c_{j.d}\overset{L}{c}{}^{c}_{1.}\overset{L}{c}{}^{d}_{i.} + \epsilon_{ijk.}\overset{L}{c}{}^{a}_{j.}\overset{L}{d}_{k.} \quad , \tag{5-31a}$$

which become

$$\overset{L}{c}{}^{a(1)}_{i.} + \Gamma^{a}_{cd}\overset{L}{c}{}^{c}_{1.}\overset{L}{c}{}^{d}_{i.} = \epsilon_{ijk.}\overset{L}{c}{}^{a}_{j.}\overset{L}{d}_{k.} \, , \quad \overset{L}{c}{}^{c}_{1.} = q^{c(1)} \tag{5-31b}$$

by substituting the contravariant differential equation of (5-10d) for the first term on the right. (5-31b) is the more general form of the contravariant equations of Frenet. It follows from (5-31b) that the values of

$$\overset{L}{c}{}^{a(n)}_{i.} := d^{n}\overset{L}{c}{}^{a}_{i.}/ds^{n} \quad \text{for} \quad n \geq 1 \tag{5-32a}$$

can be easily determined for every point P on a space curve C with the functions

$$g_{ab} = g_{ab}(q^{c}) \; , \quad \kappa = \kappa(s) \; , \quad \tau = \tau(s) \; , \tag{5-32b}$$

and the initial values at P, i.e.,

$$\overset{L}{c}{}^{a}_{i.} = e_{ji.L}c^{a}_{j.} \tag{5-32c}$$

are given. Besides (5-31), the equations of Frenet of (2-84) are also valid for <u>surface curves</u>, for which the statements for (2-86) replace those for (5-32).

If the <u>space curve C is a straight line</u>, i.e.,

$$\kappa \equiv 0 \; , \quad \tau = \text{undetermined} \; , \tag{5-33a}$$

only those equations of Frenet of (5-30b) with $j = 1$ and those equations of (5-31) with $i = 1$ are relevant. Based on (5-28a) or (5-29d) and in accordance with (5-19a,b) we obtain

$$y^{(2)}_{i.} = d^{2}y_{i.}/ds^{2} = 0_{i.} \quad , \tag{5-33b}$$

$$q^{a(2)} = d^{2}q^{a}/ds^{2} = c^{a}_{j.c}c_{j.d}q^{c(1)}q^{d(1)} = -\Gamma^{a}_{cd}q^{c(1)}q^{d(1)} \quad . \tag{5-33c}$$

These equations are the <u>differential equations of geodesic (straight) lines in Euclidean space</u> for Cartesian and curvilinear coordinates. The second form of (5-33c) is identical with (5-19b) and is also true for non-Euclidean space, e.g., for geodesics on surfaces [see Chapter 2.1.9].

Let P_o and P be two points on a space curve C separated by the distance s along C and with the coordinates and coordinate differences

$$(q^a)_o, \; q^a, \qquad \Delta q^a = q^a - (q^a)_o . \tag{5-34a}$$

A <u>series expansion of Δq^a with respect to the powers of s</u> reads

$$\Delta q^a = \sum_{n=1}^{n'} (1/n!)(q^{a(n)})_o s^n . \tag{5-34b}$$

$$q^{a(n)} = \overset{L}{c}{}^{a(n-1)}_{1.} \tag{5-34c}$$

follows from (5-29d), and thus for $n = 1$

$$(q^{a(1)})_o = (\overset{L}{c}{}^{a}_{1.})_o = (e_{i1.L})_o (c^a_{i.})_o . \tag{5-34d}$$

If these initial values are given, the derivatives of (5-34c) for $n > 1$ - and thus the coefficients of (5-34b) - can be calculated stepwise according to the explanation given for (5-31) and (5-32)

Of particular interest, especially in geodetic applications, are <u>straight lines C</u> [see (5-33)]. Δq^a is then determined using the three-dimensional polar coordinates of (5-16a) or the contravariant normal coordinates of P relative to P_o [see (5-18d)]. Due to (5-34d)

$$p^a = (q^{a(1)})_o s = (\overset{L}{c}{}^{a}_{1.})_o s \tag{5-35}$$

is also true for these coordinates. Transforming Δq^a into p^a involves an initial value problem of the differential equation of geodesics of (5-33c), which is referred to as the <u>direct geodetic problem</u> in three-dimensional Euclidean space analogous to (2-91). In this case the power series of (5-34b) change into Legendrean series, for which all the results derived from (3-12) to (3-16) are valid after exchanging the coordinate and index designations. The result equivalent to (3-15b,c) is already given here in the power series of (5-26c,d). This application of Legendrean series in three-dimensional space was first proposed by

A. Marussi in 1950 [see Marussi 1985]. The inverse transformation, i.e., the transformation of p^a into Δq^a, is then an inverse geodetic problem in three-dimensional Euclidean space corresponding to (2-92). This problem can be solved by means of the power series of (5-20) to (5-23) in formal accordance with (3-15e,f).

5.3 SURFACE-NORMAL COORDINATES

5.3.1 GENERAL FUNDAMENTALS

For defining surface-normal coordinates corresponding to (1-1) a

reference surface F (5-36a)

is introduced on which the points P of the three-dimensional space are projected by means of the surface normals into the

vertical foot points $Q \in F$ (5-36b)

[see Fig. 5.1]. The

Surface-normal coordinates $q^a = (u^\alpha, H)$ (5-37)

consist of

$$q^\alpha =: u^\alpha, \quad \alpha \in \{1,2\} = \text{surface coordinates of } Q \in F \, , \tag{5-37a}$$

$$q^3 =: H = \text{height coordinates of } P = \text{distances } P - Q \, . \tag{5-37b}$$

The applicability of surface-normal coordinates generally presupposes that there is only one foot point Q for every point P to be projected. With

$$y_{i.} = y_{i.}(q^a) = \text{position vector of } P \, , \tag{5-38a}$$

$$x_{i.} = x_{i.}(u^\alpha) = \text{position vector of } Q \, , \tag{5-38b}$$

$$n_{i.} = n_{i.}(u^\alpha) = \text{normal vector of } F \text{ at } Q \, , \tag{5-38c}$$

$$|n_{i.}| = 1 \, .$$

The transformation equations between surface-normal coordinates and Cartesian coordinates become

$$y_{i.}(q^a) = x_{i.}(u^\alpha) + H \, n_{i.}(u^\alpha) \, . \tag{5-38d}$$

If the position vector of F [see (5-38b)] is given as a function of surface coordinates u^α, then the normal vector of (5-38c) is also given according to (2-25). Hence, it follows that the geometry of the surface-normal coordinates represented by (5-38d) is fully determined by the intrinsic and extrinsic geometry of the reference surface F. In terms of differential geometry this means that the metric tensor of surface-normal coordinates, for which the expression

$$g'_{ab} = g'_{ab}(q^c) \tag{5-39a}$$

is chosen here, must be a function of the first and second fundamental tensors of the reference surface of (2-17b) and (2-47b), i.e.,

$$g_{\alpha\beta} = g_{\alpha\beta}(u^\gamma) , \qquad L_{\alpha\beta} = L_{\alpha\beta}(u^\gamma) . \tag{5-39b}$$

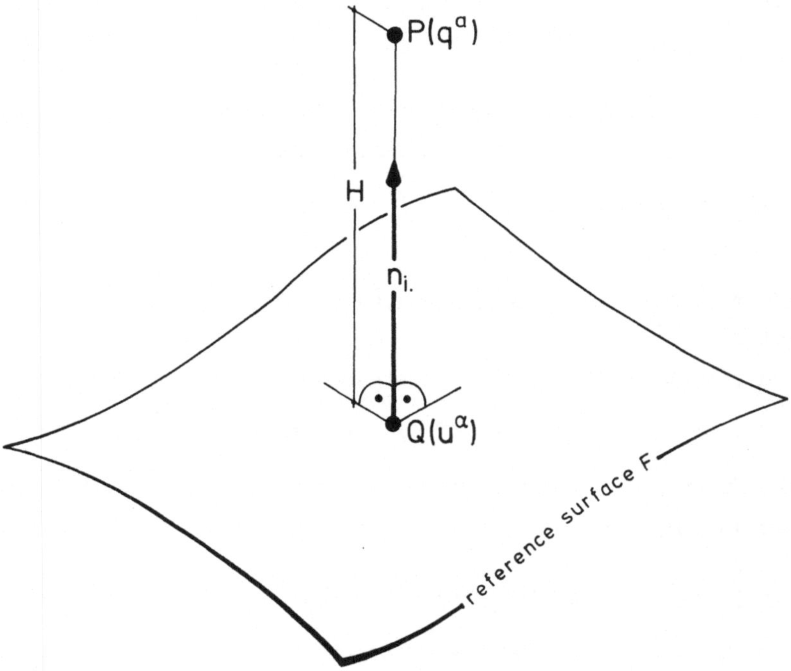

Fig. 5.1. Surface-normal coordinates $q^a = (u^\alpha, H)$

Accordingly, the differential geometry of q^a can be founded completely on (5-39b), for which the most important fundamentals in geodetic applications are dealt with in the following.

In the subsequent representations it will frequently be purposeful to combine the covariant and contravariant <u>moving trihedrons of the reference surface F</u> [see (2-5) and (2-32)] into basis vectors, i.e.,

$$b_{i.a} := b_{i.\alpha}\delta^{\alpha}_{a} + n_{i.}\delta^{3}_{a} , \qquad (5\text{-}40a)$$

$$b^{a}_{i.} := b^{\alpha}_{i.}\ \delta^{a}_{\alpha} + n_{i.}\delta^{a}_{3} . \qquad (5\text{-}40b)$$

Herein,

$b_{i.\alpha}$, $b^{\alpha}_{i.}$ = covariant and contravariant basis vectors of u^{γ} of F after (2-32), (5-40c)

and

$$n_{i.}b_{i.\alpha} = 0 , \quad n_{i.}b^{\alpha}_{i.} = 0 , \quad n_{i.}n_{i.} = 1 . \qquad (5\text{-}40d)$$

Based on (5-38d) and observing (2-46), (2-47), and (5-40a) the <u>covariant basis vectors of q^a</u> become

$$c_{i.a} := y_{i,a} = (\delta^{\beta}_{\alpha} - H\,L^{\beta}_{\alpha})b_{i.\beta}\delta^{\alpha}_{a} + n_{i.}\delta^{3}_{a} =: h^{b}_{a}\,b_{i.b} , \qquad (5\text{-}41a)$$

and

$$c_{i.a} = b_{i.a} \quad \text{for} \quad H = 0 \rightarrow P \equiv Q . \qquad (5\text{-}41b)$$

The transformation matrices h^{b}_{a} between the covariant basis vectors in Q and P which were introduced into (5-41a) have the following components:

$$h^{b}_{a} = h^{\beta}_{\alpha}\ \delta^{\alpha}_{a}\ \delta^{b}_{\beta} + \delta^{3}_{a}\ \delta^{b}_{3} , \qquad h^{\beta}_{\alpha} = \delta^{\beta}_{\alpha} - H\,L^{\beta}_{\alpha} . \qquad (5\text{-}41c)$$

For the <u>covariant metric tensor of q^a</u> [see (5-39a)] we obtain

$$\begin{aligned} g'_{ab} &= h^{c}_{a}\ h^{d}_{b}\ b_{i.c}b_{i.d} = h^{c}_{a}\ h^{d}_{b}(g_{\gamma\delta}\delta^{\gamma}_{c}\ \delta^{\delta}_{d} + \delta^{3}_{c}\ \delta^{3}_{d}) \\ &= g_{\gamma\delta}(\delta^{\gamma}_{\alpha} - H\,L^{\gamma}_{\alpha})(\delta^{\delta}_{\beta} - H\,L^{\delta}_{\beta})\delta^{\alpha}_{a}\ \delta^{\beta}_{b} + \delta^{3}_{a}\ \delta^{3}_{b} \end{aligned} \qquad (5\text{-}42a)$$

or

$$g'_{ab} = (g_{\alpha\beta} - 2H\,L_{\alpha\beta} + H^{2}L_{\alpha\gamma}L^{\gamma}_{\beta})\delta^{\alpha}_{a}\ \delta^{\beta}_{b} + \delta^{3}_{a}\ \delta^{3}_{b} . \qquad (5\text{-}42b)$$

From this we obtain in particular

$$g'_{\alpha 3} = 0 , \qquad g'_{33} = 1 , \qquad (5\text{-}42c)$$

so that the H-lines are thus perpendicular to the u^{α}-lines at all points P of the space. Due to (5-42c) the determinant of the metric tensor has the form of

$$g' = |g'_{ab}| = g'_{11}g'_{22} - g'^{2}_{12} . \qquad (5\text{-}42d)$$

<u>Completely orthogonal surface-normal coordinates q^a</u> exist when

$$g'_{ab} = 0 \quad \text{for} \quad a \neq b . \qquad (5\text{-}43a)$$

It follows from (5-42b) that this condition is met by

$$g_{12} = 0 , \qquad L_{12} = 0 \qquad (5\text{-}43b)$$

because then $g^{12} = 0$ as well, and consequently

$$L_1^2 = L_2^1 = 0 , \qquad L_{1\gamma}L_2^\gamma = L_{2\gamma}L_1^\gamma = 0 . \tag{5-43c}$$

In this case we obtain the following for (5-42d):

$$\begin{aligned} g' = g'_{11}g'_{22} &= g_{11}g_{22} - 2H(g_{11}L_{22} + g_{22}L_{11}) \\ &+ H^2((g_{11}/g_{22})L_{22}^2 + 4L_{11}L_{22} + (g_{22}/g_{11})L_{11}^2) \\ &+ 2H^3L_{11}L_{22}(L_{11}/g_{11} + L_{22}/g_{22}) + H^4(L_{11}^2/g_{11})(L_{22}^2/g_{22}) . \end{aligned} \tag{5-43d}$$

Proceeding from the covariant basis vectors of (5-41) the <u>contravariant basis vectors of q^a</u> can be calculated in accordance with (5-14b) as follows:

$$\delta_{ij.} = c_{i.a}c^a_{j.} = h^b_a \, b_{i.b}c^a_{j.} = h^\beta_\alpha \, b_{i.\beta}c^\alpha_{j.} + n_{i.}c^3_{j.} , \tag{5-44a}$$

whereby

$$h^\beta_\alpha = \delta^\beta_\alpha - H\, L^\beta_\alpha \quad \text{after (5-41c)} . \tag{5-44b}$$

Due to (5-40d) inner multiplication of (5-44a) by $b^\gamma_{i.}$ or $n_{i.}$ following the exchange of indices yields

$$c^\alpha_{i.} h^\beta_\alpha = b^\beta_{i.} , \qquad c^3_{i.} = n_{i.} , \tag{5-44c}$$

respectively. The inversion of the first tensor equation gives

$$c^\alpha_{i.} = \bar{h}^\alpha_\beta \, b^\beta_{i.} \tag{5-44d}$$

with the following components for the transformation matrix:

$$\begin{aligned} &\bar{h}^1_1 = h^2_2/D , \quad \bar{h}^1_2 = -h^1_2/D , \quad \bar{h}^2_1 = -h^2_1/D , \quad \bar{h}^2_2 = h^1_1/D , \\ &D = h^1_1\, h^2_2 - h^2_1\, h^1_2 . \end{aligned} \tag{5-44e}$$

The representation of the contravariant basis vectors corresponding to (5-41a) reads

$$c^a_{i.} = \bar{h}^\alpha_\beta \, b^\beta_{i.} \, \delta^a_\alpha + n_{i.} \, \delta^a_3 =: \bar{h}^a_b \, b^b_{i.} \tag{5-44f}$$

with

$$\bar{h}^a_b = \bar{h}^\alpha_\beta \, \delta^a_\alpha \, \delta^\beta_b + \delta^a_3 \, \delta^3_b \tag{5-44g}$$

corresponding to (5-41c). Analogous to (5-41b)

$$c^a_{i.} = b^a_{i.} \quad \text{for} \quad H = 0 \;\rightarrow\; P \equiv Q . \tag{5-44h}$$

In the case of orthogonal surface-normal coordinates [see (5-43)]

$$h_2^1 = h_1^2 = 0 \quad \text{and thus} \quad \bar{h}_2^1 = \bar{h}_1^2 = 0 \,. \tag{5-45a}$$

Thereby, (5-44e) yields

$$\bar{h}^{(\alpha)}_{(\alpha)} = 1/h^{(\alpha)}_{(\alpha)} \,, \tag{5-45b}$$

and (5-44f) becomes

$$c^a_{i.} = (1/h^{(\alpha)}_{(\alpha)}) b^{(\alpha)}_{i.} \delta^a_{(\alpha)} + n_{i.} \delta^a_3 \,. \tag{5-45c}$$

Proceeding from (5-44f,g) the components of the contravariant metric tensor g'^{ab} [see (5-14c)] can be calculated analogous to (5-42a), i.e.,

$$g'^{ab} = \bar{h}^a_c \, \bar{h}^b_d \, b^c_{i.} b^d_{i.} = g^{\gamma\delta} \, \bar{h}^\alpha_\gamma \, \bar{h}^\beta_\delta \, \delta^a_\alpha \, \delta^b_\beta + \delta^a_3 \, \delta^b_3 \,. \tag{5-46a}$$

On the other hand, we can also begin with the components of the covariant metric tensor of (5-42a). Based on

$$g'_{ab} g'^{bc} = \delta^c_a \tag{5-46b}$$

and observing (5-42c) we obtain

$$\begin{aligned} g'^{11} &= g'_{22}/g', \quad g'^{12} = -g'_{12}/g', \quad g'^{22} = g'_{11}/g' \,, \\ g'^{13} &= g'^{23} = 0 \,, \qquad\qquad\qquad g'^{33} = 1 \,. \end{aligned} \tag{5-46c}$$

For orthogonal q^a [see (5-43)]

$$\begin{aligned} g'^{11} &= 1/g'_{11}, \quad g'^{22} = 1/g'_{22}, \quad g'^{33} = 1 \,, \\ g'^{12} &= g'^{13} = g'^{23} = 0 \end{aligned} \tag{5-46d}$$

follows. Thus, the representation of the basis vectors and metric tensors of the surface-normal coordinates of (5-37) as functions of the moving trihedrons of (5-40) and the fundamental tensors of (5-39b) of the reference surface F is complete. In the following we will go into tracing back the differential equations of (5-10d) to those of Gauss and Weingarten [see (2-46) and (2-49)] for the trihedrons of (5-40) of the reference surface F. This can be done by substituting (5-41a) or (5-44f) for the right side of the equations of (5-10d). We then obtain the following results:

$$c_{i.a,b} = G^c_{ab} b_{i.c} \,, \qquad G^c_{ab} := \Gamma'^d_{ab} \, h^c_d \,, \tag{5-47a}$$

$$c^a_{i,b} = \bar{G}^a_{bc} b^c_{i.} \,, \qquad \bar{G}^a_{bc} := -\Gamma'^a_{bd} \, \bar{h}^d_c \,, \tag{5-47b}$$

h^c_d, $\bar{h}^d_c$ after (5-41c), (5-44g).

The Christoffel's symbols of (5-10e) can be calculated based on (5-42) and (5-46). Another possibility of deducing the differential equations of (5-47a,b) consists of calculating the partial derivatives of (5-41a) and (5-44f) in consideration of the differential equations of (2-46) and (2-49). In this way we obtain the following for covariant differential equations of (5-10d) in accordance with (5-41c):

$$c_{i.a,b} = c_{i.b,a} = [((h^{\delta}_{\alpha}\, \Gamma^{\gamma}_{\beta\delta} + h^{\gamma}_{\alpha,\beta})\delta^{\alpha}_{a} - L^{\gamma}_{\beta}\, \delta^{3}_{a})\delta^{\beta}_{b} + h^{\gamma}_{\alpha,3}\delta^{\alpha}_{a}\, \delta^{3}_{b}]\, b_{i.\gamma} + h^{\delta}_{\alpha}\, L_{\delta\beta}\delta^{\alpha}_{a}\, \delta^{\beta}_{b}\, n_{i.} \tag{5-48a}$$

$$h^{\gamma}_{\alpha} = \delta^{\gamma}_{\alpha} - H\, L^{\gamma}_{\alpha}\,, \quad h^{\gamma}_{\alpha,\beta} = -H\, L^{\gamma}_{\alpha,\beta}\,, \quad h^{\gamma}_{\alpha,3} = -L^{\gamma}_{\alpha}\,. \tag{5-48b}$$

The symmetry in a and b also requires symmetry in α and β. By substituting (5-48b) in (5-48a) the restriction of

$$\Gamma^{\gamma}_{\beta\delta}L^{\delta}_{\alpha} + L^{\gamma}_{\alpha,\beta} = \Gamma^{\gamma}_{\alpha\delta}L^{\delta}_{\beta} + L^{\gamma}_{\beta,\alpha} \tag{5-48c}$$

results, which is a special form of the equations of Mainardi and Codazzi of (2-50c) [see also Blaschke & Leichtweiss 1973, § 58]. The relationships of (5-43) are valid for orthogonal surface-normal coordinates q^{a}. With these relationships and based on (5-48a-c) we obtain

$$\begin{aligned}
c_{i.1,1} &= ((1/2)(g_{11,1}/g_{11})(1-H\, L^{1}_{1}) - H\, L^{1}_{1,1})\, b_{i.1} \\
&\quad -(1/2)(g_{11,2}/g_{22})(1-H\, L^{1}_{1})\, b_{i.2} + L_{11}(1-H\, L^{1}_{1})\, n_{i.}\,, \\
c_{i.1,2} &= (1/2)((g_{11,2}/g_{11})(1-H\, L^{2}_{2})\, b_{i.1} + (g_{22,1}/g_{22})(1-H\, L^{1}_{1})\, b_{i.2})\,, \\
c_{i.2,1} &= c_{i.1,2}\,, \\
c_{i.2,2} &= ((1/2)(g_{22,2}/g_{22})(1-H\, L^{2}_{2}) - H\, L^{2}_{2,2})\, b_{i.2} \\
&\quad -(1/2)(g_{22,1}/g_{11})(1-H\, L^{2}_{2})\, b_{i.1} + L_{22}(1-H\, L^{2}_{2})\, n_{i.}\,, \\
c_{i.\alpha,3} &= c_{i.3,\alpha} = -L^{(\alpha)}_{(\alpha)}b_{i.\alpha}\,, \quad c_{i.3,3} = 0\,.
\end{aligned} \tag{5-48d}$$

Analogous to (5-48a,b) the contravariant differential equations of (5-10d) become

$$c^{a}_{i,b} = [((-\bar{h}^{\alpha}_{\delta}\, \Gamma^{\delta}_{\beta\gamma} + \bar{h}^{\alpha}_{\gamma,\beta})\delta^{a}_{\alpha} - L_{\beta\gamma}\delta^{a}_{3})\delta^{\beta}_{b} + \bar{h}^{\alpha}_{\gamma,3}\delta^{a}_{\alpha}\, \delta^{3}_{b}]\, b^{\gamma}_{i.} - \bar{h}^{\alpha}_{\delta}\, L^{\delta}_{\beta}\, \delta^{a}_{\alpha}\, \delta^{\beta}_{b}\, n_{i.} \tag{5-49a}$$

with $\bar{h}^{\alpha}_{\gamma}$ according to (5-44e). The partial derivatives of $\bar{h}^{\alpha}_{\gamma,\beta}$ and $\bar{h}^{\alpha}_{\gamma,3}$ are also calculated on the basis of (5-44e). For orthogonal surface-normal coordinates of q^a [see (5-43)] we obtain

$$
\begin{aligned}
c^1_{i,1} &= ((-(1/2)(g_{11,1}/g_{11}) + H\,L^1_{1,1}/(1-H\,L^1_1))\,b^1_{i.} \\
&\quad -(1/2)(g_{11,2}/g_{11})\,b^2_{i.} + L^1_1\,n_{i.})/(1-H\,L^1_1)\,, \\
c^1_{i,2} &= ((-(1/2)(g_{11,2}/g_{11}) + H\,L^1_{1,2}/(1-H\,L^1_1))\,b^1_{i.} \\
&\quad +(1/2)(g_{22,1}/g_{11})\,b^2_{i.})/(1-H\,L^1_1)\,, \\
c^2_{i,1} &= ((-(1/2)(g_{22,1}/g_{22}) + H\,L^2_{2,1}/(1-H\,L^2_2))\,b^2_{i.} \\
&\quad +(1/2)(g_{11,2}/g_{22})\,b^1_{i.})/(1-H\,L^2_2)\,, \\
c^2_{i,2} &= ((-(1/2)(g_{22,2}/g_{22}) + H\,L^2_{2,2}/(1-H\,L^2_2))\,b^2_{i.} \\
&\quad -(1/2)(g_{22,1}/g_{22})\,b^1_{i.} + L^2_2\,n_{i.})/(1-H\,L^2_2)\,, \\
c^{\alpha}_{i,3} &= (L^{(\alpha)}_{(\alpha)}/(1-H\,L^{(\alpha)}_{(\alpha)})^2)b^{\alpha}_{i.}\,, \quad c^3_{i,\alpha} = -L_{\alpha(\alpha)}b^{(\alpha)}_{i.}\,, \quad c^3_{i,3} = 0\,.
\end{aligned}
\tag{5-49b}
$$

If higher partial derivatives with respect to q^a are calculated based on (5-48) or (5-49), then for their representation in the trihedrons of (5-40a,b) $b_{i.\gamma,\delta}$ and $b^{\gamma}_{i,\delta}$ are to be eliminated stepwise using the equations of Gauss and Weingarten of (2-46) and (2-49).

5.3.2 REPRESENTING TRANSFORMATIONS BETWEEN SURFACE-NORMAL COORDINATES AND CARTESIAN COORDINATES BY POWER SERIES

The following considerations are based on the results of Chapter 5.2.2. Moreover, the characteristics of special curvilinear coordinates which result when surface-normal coordinates q^a are concerned will be pointed out here.

We will first go into the power series (5-20) and (5-21) for <u>transforming q^a into $y_{i.}$</u>. In this case the transformation equations are the same as (5-38d), from which the following partial derivatives of $y_{i.}$ with respect to q^a are obtained observing (5-4a,b):

$$y_{i,\alpha_1\alpha_2\ldots\alpha_n} = x_{i,\alpha_1\alpha_2\ldots\alpha_n} + H\, n_{i,\alpha_1\alpha_2\ldots\alpha_n}\,, \qquad (5\text{-}50a)$$
$$y_{i,3\,\alpha_2\ldots\alpha_n} = n_{i,\alpha_2\alpha_3\ldots\alpha_n}\,, \quad y_{i,33\ldots} = 0\,.$$

The differences in the values of q^a [see (5-15b)] are

$$\Delta q^a = (\Delta u^\alpha,\ \Delta H)\,, \quad \Delta u^\alpha = u^\alpha - (u^\alpha)_o\,, \quad \Delta H = H - H_o\,. \qquad (5\text{-}50b)$$

The power series of (5-20) together with (5-50a,b) can be written as follows:

$$\Delta y_{i.} = \sum_{n=1}^{n'} (1/n!)(x_{i,\alpha_1\ldots\alpha_n} + H_o n_{i,\alpha_1\ldots\alpha_n})_o \Delta u^{\alpha_1} \Delta u^{\alpha_2} \ldots \Delta u^{\alpha_n}$$
$$+ \Delta H(n_{i.o} + \sum_{n=1}^{n''} (1/n!)(n_{i,\alpha_2\ldots\alpha_n})_o \Delta u^{\alpha_2} \Delta u^{\alpha_3} \ldots \Delta u^{\alpha_n})\,. \qquad (5\text{-}50c)$$

n' and n'' are generally chosen in dependency on Δu^α and ΔH. For

$$H_o = H = 0 \rightarrow \Delta H = 0 \qquad (5\text{-}51a)$$

$\Delta y_{i.} =: \Delta x_{i.}$ represents the differences in Cartesian coordinates of the foot points Q_o and Q of the reference surface F. In return, (5-50c) becomes

$$\Delta x_{i.} = \sum_{n=1}^{n'} (1/n!)(x_{i,\alpha_1\alpha_2\ldots\alpha_n})_o \Delta u^{\alpha_1} \Delta u^{\alpha_2} \ldots \Delta u^{\alpha_n}\,. \qquad (5\text{-}51b)$$

The case of (5-22) is explained here in such a way that instead of the position vector of (5-38d) the first and second fundamental tensors of the reference surface F [see (5-39b)] are given as functions of the surface coordinates u^γ. The partial derivatives of (5-50a) are then calculated based on the covariant base vectors of (5-41) and the accompanying differential equations of (5-48). In the second and higher derivatives of $c_{i.a}$ the equations of Gauss and Weingarten of (2-46) are substituted for $b_{i.\gamma,\delta}$ and $n_{i,\delta}$. The result corresponding to (5-22b) together with the moving trihedrons of (5-40a) can be written in the following form:

$$\Delta y_{i.} = \sum_{n=1}^{n'} ((\bar{d}^d_{\alpha_1\alpha_2\ldots\alpha_n} \Delta u^{\alpha_1} + \bar{d}^d_{\alpha_2\ldots 3\ldots\alpha_n} \Delta H)\Delta u^{\alpha_2} \Delta u^{\alpha_3} \ldots \Delta u^{\alpha_n})(b_{i.d})_o\,. \qquad (5\text{-}52a)$$

In this manner $\Delta y_{i.}$ is spanned by the covariant basis at $Q_o \in$ F [see (5-40a)], i.e.,

$$(b_{i.d})_o = (b_{i.\alpha})_o\delta^{\alpha}_{d} + n_{i.o}\delta^{3}_{d} \ . \qquad (5\text{-}52b)$$

The coefficients of (5-52a) are functions of the fundamental tensors of (5-39b) and their partial derivatives.

Inner multiplication of (5-52a) by $(b^a_{i.})_o$ [see (5-40b)] yields the following Legendrean inverse series for solving the inverse geodetic problem in three-dimensional Euclidean space:

$$\begin{aligned} p^a &:= (b^a_{i.})_o \Delta y_{i.} \\ &= \sum_{n=1}^{n'} (\bar{d}^a_{\alpha_1 a_2 \ldots \alpha_n} \Delta u^{\alpha_1} + \bar{d}^a_{\alpha_2 \ldots 3 \ldots \alpha_n} \Delta H) \Delta u^{\alpha_2} \Delta u^{\alpha_3} \ldots \Delta u^{\alpha_n} \ . \end{aligned} \qquad (5\text{-}53a)$$

The quantities of p^a introduced here involve the contravariant components of the <u>normal coordinates (5-18d) related to the covariant basis of (5-40a) at Q_o</u>. When transforming p^a into polar coordinates corresponding to (5-18e), the metric tensor g'_{ab} at Q_o is necessary, for which we obtain

$$(g_{ab})_o := (g'_{ab})_{H=0} = (g_{\alpha\beta})_o \delta^{\alpha}_{a}\, \delta^{\beta}_{b} + \delta^{3}_{a}\, \delta^{3}_{b} \qquad (5\text{-}53b)$$

according to (5-42b). Herewith,

$$s^2 = (g_{ab})_o p^a p^b = (g_{\alpha\beta})_o p^{\alpha} p^{\beta} + (p^3)^2 \qquad (5\text{-}53c)$$

and

$$r^a := (b^a_{i.})_o r_{i.} = p^a / s \ . \qquad (5\text{-}53d)$$

The subject of the following considerations is <u>transforming $y_{i.}$ into q^a</u>:

$$q^a = q^a(y_{i.}) = (u^{\alpha}(y_{i.}),\ H(y_{i.})) \ , \qquad (5\text{-}54a)$$

whose components for $a = 1,2$ represent the <u>mapping of three-dimensional space on the reference surface F</u>:

$$q^{\alpha} = u^{\alpha}(y_{i.}) \ . \qquad (5\text{-}54b)$$

The power series for (5-54a,b) generally have the forms of (5-25) and (5-26). The method for determining their coefficients described for (5-25) and (5-26) presumes that the metric tensor g'_{ab} of q^a [see (5-39a)] is given. Here, the metric tensor is replaced by first and second fundamental tensors of the reference surface F [see (5-39b)], as done for (5-52) and (5-53). Proceeding from (5-44f) and the accompanying differential equations of (5-47b) and (5-49) the partial derivatives of

q^a with respect to $y_{i.}$ are determined completely analogous to (5-26a). The resulting power series corresponding to (5-26c) has the form of

$$\Delta q^a = (\Delta u^\alpha, \Delta H) = \sum_{n=1}^{n'} d^a_{a_1 a_2 \ldots a_n} p^{a_1} p^{a_2} \ldots p^{a_n} , \tag{5-54c}$$

$$p^a = (b^a_{i.})_o \Delta y_{i.} \quad \text{after (5-53)} .$$

The components for $a = 1,2$ of this series involve the mapping of (5-54b). Only these two components are relevant in the case of (5-51a). (5-54c) can then be written as

$$\Delta u^\alpha = p^\alpha + \sum_{n=2}^{n'} d^\alpha_{a_1 a_2 \ldots a_n} p^{a_1} p^{a_2} \ldots p^{a_n} , \tag{5-54d}$$

$$p^a = (b^a_{i.})_o \Delta x_{i.} , \quad \text{indices after (5-4a,b)} ,$$

whereby $\Delta x_{i.}$ is the difference in position vectors between the points $P_o \equiv Q_o$ and $P \equiv Q$ on the reference surface F. (5-54c) and (5-54d) are the solution to the direct geodetic problem in three-dimensional space by means of Legendrean series corresponding to (3-14) and (3-15).

5.3.3 TRANSFORMATIONS BETWEEN THREE-DIMENSIONAL POLAR COORDINATES AND POLAR COORDINATES ON THE REFERENCE SURFACE

The <u>three-dimensional polar coordinates of a point P relative to a pole P_o</u> of (5-16a) are designated in the following by

$$r'_{i.} := \Delta y_{i.} / |\Delta y_{j.}| , \quad s' := |\Delta y_{i.}| , \tag{5-55a}$$

and their contravariant components of (5-18c) by

$$(r'^a)_o = (q^{a(1')})_o = (c^a_{i.})_o r'_{i.} , \quad s' , \tag{5-55b}$$

whereby $f^{(n')} := d^n f / ds'^n$ corresponding to (5-17). The accompanying contravariant normal coordinates relative to the covariant basis at P_o are calculated in accordance with (5-18d) as follows:

$$p'^a := (r'^a)_o s' = (c^a_{i.})_o \Delta y_{i.} . \tag{5-55c}$$

In addition to these coordinates, the contravariant normal coordinates relative to the covariant basis at Q_o, i.e., the plumb foot point of P_o on the reference surface F, which were introduced in (5-54a), can also be used in the following way:

$$p^a := (b^a_{i.})_o \Delta y_{i.} = (r^a)_o s', \quad (r^a)_o = (b^a_{i.})_o r'_{i.} \quad . \tag{5-55d}$$

<u>Polar coordinates on F of the plumb foot points Q relative to the pole Q_o</u> [see Fig. 5.2] are generally defined corresponding to (2-112). (2-112c,d) is replaced here by the equivalent definition analogous to (5-55a), i.e.,

$$(x^{(1)}_{i.})_o := (dx_{i.}/ds)_o = (b_{i.\alpha})_o (u^{\alpha(1)})_o \tag{5-56a}$$

$= $ tangent vector on the geodesic line $Q_o - Q$ in Q_o ,

$$s_o \quad = \text{distance } Q_o - Q \text{ along the geodesic line} \ . \tag{5-56b}$$

The accompanying contravariant components are

$$(u^{\alpha(1)})_o := (du^\alpha/ds)_o, \qquad s_o \tag{5-56c}$$

with which the contravariant normal coordinates relative to Q_o become

$$v^\alpha := (u^{\alpha(1)})_o s_o \ . \tag{5-56d}$$

This definition agrees with (2-131j) and (3-15a), but the crossbar over v is omitted here.

Representing transformation equations between p'^a or p^a, on one hand, and v^α, on the other, by power series in the coordinates can generally be done with the method described in Chapter 5.3.2, whereby the surface-normal coordinates are given in the form of

$$q^a = (v^\alpha, H) \ . \tag{5-57}$$

Then, in (5-50b)ff only u^α is replaced by the special surface coordinates v^α [see (5-56d)] in a purely formal way, and it is presumed that the first and second fundamental tensors of (5-39b) as functions of v^α are known. Since the coordinates v^α can, as a rule, only be obtained based on other surface coordinates u^α, this is also true for the fundamental tensors of v^α.

The previously described method will not be applied further here. Instead, we will proceed by solving the geodetic problems for

$P_o - P$ = geodesic (straight) line in three-dimensional Euclidean space , (5-58a)

$Q_o - Q$ = geodesic line on the reference surface (5-58b)

by means of Legendrean series. For the inverse geodetic problem for (5-58a), (5-53a) yields

$$p^a = \sum_{n=1}^{n'} (\bar{d}^a_{\alpha_1\alpha_2\ldots\alpha_n} \Delta u^{\alpha_1} + \bar{d}^a_{3\ \alpha_2\ldots\alpha_n} \Delta H)\Delta u^{\alpha_2} \Delta u^{\alpha_3} \ldots \Delta u^{\alpha_n} \tag{5-59a}$$

with

$$\bar{d}^a_{a_1a_2\ldots a_n} = (1/n!)(c_{i.a_1,a_2\ldots a_n})_o (b^a_{i.})_o \ , \tag{5-59b}$$

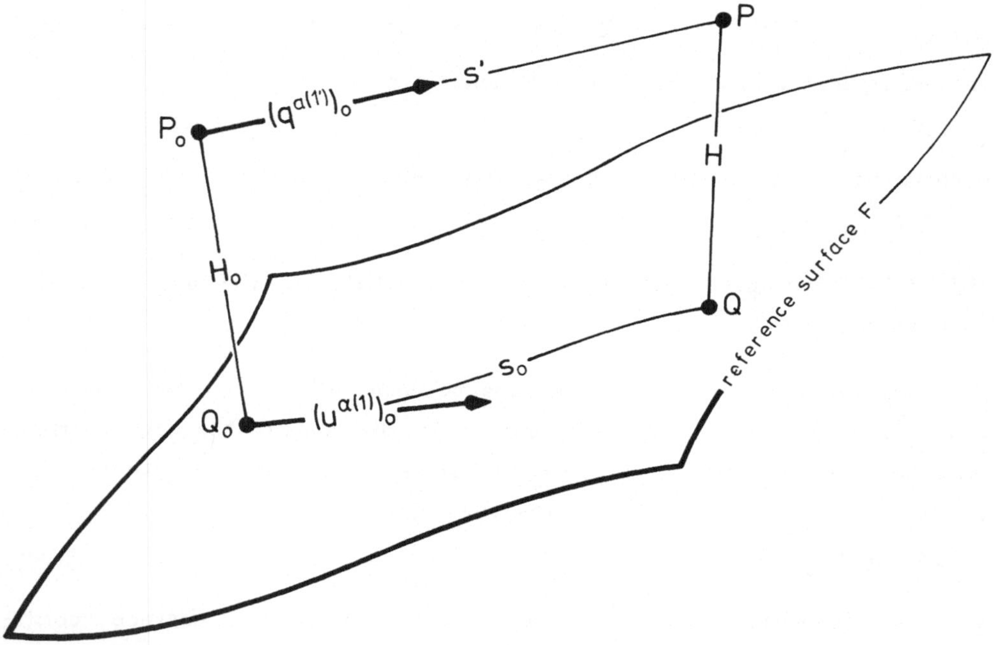

Fig. 5.2. Polar coordinates $(q^{\alpha(1')})_o$, s' and $(u^{\alpha(1)})_o$, s_o

in which the basis vectors and their partial derivatives are calculated based on (5-40), (5-41), and (5-48). Δu^α on the right side of (5-59a) can be represented as a function of the contravariant normal coordinates of (5-56d) by solving the direct geodetic problem for (5-58b) in the form of the Legendrean series of (3-15b), i.e.,

$$\Delta u^\alpha = v^\alpha + \sum_{n=2}^{n'} a^\alpha_{\beta_1\beta_2\ldots\beta_n} v^{\beta_1} v^{\beta_2} \ldots v^{\beta_n} . \tag{5-60}$$

Inserting (5-60) into (5-59) leads to the desired transformation equations between v^α and p^a with given heights H. If the normal coordinates p^a are calculated, then the polar coordinates of (5-55b) can be obtained from them as follows:

$$s' = ((g_{ab})_o p^a p^b)^{1/2} , \qquad (r^a)_o = p^a/s' . \tag{5-61}$$

We will now go into

P_o, $P \in$ reference surface F, i.e., (5-62)

$$P_o \equiv Q_o, \quad P \equiv Q \rightarrow H_o = H = 0, \quad y_{i.} \equiv x_{i.}$$

in more detail, for which (5-59) can be greatly simplified as follows:

$$p^a_o = (r^a_o) s'_o = \sum_{n=1}^{n'} \overset{o}{d}{}^a_{\alpha_1\alpha_2\ldots\alpha_n} \Delta u^{\alpha_1} \Delta u^{\alpha_2} \ldots \Delta u^{\alpha_n} , \tag{5-63a}$$

$$\overset{o}{d}{}^a_{\alpha_1\alpha_2\ldots\alpha_n} = (1/n!)(b_{i.\alpha_1,\alpha_2\ldots\alpha_n})_o (b^a_{i.})_o . \tag{5-63b}$$

If (5-60) is inserted here, we then obtain the power series

$$p^a_o = (r^a_o) s'_o = \sum_{n=1}^{n'} p^a_{\alpha_1\alpha_2\ldots\alpha_n} v^{\alpha_1} v^{\alpha_2} \ldots v^{\alpha_n} , \tag{5-64a}$$

for which the coefficients up to $n' = 3$ become

$$p^a_{\alpha_1} = \delta^a_{\alpha_1} , \qquad p^a_{\alpha_1\alpha_2} = (1/2)(L_{\alpha_1\alpha_2})_o \delta^a_3 ,$$
$$p^a_{\alpha_1\alpha_2\alpha_3} = (1/6)(-L_{\alpha_1\alpha_2} L^\beta_{\alpha_3} \delta^a_\beta + (L_{\alpha_1\alpha_2,\alpha_3} - 2L_{\alpha_2\beta}\Gamma^\beta_{\alpha_1\alpha_3})\delta^a_3) . \tag{5-64b}$$

The normal curvature of the reference surface F [see (2-63)] and its derivative with respect to s at Q_o are given as follows with the changed signs in accordance with (4-17a)

$$(\kappa_n)_o = -(L_{\alpha_1\alpha_2} u^{\alpha_1(1)} u^{\alpha_2(1)})_o ,$$
$$(\kappa^{(1)}_n)_o = -(L_{\alpha_1\alpha_2,\alpha_3} - 2L_{\alpha_2\beta}\Gamma^\beta_{\alpha_1\alpha_3})_o (u^{\alpha_1(1)} u^{\alpha_2(1)} u^{\alpha_3(1)})_o . \tag{5-64c}$$

Thus, (5-64a) up to $n' = 3$ can also be written as

$$p^a_o = v^\alpha \delta^a_\alpha + \delta p^a , \qquad p^3_o \equiv \delta p^3 , \tag{5-65a}$$

$$\delta p^a = (1/6)((\kappa_n L^\alpha_\beta u^{\beta(1)})_o s_o^3 + \ldots)\delta^a_\alpha \\ -(1/2)((\kappa_n)_o s_o^2 + (1/3)(\kappa_n^{(1)})_o s_o^3 + \ldots)\delta^a_3 . \tag{5-65b}$$

Corresponding to (5-61)

$$s'_o = ((g_{ab})_o p_o^a\, p_o^b)^{1/2} , \qquad s_o = ((g_{\alpha\beta})_o v^\alpha v^\beta)^{1/2} \tag{5-66a}$$

with

$$(g_{ab})_o = (g_{\alpha\beta})_o \delta^\alpha_a\, \delta^\beta_b + \delta^3_a\, \delta^3_b \tag{5-66b}$$

according to (5-42b). Inserting (5-65b) and (5-66b) into the first equation of (5-66a) produces

$$s'_o = s_o(1+2(g_{\alpha\beta})_o(u^{\alpha(1)})_o \delta p^\beta/s_o+(p_o^3/s_o)^2+(g_{\alpha\beta})_o \delta p^\alpha \delta p^\beta/s_o^2)^{1/2}. \tag{5-66c}$$

Together with (5-65b) and neglecting terms of s_o^n for $n > 3$ we obtain the following expression for the difference in distance:

$$\delta s := s_o - s'_o = (1/24)(\kappa_n)_o^2 s^3 + \ldots, \quad s \approx s_o \approx s'_o . \tag{5-66d}$$

Corresponding to the second equation of (5-61)

$$(r_o^a)_o = p_o^a/s'_o \quad , \qquad (u^{\alpha(1)})_o = v^\alpha/s_o \tag{5-67a}$$

are true under the conditions of (5-62). This involves the direction vectors of both geodesic lines of (5-58a,b) in the case of (5-62). For calculating the difference between them the first equation of (5-67a) is rewritten as follows using (5-65) and (5-66d):

$$(r_o^a)_o = ((v^\alpha/s_o)\, \delta^a_\alpha + \delta p^a/s_o)(1+\delta s/s_o) \\ = ((u^{\alpha(1)})_o \delta^a_\alpha + \delta p^a/s_o)(1+\delta s/s_o) . \tag{5-67b}$$

Following insertion of the expressions of (5-65b) and (5-66d) we then obtain the desired difference in direction, i.e.,

$$\delta r^a = (u^{\alpha(1)})_o \delta^a_\alpha - (r_o^a)_o = (1/2)(\kappa_n)_o \delta^a_3\, s \\ + (1/24)(\kappa_n(4L^\alpha_\beta + \kappa_n \delta^\alpha_\beta)u^{\beta(1)}\delta^a_\alpha + 4\kappa_n^{(1)}\delta^a_3)_o s^2 + \ldots , \\ s \approx s_o \approx s'_o . \tag{5-67c}$$

All results obtained from (5-63) to (5-67) are based on the special assumptions of (5-62). For solving the problems described for (5-58) to (5-61) completely these results must be supplemented by the differences between the normal coordinates of (5-59a) and (5-63a), i.e.,

$$\delta p'^a = p^a - p_o^a , \tag{5-68a}$$

or the resulting differences between the polar coordinates, i.e.,

$$\delta s' = s' - s'_o \,, \qquad \delta r'^{a} = (r^{a})_o - (r^{a}_o)_o \,. \qquad (5\text{-}68b)$$

For (5-68a) the difference between (5-59a) and (5-63a) first yields

$$\delta p'^{a} = \sum_{n=1}^{n'} (f^{a}_{\alpha_1\alpha_2\ldots\alpha_n} \Delta u^{\alpha_1} + \bar{d}^{a}_{3\,\alpha_2\ldots\alpha_n} \Delta H) \Delta u^{\alpha_2} \Delta u^{\alpha_3} \ldots \Delta u^{\alpha_n} \,, \qquad (5\text{-}69a)$$

with

$$f^{a}_{\alpha_1\alpha_2\ldots\alpha_n} = (1/n!)(c_{i.\alpha_1,\alpha_2\ldots\alpha_n} - b_{i.\alpha_1,\alpha_2\ldots\alpha_n})_o (b^{a}_{i.})_o \,,$$
$$\bar{d}^{a}_{3\,\alpha_2\ldots\alpha_n} = (1/n!)(c_{i.3,\alpha_2\ldots\alpha_n})_o (b^{a}_{i.})_o \,. \qquad (5\text{-}69b)$$

Due to (5-38) to (5-41) we obtain

$$(c_{i.\alpha_1} - b_{i.\alpha_1})_o = (n_{i,\alpha_1})_o H_o = -(L^{\beta}_{\alpha_1} b_{i.B})_o H_o \,,$$
$$(c_{i.3})_o = (n_{i.})_o \qquad (5\text{-}70a)$$

for $n = 1$, and consequently (5-69b) can be written as

$$f^{a}_{\alpha_1\alpha_2\ldots\alpha_n} = (1/n!)(n_{i,\alpha_1\alpha_2\ldots\alpha_n})_o (b^{a}_{i.})_o H_o \,,$$
$$\bar{d}^{a}_{3\,\alpha_2\ldots\alpha_n} = (1/n!)(n_{i,\alpha_2\ldots\alpha_n})_o \; (b^{a}_{i.})_o \,. \qquad (5\text{-}70b)$$

The partial derivatives of $n_{i.}$ at P_o are calculated considering the equations of Gauss and Weingarten (2-46). Up to $n = 2$ we have

$$f^{a}_{\alpha_1} = -(L^{\alpha}_{\alpha_1})_o \delta^{a}_{\alpha} H_o \,,$$
$$f^{a}_{\alpha_1\alpha_2} = -(1/2)((L^{\alpha}_{\alpha_1,\alpha_2} + L^{\beta}_{\alpha_1} \Gamma^{\alpha}_{\beta\alpha_2})\delta^{a}_{\alpha} - L^{\alpha}_{\alpha_1\alpha_2} \delta^{a}_{3})_o H_o \,, \qquad (5\text{-}70c)$$
$$\bar{d}^{a}_{3} = \delta^{a}_{3} \,, \qquad \bar{d}^{a}_{3\alpha_2} = -(1/2)(L^{\alpha}_{\alpha_2})_o \delta^{a}_{\alpha} \,.$$

If these results and (5-60) are substituted for Δu^{α} in (5-69a), then the series

$$\delta p'^{a} = f^{a}_{\alpha} v^{\alpha} + (f^{a}_{\alpha\beta} - f^{a}_{\gamma}(\Gamma^{\gamma}_{\alpha\beta})_o/2) v^{\alpha} v^{\beta} + (\delta^{a}_{3} + \bar{d}^{a}_{3\alpha} v^{\alpha}) \Delta H + \ldots \,, \qquad (5\text{-}71a)$$

is produced, in which

$$v^{\alpha} = (u^{\alpha(1)})_o s_o \approx (u^{\alpha(1)})_o s'_o \qquad (5\text{-}71b)$$

is accurate enough for the normal coordinates of (5-56d). (5-71) is the solution to (5-68a). To arrive at the differences in polar coordinates of (5-68b), (5-68a,b) will first be inserted into (5-61). By doing so we obtain

$$s' = (s_o'^2 + 2(g_{ab})_o(r_o^a)_o s_o'\delta p'^{,b} + (g_{ab})_o\delta p'^{,a}\delta p'^{,b})^{1/2},$$

$$(r^a)_o = (p_o^a + \delta p'^{,a})/(s_o' + \delta s') \tag{5-72a}$$

$$= (r_o^a)_o - (r_o^a)_o\delta s'/s_o' + \delta p'^{,a}/s_o' + \dots .$$

Herein, according to (5-42b) and due to $H = 0$ at Q_o

$$(g_{ab})_o = (g_{\alpha\beta})_o\delta_a^\alpha\,\delta_b^\beta + \delta_a^3\,\delta_b^3\,, \tag{5-72b}$$

and corresponding to (5-63)

$$(r_o^a)_o = p_o^a/s_o'\,. \tag{5-72c}$$

If s_o' and/or H_o and ΔH are relatively small compared with the normal curvature radius of the reference surface F, then (5-72c) can be approximated by

$$(r_o^a)_o = (u^{\alpha(1)})_o\delta_\alpha^a - (\kappa_n/2)_o s_o'\delta_3^a \tag{5-72d}$$

based on (5-64). Thereby and together with (5-71a) the first equation of (5-72a) can be written as

$$s' = s_o'(1+(\Delta H/s_o')^2 + (\kappa_n)_o(H_o+ H))^{1/2}\,. \tag{5-72e}$$

For

$$\Delta H/s_o'\,,\ (\kappa_n)_o(H_o+ H) \ll 1 \tag{5-73a}$$

the resulting <u>difference in distance</u> of (5-68b) is

$$\delta s' = s'- s_o' = (1/2)(\Delta H^2/s_o' + (\kappa_n)_o s_o'(H_o+ H))\,. \tag{5-73b}$$

Based on (5-72) the <u>difference in direction</u> of (5-68b) becomes

$$\begin{aligned} \delta r'^{,\alpha} &= -(L_\beta^\alpha + \kappa_n\delta_\beta^\alpha)_o(r_o^\beta)_o H - (1/2)(\Delta H/s_o'^2-(\kappa_n)_o)(r_o^\alpha)_o\Delta H\,, \\ \delta r'^{,3} &= \Delta H/s_o' \end{aligned} \tag{5-73c}$$

with equal accuracy. If the expression of (5-64c) is inserted for κ_n into the first term of the first equation of (5-73c), we then obtain

$$\delta r'^{,\alpha} = -(g^{\alpha\beta}- r_o^\alpha\,r_o^\beta)_o(r_o^\gamma)_o(L_{\beta\gamma})_o H-(1/2)(\Delta H/s_o'^2- (\kappa_n)_o)(r_o^\alpha)_o\Delta H, \tag{5-73d}$$

in which the first vector dependent on the target height H is perpendicular to $(r_o^\alpha)_o$, and the second, ΔH-dependent vector points in the direction of $(r_o^\alpha)_o$.

Based on the results of the special cases of (5-58) to (5-61) treated for (5-62) to (5-67) and (5-68) to (5-73) the complete solution reads

$$p^a = p_o^a + \delta p'^a = v^\alpha \delta_\alpha^a + \delta p^a + \delta p'^a ; \qquad (5\text{-}74a)$$

δp^a, $\delta p'^a$ after (5-65b), (5-71a).

For the accompanying polar coordinates of (5-55) and (5-56) we obtain

$$s' = s_o - \delta s + \delta s' , \qquad (r^a)_o = (u^{\alpha(1)})_o \delta_\alpha^a + \delta r^a - \delta r'^a ; \qquad (5\text{-}74b)$$

δs , δr^a after (5-66d), (5-67c) , $\delta s'$, $\delta r'^a$ after (5-73)

5.4 GEODETIC COORDINATES

5.4.1 PRELIMINARY REMARKS

The most important geodetic example of the application of surface-normal coordinates is

Geodetic coordinates (5-75)

$q^a = (u^\alpha, H)$, (5-75a)

u^α = surface coordinates of the plumb foot points Q on the reference ellipsoid F , (5-75b)

H = ellpsoidal heights of the points P after (5-37b) , (5-75c)

which have already been defined for (1-2) with regard to an ellipsoid of revolution as a reference surface F [see Fig. 5.3]. The reference surfaces, also called reference ellipsoids F here, are primarily represented by geographic coordinates, for which the most important fundamentals are summarized in Chapter 4.2.

Chapters 5.4.2 and 5.4.3 will essentially deal with transformations between three-dimensional Cartesian coordinates and polar coordinates based on geographic and Gaussian coordinates as well as polar coordinates as surface coordinates [see (5-75b)]. Transformations between two systems of geodetic coordinates will be discussed in Chapter 5.5.

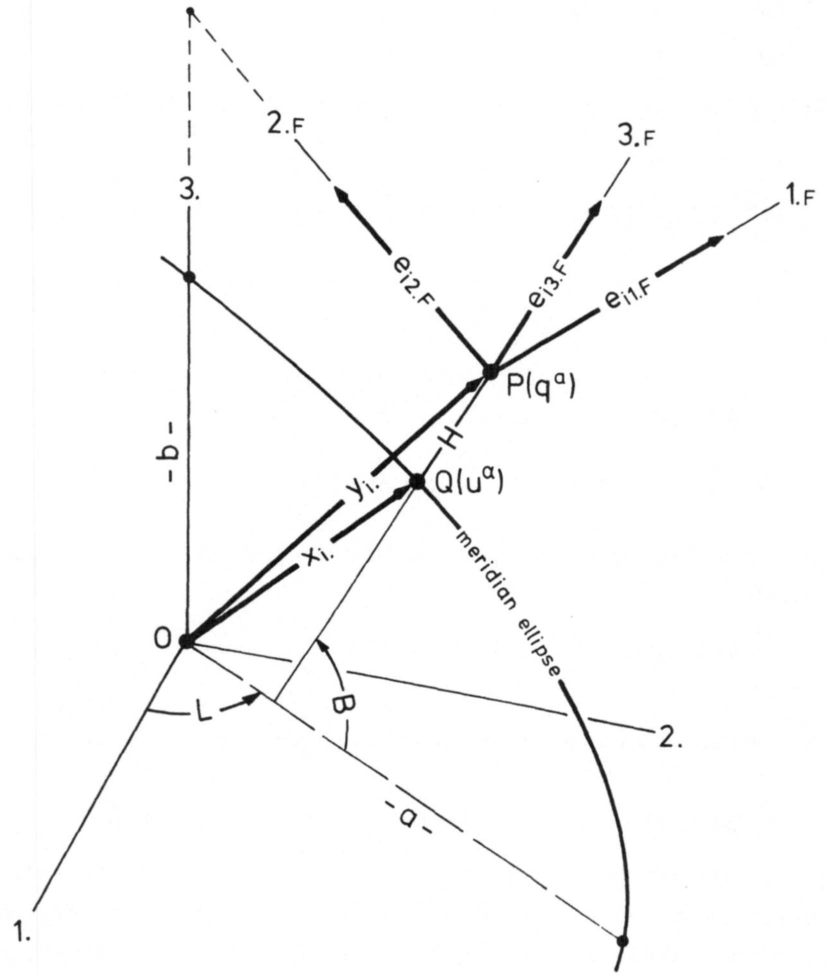

Fig. 5.3. Geodetic coordinates $q^a = (u^\alpha, H)$

5.4.2 GEOGRAPHICALLY GEODETIC COORDINATES

If the geographic coordinates (4-7), i.e.,

$$u^\alpha = (L, B) \ , \tag{5-76a}$$

are used as the surface coordinates of (5-75b), the geodetic coordinates of (5-75a) formed in this way, i.e.,

$$q^a = (u^\alpha,\ H) = (L,\ B,\ H)\ , \tag{5-76b}$$

are then referred to as <u>geographically geodetic coordinates</u>.

For solving the transformation problems discussed in Chapter 5.3 the intrinsic and extrinsic geometry of the reference ellipsoid F as functions of the geographic coordinates of (5-76a) are necessary. This has already been explained in Chapter 4.2 in great detail so that only little needs to be said in the following.

The <u>transformation equations between geographically geodetic coordinates and Cartesian coordinates</u>, the latter referring to the ellipsoid axes S given in Fig. 5.3, are written as follows due to (5-38d) with $x_{i.}$ and $n_{i.}$ according to (4-8) and (4-10a):

$$y_{i.}(L,B,H) = x_{i.}(L,B) + H\, n_{i.}(L,B) \tag{5-77a}$$

$$= ((c/V+H)\cos B\ \cos L,\ (c/V+H)\cos B\ \sin L,\ (b/V+H)\sin B)\ .$$

For tracing this equation back to the independent ellipsoid parameters c and e''

$$b/V = (c/V)/(1+e''^2) \tag{5-77b}$$

is to be inserted into the third component corresponding to (4-6f). By inverting the transformation equations of (5-77a) we directly obtain

$$L = \arctan(y_{2.}/y_{1.})\ . \tag{5-77c}$$

For calculating B

$$B = \arctan((y_{3.} + ((c-b)/V)\sin B)/(y_{1.}^2 + y_{2.}^2)^{1/2})\ , \tag{5-77d}$$

$$c-b = e''^2 c/(1+e''^2)$$

is a very appropriate iteration formula whereby we can begin with the first approximation, i.e.,

$$c-b = 0 \ \rightarrow\ B_1 = \arctan(y_{3.}/(y_{1.}^2 + y_{2.}^2)^{1/2})\ . \tag{5-77e}$$

H can then be calculated as follows:

$$H = y_{3.}/\sin B - b/V = (y_{1.}^2 + y_{2.}^2)^{1/2}/\cos B - c/V\ . \tag{5-77f}$$

<u>Local systems S_F</u> (5-78)

with the rotation matrices relative to S, i.e.,

$$e_{ij.F} = (n_{i,1}/\cos B)\delta_{j1.} + n_{i,2}\delta_{j2.} + n_{i.}\delta_{j3.} \tag{5-78a}$$

$$= R_{ik.3}(\pi/2+L)R_{kj.1}(\pi/2-B) \tag{5-78b}$$

are spanned by the orthogonal trihedrons of (4-10a-c) [see Fig. 5.3, and regarding (5-78b) see Heitz 1980-1983, (02-63)ff]. The rotation matrices $e_{ij.F}$ are orthonormal bases with which the covariant and contravariant basis vectors of (5-40a,b) in the plumb foot points $Q \in F$ are written as follows:

$$b_{i.a} = e_{ia.F} \; (R_1 \cos B \; , R_2 \, , 1)^{(a)} \; , \tag{5-79a}$$

$$b^a_{i.} = e_{i(a).F}((R_1 \cos B)^{-1}, \, R_2^{-1}, \, 1)^a \tag{5-79b}$$

with

$$R_1 = c/V \; , \quad R_2 = c/V^3 , \quad V = (1+e''^2 \cos^2 B)^{1/2} \tag{5-79c}$$

according to (4-18d). Corresponding to (5-41a) and (5-44f,g) the basis vectors for $H \neq 0$ are

$$c_{i.a} = h^b_a \, b_{i.b} \; , \qquad c^a_{i.} = \bar{h}^a_b \, b^b_{i.} \tag{5-79d}$$

with

$$\begin{aligned} &h^1_1 = 1/\bar{h}^1_1 = 1 + H/R_1 \; , \qquad h^2_2 = 1/\bar{h}^2_2 = 1 + H/R_2 \\ &h^\beta_\alpha = \bar{h}^\beta_\alpha = 0 \quad \text{for} \quad \alpha \neq \beta \; , \\ &h^3_\alpha = h^\alpha_3 = \bar{h}^3_\alpha = \bar{h}^\alpha_3 = 0 \; , \qquad h^3_3 = \bar{h}^3_3 = 1 \; . \end{aligned} \tag{5-79e}$$

Thereby, (5-79d) can be written in detail, i.e.,

$$\begin{aligned} c_{i.1} &= h^1_1 \, b_{i.1} = (1+H/R_1) \;\; b_{i.1} \; , \\ c_{i.2} &= h^2_2 \, b_{i.2} = (1+H/R_2) \;\; b_{i.2} \; , \\ c_{i.3} &= h^3_3 \, b_{i.3} = b_{i.3} = n_{i.} \; , \end{aligned} \tag{5-79f}$$

$$\begin{aligned} c^1_{i.} &= \bar{h}^1_1 \, b^1_{i.} = (1+H/R_1)^{-1} b^1_{i.} \; , \\ c^2_{i.} &= \bar{h}^2_2 \, b^2_{i.} = (1+H/R_2)^{-1} b^2_{i.} \; , \\ c^3_{i.} &= \bar{h}^3_3 \, b^3_{i.} = b^3_{i.} = n_{i.} \; . \end{aligned} \tag{5-79g}$$

The covariant metric tensor of (5-42) has the form of

$$g'_{ab} = (R_1 + H)^2 \cos^2 B \; \delta^1_a \, \delta^1_b + (R_2 + H)^2 \delta^2_a \, \delta^2_b + \delta^3_a \, \delta^3_b \; . \tag{5-80a}$$

(5-43a) is true here, i.e.,

$$g'_{ab} = 0 \quad \text{for} \quad a \neq b \; , \tag{5-80b}$$

so that the components of the contravariant metric tensor are obtained in accordance with (5-46d). At the points $Q \in F$ we have

$$g_{ab} = g'_{ab}(H = 0) \; , \qquad g^{ab} = g'^{ab}(H = 0) \; . \tag{5-80c}$$

Together with (4-13) and (4-15b) the equations of Gauss and Weingarten [see (2-46) and (2-49)] for the basis vectors of (5-79a,b) assume the following forms:

$$
\begin{aligned}
b_{i.1,1} &= V^2 \sin B \cos B \, b_{i.2} - R_1 \cos^2 B \, b_{i.3} \,, \\
b_{i.1,2} &= b_{i.2,1} = -(1/V^2) \tan B \, b_{i.1} \,, \\
b_{i.2,2} &= 3(e''/V)^2 \sin B \cos B \, b_{i.2} - R_2 \, b_{i.3} \,, \\
b_{i.3,1} &= (1/R_1) \, b_{i.1} \,, \qquad b_{i.3,2} = (1/R_2) \, b_{i.2} \,;
\end{aligned}
\tag{5-81a}
$$

$$
\begin{aligned}
b^1_{i,1} &= (1/V^2) \tan B \, b^2_{i.} - (1/R_1) \, b^3_{i.} \,, \\
b^1_{i,2} &= (1/V^2) \tan B \, b^1_{i.} \,, \quad b^2_{i,1} = -V^2 \sin B \cos B \, b^1_{i.} \,, \\
b^2_{i,2} &= -3(e''/V)^2 \sin B \cos B \, b^2_{i.} - (1/R_2) \, b^3_{i.} \,, \\
b^3_{i,1} &= R_1 \cos^2 B \, b^1_{i.} \,, \qquad b^3_{i,2} = R_2 \, b^2_{i.}
\end{aligned}
\tag{5-81b}
$$

with

$$
b_{i.3} = b^3_{i.} = n_{i.} \,. \tag{5-81c}
$$

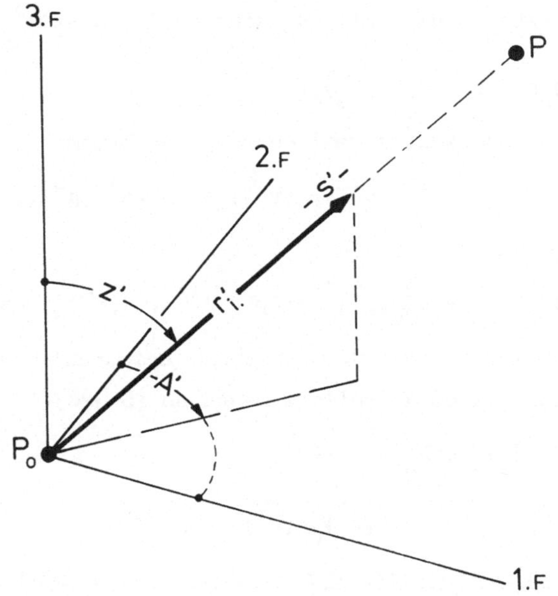

Fig. 5.4. Ellipsoidal polar coordinates A',z',s' in S_F

Now that we have taken a general look at the geometry of geographically geodetic coordinates, we will go into transforming them into Cartesian coordinates using power series corresponding to (5-50) to (5-53). Here, the coordinate differences of (5-51b) have the form of

$$\Delta q^a = (\Delta u^\alpha, \Delta H) = (\Delta L, \Delta B, \Delta H) \ . \qquad (5\text{-}82a)$$

For representing contravariant normal coordinates of p'^a and p^a, which are generally defined by (5-18d) and (5-53a) and (5-55c,d) respectively, we usually begin with ellipsoidal polar coordinates in S_F [see (5-78)] in the form of

$$\begin{aligned} A' &= \text{ellipsoidal azimuth } P_o \rightarrow P \ , \\ z' &= \text{ellipsoidal zenith distance } P_o \rightarrow P \ , \\ s' &= \text{distance } P_o - P \end{aligned} \qquad (5\text{-}82b)$$

corresponding to Fig. 5.4. Herewith, the Cartesian direction vector in S_F and S becomes

$$\overset{F}{r}{}'_{i.} = (\sin z'\sin A', \ \sin z'\cos A', \ \cos z'), \quad r'_{i.} = e_{ij.F}\overset{F}{r}{}'_{j.} \ . \qquad (5\text{-}82c)$$

According to (5-55) and observing (5-79) the results of this are

$$\begin{aligned} (r^a)_o &= (b^a_{i.})_o e_{ij.F} \overset{F}{r}{}'_{j.} \\ &= ((R_1\cos B)^{-1}_o \sin z'\sin A', \ R^{-1}_{2o}\sin z'\cos A', \ \cos z')^a \ , \end{aligned} \qquad (5\text{-}82d)$$

$$(r'^a)_o = (c^a_{i.})_o e_{ij.F} \overset{F}{r}{}'_{j.} = (\bar{h}^a_b)_o (r^b)_o \ , \qquad (5\text{-}82e)$$

with which the contravariant normal coordinates become

$$p^a = (r^a)_o s' \ , \qquad p'^a = (r'^a)_o s' = (\bar{h}^a_b)_o p^b \ . \qquad (5\text{-}82f)$$

If p^a are given, we first obtain

$$s' = ((R_1\cos B)^2_o (p^1)^2 + R^2_{2o}(p^2)^2 + (p^3)^2)^{1/2}, \quad (r^a)_o = p^a/s' \qquad (5\text{-}82g)$$

corresponding to (5-61), and the direction components A' and z' [see (5-82b)] can be calculated as follows based on (5-82d):

$$\begin{aligned} A' &= \arctan[(R_1/R_2)_o \cos B_o (r^1)_o/(r^2)_o] \ , \\ z' &= \arccos((r^3)_o) \ , \qquad (r^1)_o/(r^2)_o = p^1/p^2 \ . \end{aligned} \qquad (5\text{-}82h)$$

For determining the partial derivatives of (5-50a) the procedure described for (5-52) is chosen, according to which

$$(y_{i,a_1a_2\ldots a_n})_o = (c_{i.a_1,a_2\ldots a_n})_o = n! \ \bar{d}^d_{a_1a_2\ldots a_n}(b_{i.d})_o \qquad (5\text{-}83a)$$

is true. Then, it follows that

$$\bar{d}^a_{a_1a_2\ldots a_n} = (1/n!)(c_{i.a_1,a_2\ldots a_n})_o(b^a_{i.})_o \;. \qquad (5\text{-}83b)$$

It ensues from (5-50c) and (5-52a) that only one of the covariant indices of the nonvanishing coefficients of (5-83b) can be equal to three. With (5-79d,f) and for $n = 1$ we obtain

$$c_{i.a_1} = h^b_{a_1} b_{i.b} = (1+H/R_1,\ 1+H/R_2,\ 1)^b b_{i.b} \;. \qquad (5\text{-}83c)$$

Based on (5-83c) and observing (5-81a) the partial derivatives of $c_{i.a}$ can be calculated stepwise. The differential equations of (5-48d) with the first and second fundamental tensors corresponding to (4-12) and (4-13) are valid for the first derivatives of (5-83c). With these results the coefficients of (5-83b) of series (5-52a) and (5-53a) can then be calculated. Here we will only pursue series of (5-53a), which can also be written as follows:

$$p^a = \sum_{n=1}^{n'} \bar{e}^a_{n_1n_2n_3} \Delta L^{n_1} \Delta B^{n_2} \Delta H^{n_3} \;, \quad n_1+n_2+n_3 = n \;, \quad n_3 \in \{0,1\} \;. \qquad (5\text{-}84a)$$

The coefficients occurring in (5-84a) generally become

$$\bar{e}^a_{n_1n_2n_3} = \sum \bar{d}_{a_1a_2\ldots a_n} \;, \qquad (5\text{-}84b)$$

$$n_1+n_2+n_3 = n \;, \qquad a_1a_2\ldots a_n = n_1{\times}1\ n_2{\times}2\ n_3{\times}3$$

due to the coefficients of (5-83b). With the abbreviations

$$co := \cos B_o,\ t := \tan B_o,\ V := V_o = (1+e''^2\cos^2 B_o)^{1/2} \;, \qquad (5\text{-}84c)$$
$$h^{(\alpha)}_{(\alpha)} := (1+H/R_\alpha)_o \;, \quad h^3_3 = 1 \;, \quad R_\alpha := (R_\alpha)_o$$

we obtain the following up to $n' = 3$:

$$\bar{e}^a_{100} = h^1_1\,\delta^a_1 \;, \quad \bar{e}^a_{010} = h^2_2\,\delta^a_2 \;, \quad \bar{e}^a_{001} = \delta^a_3 \;;$$

$$\bar{e}^a_{200} = (1/2)co^2 h^1_1(\ 0\ ,\ V^2 t,\ -R_1) \;,$$
$$\bar{e}^a_{110} = (-(1/V^2)t\ h^2_2,\ 0\ ,\ 0\ ) \;,$$
$$\bar{e}^a_{020} = -(1/2)(\ 0\ ,\ (3/V^2)(1-V^2)t,\ R_2 h^2_2) \;,$$
$$\bar{e}^a_{101} = (\ 1/R_1,\ 0\ ,\ 0\ ) \;,$$
$$\bar{e}^a_{011} = (\ 0\ ,\ 1/R_2,\ 0\ ) \;,$$
$$\bar{e}^a_{002} = (\ 0\ ,\ 0\ ,\ 0\ ) \;;$$

$$\bar{e}^a_{300} = (-(1/6)co^2 h^1_1(1+t^2),\ 0\ ,\ 0\ ) \;, \qquad (5\text{-}84d)$$

$$\bar{e}^a_{210} = -(1/2)\mathrm{co}^2 t\ h^2_2(\ 0\ ,\ t\ ,\ -R_2)\ , \tag{5-84d}$$
$$\bar{e}^a_{120} = (-(1/(2V^2))(h^2_2-(3/V^2)(1-V^2)t^2),\ 0\ ,\ 0\ )\ ,$$
$$\bar{e}^a_{030} = (\ 0\ ,\ -(1/6)(h^2_2+(3/V^4)(1-V^2)(V^2-(1-V^2)t^2)),\ (R_2/V^2)t(1-V^2)),$$
$$\bar{e}^a_{201} = (1/2)\mathrm{co}^2(\ 0\ ,\ (1/R_2)t,\ -1)\ ,$$
$$\bar{e}^a_{111} = (-(1/R_1)t,\ 0\ ,\ 0\ )\ ,$$
$$\bar{e}^a_{021} = (\ 0\ ,\ 0\ ,\ -1/2)\ ,$$
$$\bar{e}^a_{102} = \bar{e}^a_{012} = \bar{e}^a_{003} = (\ 0\ ,\ 0\ ,\ 0\ )\ .$$

For <u>transforming $\Delta y_{i.}$ and p^a into Δq^a</u>, i.e., the normal coordinates of (5-82f) into the differences in the geographically geodetic coordinates of (5-82a), by means of the power series in p^a we will use the procedure described for (5-25) to (5-26) with the modification given for (5-54). Thereby, we begin with

$$\partial q^a/\partial y_{i.} = c^a_{i.} = \bar{h}^{(a)}_{(a)} b^a_{i.} \tag{5-85a}$$

for the first derivative of (5-26a) in accordance with (5-79d,e). The second gradient of q^a becomes

$$\partial^2 q^a/(\partial y_{i.}\,\partial y_{j.}) = c^a_{i,b} c^b_{j.} = c^a_{i,b} \bar{h}^{(b)}_{(b)} b^b_{j.}\ , \tag{5-85b}$$

$c^a_{i,b}$ after (5-49b) ,

and based on these results the higher gradients are calculated observing (5-85a) and (5-81b). At P_o these gradients usually have the form of

$$(\partial^n q^a/(\partial y_{i_1.}\partial y_{i_2.}\dots\partial y_{i_n.}))_o$$
$$= n!\ d^a_{a_1 a_2\dots a_n}(b^{a_1}_{i_1.})_o(b^{a_2}_{i_2.})_o\dots(b^{a_n}_{i_n.})_o\ , \tag{5-85c}$$

from which

$$d^a_{a_1 a_2\dots a_n} = (1/n!)(\partial^n q^a/(\partial y_{i_1.}\partial y_{i_2.}\dots\partial y_{i_n.}))_o$$
$$\cdot\ (b_{i_1.a_1})_o(b_{i_2.a_2})_o\dots(b_{i_n.a_n})_o \tag{5-85d}$$

ensues. These are the coefficients of the series of (5-54c), which can also be written in the following way:

$$\Delta q^a = (\Delta L,\ \Delta B,\ \Delta H) = \sum_{n=1}^{n'} e^a_{n_1 n_2 n_3}(p^1)^{n_1}(p^2)^{n_2}(p^3)^{n_3}\ . \tag{5-86a}$$

The coefficients $e^a_{n_1 n_2 n_3}$ are calculated from the coefficients of (5-85d) analogous to (5-84b). With the abbreviations of (5-84c) and

$$\bar{h}^{(\alpha)}_{(\alpha)} := 1/(1+H/R_{(\alpha)})_o, \qquad \bar{h}^3_3 = 1 \tag{5-86b}$$

we obtain the following up to $n' = 2$:

$$e^a_{100} = \bar{h}^1_1\,\delta^a_1\,, \quad e^a_{101} = \bar{h}^2_2\,\delta^a_2\,, \quad e^a_{001} = \delta^a_3\,;$$

$$\begin{aligned}
e^a_{200} &= (1/2)co^2\bar{h}^1_1(\ 0\ ,\ -V^2t\bar{h}^2_2,\ R_1)\ ,\\
e^a_{110} &= ((1/2V^2)t\bar{h}^1_1(\bar{h}^1_1+\bar{h}^2_2(1-(1-V^2)(H/R_1)\bar{h}^1_1)),\ 0\ ,\ 0\ )\ ,\\
e^a_{020} &= (1/2)\bar{h}^2_2(\ 0\ ,\ (3/V^2)t(1-V^2)\bar{h}^2_2(1-(H/R_2)\bar{h}^2_2),\ R_2)\ ,\\
e^a_{101} &= (-(\bar{h}^1_1)^2/R_1,\ 0\ ,\ 0\ )\ ,\\
e^a_{011} &= (\ 0\ ,-(\bar{h}^2_2)^2/R_2,\ 0\ )\ ,\\
e^a_{002} &= (\ 0\ ,\ 0\ ,\ 0\ )\ .
\end{aligned} \tag{5-86c}$$

For $n = 3$ the coefficients will only be approximated with the assumptions of

$$V^2 = 1 \rightarrow R_1 = R_2 = R = \text{constant},\ H/R = 0\ . \tag{5-86d}$$

The results are:

$$\begin{aligned}
e^a_{300} &= (-(1/3)co^2(1+t^2),\ 0\ ,\ 0\ )\ ,\\
e^a_{210} &= (\ 0\ ,\ -(1/2)co^2(1+t^2),\ 0\ )\ ,\\
e^a_{120} &= (\ t^2\ ,\ 0\ ,\ 0\ )\ ,\\
e^a_{030} &= (\ 0\ ,\ -1/3,\ 0\ )\ ,\\
e^a_{201} &= (1/2)co^2(\ 0\ ,\ (2/R)t,\ -1\ )\ ,\\
e^a_{111} &= (-(2/R)t,\ 0\ ,\ 0\ )\ ,\\
e^a_{021} &= (\ 0\ ,\ 0\ ,-1/2\ )\ ,\\
e^a_{102} &= (\ 1/R^2,\ 0\ ,\ 0\ )\ ,\\
e^a_{012} &= (\ 0\ ,\ 1/R^2,\ 0\ )\ ,\\
e^a_{003} &= (\ 0\ ,\ 0\ ,\ 0\ )\ .
\end{aligned} \tag{5-86e}$$

For

$B \approx 50°$, $\qquad z' = 90° \pm 4°$

the series of (5-84) and (5-86) have the following errors:

$$\begin{aligned}
&< 1\ \text{mm} \quad \text{for up to} \quad s' \approx 20\ \text{km}\ ,\\
&< 6\ \text{mm} \quad \text{for up to} \quad s' \approx 40\ \text{km}\ ,\\
&< 30\ \text{mm} \quad \text{for up to} \quad s' \approx 60\ \text{km}\ .
\end{aligned} \tag{5-87}$$

5.4.3 GAUSSIAN GEODETIC COORDINATES

If the Gaussian coordinates of (4-91d,e), i.e.,

$$\bar{u}^{\alpha} = (x,y)^{\alpha}, \tag{5-88a}$$

are used as the surface coordinates of (5-75b), the geodetic coordinates of (5-75a) obtained in this way, i.e.,

$$\bar{q}^{a} = (\bar{u}^{\alpha}, H) = (x,y,H) , \tag{5-88b}$$

are referred to as <u>Gaussian geodetic coordinates</u>.

For solving the transformation problems discussed in Chapter 5.3 we can begin with the intrinsic and extrinsic geometries of the reference ellipsoid F as functions of the Gaussian coordinates of (5-88a). The transformation equations between the contravariant normal coordinates q^a [see (5-53a) and (5-55d)] or the polar coordinates of (5-82b) and the Gaussian geodetic coordinates of (5-88b) could be obtained in the way described for (5-78) to (5-86) with the metric tensor of (4-101) and the second fundamental tensor of the Gaussian coordinates, which has not yet been determined. Since we already have the transformation equations between geographic and Gaussian coordinates on the reference ellipsoid in (4-93) and (4-95), it is also possible to exchange the geographic coordinates for the Gaussian coordinates in (5-84) and (5-86) using these equations. This will be described in more detail in the following.

It is first necessary to partially change and broaden the designations for the points and coordinates. The point pairs of P_o and P as well as Q_o and Q used in Chapter 5.2.2 and following are designated here by

$$P_1, P_2 \text{ and } Q_1, Q_2 , \tag{5-89a}$$

and the accompanying coordinates are provided with the same indices. In addition to (5-89a) the following points are introduced:

Q_o = reference point of the Gaussian coordinates after (4-75a) and (4-91a) , (5-89b)

Q_f = foot point of the ordinate after (4-96a) , (5-89c)

and for representing the transformation equations we use the coordinate differences, i.e.,

$$\Delta q^a = (\Delta u^\alpha,\ \Delta H) = (\Delta L,\ \Delta B,\ \Delta H)\ ,$$
$$\Delta L = L_2 - L_1\ , \quad \Delta B = B_2 - B_1\ , \quad \Delta H = H_2 - H_1\ , \tag{5-89d}$$

$$\Delta L_n = L_n - L_o\ , \quad \Delta B_n = B_n - B_o\ , \quad n \in \{1,2\}\ , \tag{5-89e}$$

$$\Delta \bar{q}^a = (\Delta \bar{u}^\alpha,\ \Delta H) = (\Delta x,\ \Delta y,\ \Delta H)\ ,$$
$$\Delta x = x_2 - x_1\ , \quad \Delta y = y_2 - y_1\ , \quad \Delta H = H_2 - H_1\ . \tag{5-89f}$$

p^a are the normal coordinates of P_2 with respect to P_1 so that according to (5-82d-f)

$$(r^a)_1 = (b^a_{i.})_1 e_{ij.F}\ \overset{F}{r'_{j.}},\quad \overset{F,}{r_{j.}}\ \text{after (5-82c)}\ ,$$
$$= ((R_1\cos B)_1^{-1}\sin z'\sin A',\ R_{21}^{-1}\sin z'\cos A',\ \cos z')^a, \tag{5-90a}$$

$$p^a = (r^a)_1 s'\ . \tag{5-90b}$$

Instead of the azimuth A', which is related to $e_{i2.F}$ and thus to the meridian through P_1, we use the angle

β' = horizontal direction angle of $(r^a)_1$ relative to the x-line through P_1 (5-90c)

in Gaussian coordinates. Due to A' and with the help of the meridian convergence γ [see (4-102b)] this angle becomes

$$\beta' = A' - \gamma\ . \tag{5-90d}$$

The geodetic polar coordinates of (5-82b) are calculated according to (5-82g,h) as follows:

$$s' = ((R_1\cos B)_1^2(p^1)^2 + R_{21}^2(p^2)^2 + (p^3)^2)^{1/2},\ (r^a)_1 = p^a/s',$$
$$A' = \beta' + \gamma = \arctan\,(V_1^2\cos B_1(r^1)_1/(r^2)_1)\ , \tag{5-90e}$$
$$z' = \arccos((r^3)_1)\ .$$

The <u>transformation of $\Delta\bar{q}^a$ [see (5-89f)] into the contravariant normal coordinates p^a</u> can be carried out by means of (5-84) after ΔL and ΔB as functions of Δx and Δy have been calculated. For this, the transformation equations of (4-95) are available. When they are applied to Q_2 assuming

$B_o = B_{f1}$ = latitude of the foot point $Q_1 \rightarrow x_1 = 0$, according to (4-96a), (5-91a)

then together with (5-89d-f) and observing (4-96b) we obtain

$$(\Delta B,\ \Delta L)^{\alpha} = \sum_{n=1}^{n'} a^{\alpha}_{n_1 n_2} (\Delta x^{n_1}\ y_2^{n_2} - \upsilon\ y_1^{n_2})\ , \tag{5-91b}$$

$$\upsilon = \begin{cases} 1 \\ 0 \end{cases} \text{for} \quad \begin{matrix} n_1 = 0 \\ n_1 > 0 \end{matrix}\ . \tag{5-91c}$$

For <u>transforming the contravariant normal coordinates p^a into $\Delta\bar{q}^a$</u> [see (5-89f)] p^a is first transformed into Δq^a [see (5-89d)] using (5-86). Δx and Δy can then be calculated using (4-93) based on ΔL and ΔB. If the series of (4-93) is applied to Q_2 assuming

$$B_o = B_1 \ \rightarrow\ \Delta B_2 = \Delta B\ , \tag{5-92a}$$

then with (5-89d-f) and observing (4-94b) we yield

$$(\Delta x,\ \Delta y)^{\alpha} = \sum_{n=1}^{n'} a^{\alpha}_{n_1 n_2} (\Delta B^{n_1} \Delta L_2^{n_2} - \upsilon\ \Delta L_1^{n_2})\ , \quad \upsilon \text{ after (5-91c)}. \tag{5-92b}$$

Instead of the two-step transformations given for (5-91) and (5-92) we naturally could give the direct transformation functions for p^a and $\Delta\bar{q}^a$. To do this, the series of (5-91b) would have to be substituted for ΔL and ΔB in (5-84a), and the series of (5-86a) for ΔL_n and ΔB in (5-92b). This method being of less importance will not be discussed further here.

5.4.4 TRANSFORMATIONS BETWEEN ELLIPSOIDAL POLAR COORDINATES AND POLAR COORDINATES ON THE REFERENCE ELLIPSOID

The transformation problems treated in Chapter 5.3.3 for general surface-normal coordinates will be applied to the geographically geodetic coordinates of (5-76) in the following. For the "space points P_o and P" the <u>ellipsoidal polar coordinates</u> of (5-82b), i.e.,

$$A',\ z',\ s', \tag{5-93a}$$

are introduced with the accompanying three-dimensional normal coordinates of (5-55d) and (5-82f), i.e.,

$$p^a = (b^a_{i.})_o \Delta y_{i.} = (r^a)_o s',\quad (r^a)_o = (b^a_{i.})_o r'_{i.}\ . \tag{5-93b}$$

The transformations between (5-93a) and (5-93b) are represented in

(5-82b-h). On the reference ellipsoid <u>geodesic polar coordinates</u>

$$A,\ s_o \tag{5-94a}$$

are defined for the "surface points Q_o and Q" with $r =: s_o$ according to (4-23d,e). The <u>normal coordinates</u> on the reference ellipsoid of (5-56d) are thus given by (4-24b) with $\bar{v}^\alpha =: v^\alpha$:

$$v^\alpha = (u^{\alpha(1)})_o s_o = ((R_1 \cos B)_o^{-1} \sin A,\ R_{2o}^{-1} \cos A)^\alpha s_o \ . \tag{5-94b}$$

With given ellipsoidal heights H we can proceed here from (5-84) and (4-25), i.e.,

$$p^a = \sum_{n=1}^{n'} \bar{e}^a_{n_1 n_2 n_3} \Delta L^{n_1} \Delta B^{n_2} \Delta H^{n_3} \ , \tag{5-95a}$$

$$\Delta u^\alpha = (\Delta L,\ \Delta B)^\alpha = v^\alpha + \sum_{n=2}^{n'} b^\alpha_{n_1 n_2} (v^1)^{n_1} (v^2)^{n_2} \ , \tag{5-95b}$$

for the transformation equations between v^α and p^α resulting from (5-59) and (5-60). (5-95b) is to be substituted for ΔL and ΔB in (5-95a) whereby we would obtain the result directly corresponding to (5-74a). On the other hand, we can also make use of the partial solutions treated for (5-62)ff and (5-68)ff, which is done in the following. For this purpose, the formulas of (5-66d), (5-67c), and (5-73) are to be specialized for geographic coordinates on the reference ellipsoid F. Insofar as the "horizontal direction $Q_o \rightarrow Q$" is required in these formulas, we can proceed with (5-73a) accurately enough. Observing (5-94b) we yield

$$(r_o^\alpha)_o \approx (u^{\alpha(1)})_o = ((R_1 \cos B)_o^{-1} \sin A,\ R_{2o}^{-1} \cos A)^\alpha \ , \tag{5-96a}$$

$$A \approx A' \quad \text{after (5-93a)} \ . \tag{5-96b}$$

The components of the fundamental tensors are given by (4-12) and (4-13). The theorem of Euler [see (4-20b)] is used for calculating the normal curvature, i.e.,

$$(\kappa_n)_o = R_{1o}^{-1} \sin^2 A + R_{2o}^{-1} \cos^2 A \ , \tag{5-96c}$$

and its derivative with respect to s is neglected in the following:

$$(\kappa_n^{(1)})_o = 0 \ . \tag{5-96d}$$

Thus, the <u>reductions between the chord and the geodesic line $Q_o - Q$</u> in the form of the differences in distance and direction become

$$\delta s = s_o - s'_o = (1/24)(\kappa_n)_o^2\, s^3 + \dots\,, \qquad s \approx s_o \approx s'_o\,, \tag{5-97a}$$

$$\delta r^a = (1/2)(\kappa_n)_o \delta_3^a\, s \tag{5-97b}$$
$$+ (1/24)(\kappa_n)_o [4(R_1^{-1}u^{1(1)},\ R_2^{-1}u^{2(1)})^\alpha + \kappa_n u^{\alpha(1)}]_o \delta_\alpha^a\, s^2 + \dots$$

in accordance with (5-66d) and (5-67c). Now, the difference in azimuth δA corresponding to (5-97b) and the zenith distance z'_o of the chord $Q_o - Q$ in the local system S_F are calculated. δA is equal to the absolute value of the horizontal component of δr^a perpendicular to $(u^{\alpha(1)})_o$, i.e.,

$$\delta A = |(g^{\alpha\beta} - u^{\alpha(1)}u^{\beta(1)})_o (g_{\beta\gamma})_o\ \delta r^\gamma|\,, \tag{5-98a}$$

with sufficient accuracy. Together with the metric tensor of (4-12), as well as (5-96) and (5-97b), and the assumption

$(\kappa_n/R_1)_o \approx K_o$ after (4-19a)

the <u>difference in azimuth of $Q_o - Q$</u> follows from (5-98a), i.e.,

$$\delta A = A - A'_o = (1/12)(1-V^2)K_o s^2 \sin 2A\,, \tag{5-98b}$$

with A according to (5-94a), and A'_o is the azimuth of the chord $Q_o - Q$. The third component of (5-97b) directly yields the <u>zenith distance of $Q_o - Q$</u>, i.e.,

$$z'_o = -\delta r^3 = -(1/2)(\kappa_n)_o s\,. \tag{5-98c}$$

The <u>reductions between the chords $Q_o - Q$ and $P_o - P$</u> are obtained based on (5-72) and (5-73). For greater heights H_o and H and differences in heights ΔH the differences in distance are calculated based on (5-72e) as a function of s'_o or s' as follows:

$$\begin{aligned}\delta s' = s' - s'_o &= s'_o[(1+(\Delta H/s'_o)^2 + (\kappa_n)_o(2H + \Delta H))^{1/2} - 1] \\ &= s'[1-((1-(\Delta H/s')^2)(1-(\kappa_n)_o(2H + \Delta H)))^{1/2}]\,.\end{aligned} \tag{5-99a}$$

With the presupposition of (5-73a), i.e.,

$$\Delta H/s'_o,\quad \Delta H/s',\quad (\kappa_n)_o(H_o + H) \ll 1\,, \tag{5-99b}$$

we obtain in accordance with (5-73b)

$$\delta s' = s' - s'_o = (s'_o/2)[(\Delta H/s'_o)^2 + (\kappa_n)_o(H_o + H)]\,, \tag{5-99c}$$

in which $s'_o \approx s'$ can be inserted into the right side. Due to (5-73c,d), (4-12), (4-13), and (5-96) and with the assumption of (5-99b) the difference in direction becomes

$$\delta r^{,a} = [\delta r_+^{,\alpha} - (1/2)(\Delta H/s_o^{,2} - (\kappa_n)_o)(r_o^\alpha)_o]\delta_\alpha^a + (\Delta H/s_o^{,})\delta_3^a \tag{5-99d}$$

with the horizontal component of $\delta r^{,a}$ perpendicular to $(r_o^\alpha)_o$, i.e.,

$$\delta r_+^{,\alpha} = (1-V^2)R_{1o}^{-1}((R_1\cos B)_o^{-1}\cos A, -R_{2o}^{-1}\sin A)^\alpha \sin A\cos A\ H\ . \tag{5-99e}$$

The difference in azimuths $\delta A'$ corresponding to (5-99d,e) and the zenith distance z' of the chord $P_o - P$ in the local system S_F are also calculated analogous to (5-98). The <u>difference in azimuth of $P_o \rightarrow P$ and $Q_o \rightarrow Q$</u> is equal to the absolute value of (5-99e), i.e.,

$$\delta A' = A' - A_o = |\delta r_+^{,\alpha}| = (1/2)((1-V^2)/R_{1o})\sin 2A'\ H\ , \tag{5-100a}$$

with sufficient accuracy. This involves the so-called "reduction in azimuth due to the height H of the target point". The third component of (5-99d) directly yields the <u>zenith distance $P_o \rightarrow P$</u>, i.e.,

$$z' = \delta r^{,3} = \Delta H/s_o^{,}\ . \tag{5-100b}$$

As already noted, all the results of (5-99c) to (5-100b) are only true presuming (5-99b).

With the results of (5-97) to (5-100) we obtain

$$s' - s = \delta s' - \delta s\ , \qquad A' - A = \delta A' - \delta A \tag{5-101}$$

for the reductions between polar coordinates of the chord $P_o - P$ and the geodesic line $Q_o - Q$. The reduction formulas of (5-97) and (5-98b) are accurate to within <1 cm for distances up to 40 km. This is also true for the reduction formulas of (5-99a) with zenith distances of $90° \pm 5°$ and for (5-99c) with zenith distances of $90° \pm 2°$. The reduction in azimuth due to the target height of (5-100a) is dependent solely on the target height H and not the distance. This reduction formula is accurate to <0.01" for H up to about 5000 m. See Wolfrum (1984) regarding reduction formulas for greater domains of validity.

5.5 TRANSFORMATIONS BETWEEN GEOGRAPHICALLY GEODETIC COORDINATES

5.5.1 FUNDAMENTALS

The following considerations are based on a three-dimensional point field near the earth's surface which is described by two systems of geographically geodetic coordinates according to (5-76)ff. These systems will be designated in the manner introduced in Chapter 5.2.1, by which the quantities of one of the two systems are provided with a crossbar. Thus, the coordinate systems designated by GC and $\overline{GC}$, for example, are defined as follows in accordance with Chapters 4.2 and 5.4.2:

<u>GC system</u>: origin O, system of axes S , (5-102)

ϵ^{υ} , $\upsilon \in \{1,2\}$ = ellipsoid parameters of (4-5h), (5-102a)

q^{a} = (L,B,H) = geographically geodetic coordinates , (5-102b)

$y_{i.}$ = position vectors in S [see (5-77)] , (5-102c)

S_F = local systems of (5-78) , (5-102d)

A',z',s' = local polar coordinates in S_F [see (5-82b)] ; (5-102e)

<u>$\overline{GC}$ system</u>: origin $\bar{O}$, system of axes $\bar{S}$, (5-103)

$\bar{\epsilon}^{\upsilon}$, $\upsilon \in \{1,2\}$ = ellipsoid parameters of (4-5h) , (5-103a)

$\bar{q}^{a}$ = $(\bar{L},\bar{B},\bar{H})$ = geographically geodetic coordinates , (5-103b)

$\bar{y}_{i.}$ = position vectors in $\bar{S}$ [see (5-77)] , (5-103c)

$S_{\bar{F}}$ = local systems of (5-78) , (5-103d)

$\bar{A}',\bar{z}',\bar{s}' \equiv s'$ = local polar coordinates in $S_{\bar{F}}$ [see (5-82b)]. (5-103e)

As <u>ellipsoid parameters</u> we primarily use

$$\epsilon^{\upsilon} =: (c,\ e''^{2}), \qquad \bar{\epsilon}^{\upsilon} =: (\bar{c},\ \bar{e}''^{2}) , \tag{5-104a}$$

c , $\bar{c}$ = pole curvature radii of (4-5b) ,

e'', $\bar{e}''$ = second eccentricities of (4-4c) . (5-104b)

We always presume small differences in the parameters of (5-104a), i.e.,

$$\delta\epsilon^{\upsilon} := \bar{\epsilon}^{\upsilon} - \epsilon^{\upsilon} = (\delta c,\ \delta e''^{2}) , \tag{5-104c}$$

so that

$$(\delta c/c)^2 \rightarrow 0\ , \qquad (\delta e''^2)^2 \rightarrow 0 \qquad \text{(5-104d)}$$

are true accurately enough. Based on (4-5d,g) the following relationships exist between the ellipsoid parameters of (5-104a) and a and α [see (4-1c)]:

$$c = a/(1-\alpha)\ , \qquad e''^2 = \alpha(2-\alpha)/(1-\alpha)^2 \qquad \text{(5-104e)}$$

with the accompanying differential formulas of

$$dc = (c/a)(da + c\,d\alpha)\ , \qquad de''^2 = 2(c/a)^3 d\alpha\ . \qquad \text{(5-104f)}$$

Some of the most important <u>transformation problems</u> are the transformations between the geographically geodetic coordinates of (5-102b) and (5-103b) according to (5-6a), i.e.,

$$q^a = q^a(\bar{q}^b)\ , \qquad \bar{q}^a = \bar{q}^a(q^b)\ . \qquad \text{(5-105a)}$$

Another problem is the determination of the relative rotation of the local systems of (5-102d) and (5-103d), i.e., the

$$\overset{F}{e}_{ij.\bar{F}} = \text{rotation matrices of } S_{\bar{F}} \text{ relative to } S_F \qquad \text{(5-105b)}$$

for one and the same space point. These matrices are the basis for transforming local direction parameters, which are usually represented by the components of one and the same direction vector $r'_{i.}$ in the local systems S_F and $S_{\bar{F}}$ of a point as follows:

$$\overset{F}{r'_{i.}}\ , \quad \overset{\bar{F}}{r'_{i.}}\ , \qquad |r'_{i.}| = 1\ . \qquad \text{(5-105c)}$$

Together with (5-105b) the transformation equations read

$$\overset{F}{r'_{i.}} = \overset{F}{e}_{ij.\bar{F}}\ \overset{\bar{F}}{r'_{j.}}\ , \qquad \overset{\bar{F}}{r}_{i.} = \overset{\bar{F}}{e}_{ij.F}\ \overset{F}{r'_{j.}} \qquad \text{(5-105d)}$$

with whose help we also obtain, for example, the transformation equations between the local polar coordinates of (5-102e) and (5-103e), i.e.,

$$\bar{A}' = \bar{A}'(A',\ z')\ , \qquad \bar{z}' = \bar{z}'(A',z')\ . \qquad \text{(5-105e)}$$

With regard to <u>constructing the transformation equations</u> for the problems described for (5-105) it is to be noted that for the geographically geodetic coordinates both the metric tensors

$$g'_{ab}(q^c)\ , \qquad \bar{g}'_{ab}(\bar{q}^c) \qquad \text{(5-106a)}$$

and their transformation equations

$$y_{i.} = y_{i.}(q^a) \;, \qquad \bar{y}_{i.} = \bar{y}_{i.}(\bar{q}^a) \tag{5-106b}$$

into Cartesian coordinates in S and $\bar{S}$, respectively, are given according to (5-80) and (5-77a,b). For the inverse transformations for (5-106b) we can use the method of (5-77c-f). Consequently, the transformation equations ensuing from (5-106a) can be derived based on the method given for (5-5) to (5-9). On the other hand, however, the presuppositions given by (5-106b) allow the method depicted in (5-11) to (5-13) to be applied. Both procedures can practically be applied alternatively; it must be decided from case to case which is the most appropriate.

The transformations of (5-105) are uniquely determined when the ellipsoid parameters of (5-104a,b), i.e.,

$$\epsilon^{\nu}, \; \bar{\epsilon}^{\nu} \tag{5-107a}$$

are given, and the six degrees of freedom of the relative spatial movements between the two geodetic systems of GC and $\overline{GC}$ are given appropriately. This is done in the form of so-called <u>local transformation parameters</u> corresponding to (5-8) which are defined for purposes here as

$$(\delta q^a)_o := (\bar{q}^a)_o - (q^a)_o \quad \text{and} \quad (\bar{q}^a_{\;,b})_o \tag{5-107b}$$

due to the similarity between the systems of (5-102) and (5-103). Other equi-valent parameters can also be employed [see Chapter 5.5.4]. When using the transformation equations of (5-106b), the parameters of movement between S and $\bar{S}$ [see (5-12a)] may be introduced as <u>central transformation parameters</u> in place of (5-107b), i.e.,

$$p_{i.}, \; e_{ij.} \,. \tag{5-107c}$$

If one of the systems of parameters of (5-107b,c) is given, then the other can be calculated based on (5-12b) and (5-13a) in connection with (5-77)ff.

If we proceed on the presumptions of (5-106b) and (5-107c), then a <u>method of calculating the transformations of (5-105)</u> can be given directly based on (5-17) and (5-77). Assuming that both the dimensions of the ellipsoid of (5-107a) and the central transformation parameters of (5-107c) are given, the first transformation of (5-105a), in which $\bar{q}^a$ is given and q^a is required, can be performed

- by means of (5-77a) for calculating $\bar{y}_{i.}(\bar{q}^a)$, (5-108a)

- by means of (5-12b) for calculating (5-108b)

$$y_{i.} = \bar{y}_{i.}(\bar{q}^a) + p_{i.} + (e_{ij.} - \delta_{ij.})\bar{y}_{j.}(\bar{q}^a) \ ,$$

- corresponding to (5-77c-f) for calculating q^a based on $y_{i.}$. (5-108c)

Thus, the rotation matrices of the local systems S_F and $S_{\bar{F}}$ relative to S and $\bar{S}$ can then be calculated, i.e.,

$$e_{ij.F}(L,B) \ , \qquad \bar{e}_{ij.F}(\bar{L},\bar{B}) \ , \tag{5-108d}$$

and (5-105b) becomes

$$\overset{F}{e}_{ij.\bar{F}} = e_{ki.F}\, e_{kl.}\, e_{lj.\bar{F}} \ . \tag{5-108e}$$

Using these rotation matrices the transformations of the direction vectors and polar coordinates of (5-105c-e) can ultimately be calculated, which will be gone into further in Chapters 5.5.3 and 5.5.4.

The calculation methods described for (5-108) are not suitable for many applications because above of all they do not provide any direct transformation equations. Seen purely numerically, this is of disadvantage in (5-108a-c) since the always relatively small differences between $\bar{q}^a$ and q^a must then be calculated indirectly via the geocentric position vectors $\bar{y}_{i.}$ and $y_{i.}$. <u>Direct transformation equations for (5-105)</u> can only be produced by linearizing the individual transformation steps or by series expansions corresponding to (5-7) to (5-9) or based on (5-13). Linearizations are facilitated by the fact that due to (5-104d)

$$p_{i.}p_{j.}/c^2 \ \rightarrow \ 0 \ , \tag{5-109a}$$

$$e_{ij.} = \delta_{ij.} - \epsilon_{ijk.}d_{k.} \ , \qquad d_{k.}d_{l.} \ \rightarrow \ 0 \ ,$$

$d_{k.}$ = infinitesimal rotation vector of $\bar{S}$ relative to S , (5-109b)

are generally accurate enough for the central transformation parameters of (5-107c). Thereby, the following assumptions, for example, are justified:

$$\delta q^a \delta q^b \ \rightarrow \ 0 \qquad \text{with} \qquad \delta q^a = \bar{q}^a - q^a \ , \tag{5-109c}$$

$$\overset{F}{e}_{ij.\bar{F}} = \delta_{ij.} - \epsilon_{ijk.}\overset{F}{d}{}'_{k.} \ , \qquad \overset{F}{d}{}'_{k.}\overset{F}{d}{}'_{l.} \ \rightarrow \ 0 \ ,$$

$\overset{F}{d}{}'_{k.}$ = infinitesimal rotation vector of $S_{\bar{F}}$ relative to S_F , (5-109d)

$$\delta \overset{F}{r}{}'_{i.}\, \delta \overset{F}{r}{}'_{j.} \ \rightarrow \ 0 \ , \qquad \delta \overset{F}{r}{}'_{i.} = \overset{\bar{F}}{r}_{i.} - \overset{F}{r}{}'_{i.} \ , \tag{5-109e}$$

and corresponding to (5-109a) the following is true for the local transformation parameters of (5-107b):

$$(\delta q^a)_o (\delta q^b)_o \rightarrow 0 \,,$$

$$(\bar{q}^a_{,b})_o = \delta^a_b + (\delta \bar{q}^a_{,b})_o \,, \qquad (\delta \bar{q}^a_{,b})_o (\delta \bar{q}^c_{,d})_o \rightarrow 0 \,. \tag{5-109f}$$

Based on (5-104d) and (5-109) the transformation equations for (5-105) and (5-108) will be derived in linearized form and also with the help of power series expansions in the coordinate differences in Chapters 5.5.2 to 5.5.4. One of the main problems here is the determination of the transformation parameters of (5-107b) or (5-107c), of which several examples are given in Chapter 5.5.5. Finally, in Chapter 5.5.6 we will go into the determination of the dimensions and location of the so-called "mean earth ellipsoid", which involves calculating the ellipsoid and transformation parameters.

Only the transformations between the geographically geodetic coordinates of (5-102b) and (5-103b) will be treated here because, in comparison to other surface coordinates on the ellipsoid of revolution, this leads to very clear and simple results. If in place of geographic coordinates, i.e.,

$$u^\alpha = (L,B) \,, \qquad \bar{u}^\alpha = (\bar{L},\bar{B}) \,, \tag{5-110a}$$

other <u>arbitrary surface coordinates w^α and $\bar{w}^\alpha$</u>, e.g., the Gaussian or UTM coordinates of (4-91) and (4-92), are used, then the corresponding transformation equations of

$$u^\alpha = u^\alpha(w^\beta) \,, \qquad w^\alpha = w^\alpha(u^\beta) \,,$$

$$\bar{u}^\alpha = \bar{u}^\alpha(\bar{w}^\beta) \,, \qquad \bar{w}^\alpha = \bar{w}^\alpha(\bar{u}^\beta) \tag{5-110b}$$

are to be employed before or after the transformation equations between the geographically geodetic coordinates. In the case of Gaussian coordinates w^α and $\bar{w}^\alpha$ the transformation equations of (5-110b) are given by (4-93) and (4-95).

5.5.2 TRANSFORMATIONS BETWEEN CONCENTRICALLY GEODETIC COORDINATE SYSTEMS

Geodetic systems with identical ellipsoid axes are referred to as concentrically geodetic systems GC and $\overline{GC}$, i.e.,

$$S \equiv \bar{S} \ . \tag{5-111a}$$

It follows from this that

$$p_{i.} = 0_{i.} \ , \quad e_{ij.} = \delta_{ij.} \ , \quad y_{i.}(q^a) = \bar{y}_{i.}(\bar{q}^b) \tag{5-111b}$$

for the central transformation parameters of (5-107c) and the position vectors of (5-102c) and (5-103c). The coordinate differences

$$\delta q^a = (\bar{L}-L,\ \bar{B}-B,\ \bar{H}-H) =: (\delta L_\epsilon, \delta B_\epsilon, \delta H_\epsilon) \tag{5-112a}$$

are thus functions solely of the variations in the ellipsoid parameters $\delta\epsilon^\nu$ [see (5-104c)] and vanish with these variations. Together with the assumptions made for (5-104d) and (5-109) for the transformation equations between two concentrically geodetic systems q^a and $\bar{q}^a$ the following linear setup can be made:

$$\bar{q}^a = q^a + \delta q^a = q^a + (\partial q^a/\partial\epsilon^\nu)_o \delta\epsilon^\nu \tag{5-112b}$$

with

$$(\partial q^a/\partial\epsilon^\nu)_o := \partial q^a/\partial\epsilon^\nu \quad \text{for} \quad \delta\epsilon^\nu = 0^\nu \ \rightarrow \ \delta q^a = 0^a \ . \tag{5-112c}$$

Due to the third equation of (5-111b) the condition equations for these partial derivatives become

$$\partial y_{i.}/\partial\epsilon^\nu = 0 = (\partial y_{i.}/\partial\epsilon^\nu)' + c_{i.a}(\partial q^a/\partial\epsilon^\nu)_o \tag{5-113a}$$

with

$$(\partial y_{i.}/\partial\epsilon^\nu)' := \partial y_{i.}/\partial\epsilon^\nu \ \text{für} \ \partial q^a/\partial\epsilon^\nu = 0 \ . \tag{5-113b}$$

Inner multiplication of (5-113a) with $c^b_{i.}$ following the exchange of indices leads to the desired results, i.e.,

$$(\partial q^a/\partial\epsilon^\nu)_o = -c^a_{i.}(\partial y_{i.}/\partial\epsilon^\nu)' \ . \tag{5-113c}$$

Due to (5-77a,b) we obtain the following expressions for the partial derivatives of (5-113b) with respect to the ellipsoid parameters of (5-104a):

$$(\partial y_{i.}/\partial c)' = (1/V)(\cos B\cos L,\ \cos B\sin L,\ (1+e''^2)^{-1}\sin B) \ , \tag{5-113d}$$

$$(\partial y_{i.}/\partial e''^2)' = -(c/(2V^3))(\cos^3 B\cos L,\ \cos^3 B\sin L,\ -(1+e''^2)^{-2}(\sin^2 B-3V^2)\sin B) \ , \quad V^2 = 1 + e''^2\cos^2 B \ , \tag{5-113e}$$

which together with the contravariant basis vectors of (5-79g) are inserted into (5-113c). Here, this is done only under the usually sufficient, simplified assumptions

$$e''^4,\ H/c \rightarrow 0 \tag{5-114a}$$

with which we obtain

$$(\partial q^a/\partial c)_o = (0,\ (e''^2/c)\sin B\cos B,\ -1+e''^2(1-(1/2)\cos^2 B))\ , \tag{5-114b}$$

$$(\partial q^a/\partial e''^2)_o = (0,\ \sin B\cos B(1-e''^2(2-(1/2)\cos^2 B))\ ,$$
$$c(1-(1/2)\cos^2 B-e''^2(2-\cos^2 B-(1/4)\cos^4 B)))\ . \tag{5-114c}$$

The transformation equations of (5-112b) with (5-114) can be rewritten as follows:

$$\delta q^1 =:\ \delta L_\varepsilon = 0\ , \tag{5-115a}$$

$$\delta q^2 =:\ \delta B_\varepsilon = \sin B\cos B((e''^2/c)\delta c + (1-e''^2(2-(1/2)\cos^2 B))\delta e''^2), \tag{5-115b}$$

$$\delta q^3 =:\ \delta H_\varepsilon = -(1-e''^2(1-(1/2)\cos^2 B))\delta c$$
$$+c(1-(1/2)\cos^2 B-e''^2(2-\cos^2 B-(1/4)\cos^4 B))\delta e''^2\ . \tag{5-115c}$$

Due to (5-111b)

$$\bar{q}^a_{,b} := \partial\bar{q}^a/\partial q^b = \bar{c}^a_{i.}\,c_{i.b} \tag{5-116a}$$

is true for the <u>transformation matrices</u> of (5-13a), and

$$\bar{q}^a_{,b} = \delta^a_b + \delta\bar{q}^a_{,b} \tag{5-116b}$$

is established corresponding to (5-112b). Proceeding from (5-115) we have

$$\delta\bar{q}^a_{,b} = \partial(\delta q^a)/\partial q^b\ , \tag{5-116c}$$

of which only the two following components do not vanish:

$$\delta\bar{q}^2_{,2} = -(e''^2/c)(1-2\cos^2 B)\delta c - (1-2\cos^2 B-e''^2(2-(7/2)\cos^2 B))\delta e''^2,$$
$$\delta\bar{q}^3_{,2} = \sin B\cos B(e''^2\delta c + c(1-e''^2(2+\cos^2 B))\delta e''^2)\ . \tag{5-116d}$$

Due to the given central transformation parameters after (5-111b) the <u>local transformation parameters</u> of (5-107b) for arbitrary reference points P_o are uniquely determined as follows:

$$\left.\begin{array}{l}(\delta q^a)_o := \delta q^a \text{ after (5-115)}\\ (\bar{q}^a_{,b})_o := \bar{q}^a_{,b} \text{ after (5-116)}\end{array}\right\}\ \text{for}\ B = B_o\ . \tag{5-117a}$$

In this case the method applied for (5-9) and (5-13) does not have to be applied for calculating the second and higher partial derivatives of $\bar{q}^a$. This is rather done directly by stepwise differentiating (5-116b,d) with respect to L, B, and H. Thus the second partial derivatives at P_o become

$$(\bar{q}^a_{,bc})_o = (\partial(\delta\bar{q}^a_{,b})/\partial q^c)_o \tag{5-117b}$$

with the following nonvanishing components:

$$(\bar{q}^2_{,22})_o = -\sin B_o \cos B_o (4(e''^2/c)\delta c - (4-7e''^2)\delta e''^2) , \tag{5-117c}$$

$$(\bar{q}^3_{,22})_o = -e''^2(1-2\cos^2 B_o)\delta c$$
$$-c(1-2\cos^2 B_o - e''^2(2-\cos^2 B_o - 4\cos^4 B_o))\delta e''^2 . \tag{5-117d}$$

With (5-117) and

$$\Delta L = L - L_o, \quad \Delta B = B - B_o, \quad \Delta H = H - H_o ,$$
$$\Delta\bar{L} = \bar{L} - \bar{L}_o, \quad \Delta\bar{B} = \bar{B} - \bar{B}_o, \quad \Delta\bar{H} = \bar{H} - \bar{H}_o \tag{5-118a}$$

corresponding to (5-7b) the power series of (5-7c,d) for the transformations of Δq^a and $\Delta\bar{q}^a$ up to the second order assume the following special forms:

$$\Delta\bar{L} = \Delta L , \tag{5-118b}$$

$$\Delta\bar{B} = (1 + (\delta\bar{q}^2_{,2})_o)\Delta B + (1/2)(\delta\bar{q}^2_{,22})_o \Delta B^2 + \dots , \tag{5-118c}$$

$$\Delta\bar{H} = \Delta H + (\delta\bar{q}^3_{,2})_o \, \Delta B + (1/2)(\delta\bar{q}^3_{,22})_o \Delta B^2 + \dots , \tag{5-118d}$$

which have the same accuracy as (5-114a).

Now the local rotation vector $\overset{F}{d}_{i.}$ [see (5-109d)] will be calculated based on (5-116a). Due to (5-114a)

$$\bar{q}^a_{,b} = \bar{c}^a_{i.} c_{i.b} \approx \bar{b}^a_{i.} b_{i.b} \tag{5-119a}$$

and together with (5-79a,b), (5-109d), and (5-116b)

$$\delta^a_b + \delta\bar{q}^a_{,b} = (\delta_{b(a).} - \epsilon_{b(a)k.} \overset{F}{d}{}'_{k.})((\bar{R}_1 \cos\bar{B})^{-1}, \bar{R}_2^{-1}, 1)^a$$
$$\cdot (R_1 \cos B \quad , R_2 \ , 1)^{(b)} \tag{5-119b}$$

results. For $a \neq b$ we consequently obtain the rotation vector components

$$\overset{F}{d}{}'_{i.} = (-\delta\bar{q}^3_{,2}/R_2, 0, 0) = (-\delta B_\epsilon, 0, 0) , \tag{5-119c}$$
$$\delta q^2 = \bar{B} - B =: \delta B_\epsilon \text{ after (5-115b)},$$

which is obvious.

5.5.3 TRANSFORMATIONS BETWEEN ARBITRARY GEODETIC SYSTEMS BASED ON CENTRAL TRANSFORMATION PARAMETERS

The linearization of the calculations of (5-108) regarding the ellipsoid parameters, the central transformation parameters, and the coordinates corresponding to (5-104) and (5-109) begins with (5-12b) or (5-108b) with (5-109b) in the form of

$$\bar{y}_{i.}(q^a) = y_{i.}(q^a) - p_{i.} + \epsilon_{ijk.} y_{j.}(q^a) d_{k.} \quad . \tag{5-120}$$

Combined with (5-77a) this involves linearized transformation equations between geodetic and Cartesian coordinates q^a and $\bar{y}_{i.}$ related to the central transformation parameters of (5-107c). For the inverse transformation we start with a linearization of $\bar{y}_{i.}$ around the point

$$(\bar{q}^a)_o = (\bar{L}_o, \bar{B}_o, \bar{H}_o) \; , \qquad \bar{y}_{i.o} := \bar{y}_{i.}(\bar{L}_o, \bar{B}_o, \bar{H}_o) \; , \tag{5-121a}$$

which yields

$$\bar{y}_{i.} = \bar{y}_{i.o} + \delta\bar{y}_{i.} = \bar{y}_{i.o} + \bar{c}_{i.a}\delta\bar{q}^a \; , \qquad \delta\bar{q}^a = \bar{q}^a - (\bar{q}^a)_o \; , \tag{5-121b}$$

whereby

$$\delta\bar{y}_{i.} = -p_{i.} + \epsilon_{ijk.}\bar{y}_{j.o} d_{k.} + (y_{i.} - \bar{y}_{i.o}) \tag{5-121c}$$

follows from (5-120). After inner multiplication with $\bar{c}^a_{i.o}$ we obtain the following linearized transformation equations between Cartesian and geodetic coordinates $y_{i.}$ and $\bar{q}^a$:

$$\delta\bar{q}^a = \bar{q}^a - (\bar{q}^a)_o = \bar{c}^a_{i.o}(-p_{i.} + \epsilon_{ijk.}\bar{y}_{j.o} d_{k.} + (y_{i.} - \bar{y}_{i.o})) \; . \tag{5-122a}$$

The approximate values of (5-121a) can be chosen, for example, so that

$$\bar{y}_{i.o} = y_{i.} \; , \tag{5-122b}$$

whereby the last term in (5-122a) vanishes. Since $\bar{c}^3_{i.o}$ is nearly parallel to $\bar{y}_{i.o}$,

$$\delta\bar{q}^3 = \delta\bar{H} \approx -\bar{c}^3_{i.o} p_{i.} = -\bar{n}_{i.o} p_{i.} \tag{5-122c}$$

follows from (5-122a,b), i.e., $\delta\bar{H}$ is solely a function of the central parallel displacement $p_{i.}$.

If

$$(\bar{q}^a)_o = q^a \rightarrow \delta\bar{q}^a = \bar{q}^a - q^a =: \delta q^a \text{ after (5-109b)} \tag{5-123a}$$

is assumed instead of (5-122b), then the difference $y_{i.} - \bar{y}_{i.o}$ is solely a function of the differences in the ellipsoid parameters $\delta\epsilon^{\upsilon}$ [see (5-104c)], and corresponding to (5-112)

$$\bar{c}^{a}_{i.o}(y_{i.} - \bar{y}_{i.o}) = (\partial q^{a}/\partial\epsilon^{\upsilon})_{o}\delta\epsilon^{\upsilon} = (\delta L_{\epsilon},\ \delta B_{\epsilon},\ \delta H_{\epsilon}) =: (\delta q^{a})_{\epsilon} \qquad \text{(5-123b)}$$

is true with

$$(\delta q^{a})_{\epsilon} = \delta q^{a} \quad \text{after (5-112a), (5-115)} \ . \qquad \text{(5-123c)}$$

Based on these results (5-122a) can be converted into <u>linearized transformation equations between geodetic coordinates q^{a} and $\bar{q}^{a}$</u>, i.e.,

$$\delta q^{a} = (\bar{L}-L,\ \bar{B}-B,\ \bar{H}-H) = (\delta q^{a})_{\epsilon} - c^{a}_{i.o}p_{i.} + \epsilon_{ijk.}c^{a}_{i.o}y_{j.}d_{k.} \ ; \qquad \text{(5-124a)}$$

herein

$$\bar{c}^{a}_{i.o} \approx c^{a}_{i.o} \ , \qquad \bar{y}_{i.o} \approx y_{i.} \qquad \text{(5-124b)}$$

are inserted in the coefficients of $p_{i.}$ and $d_{i.}$. Together with

$$R_1 + H =: R_1' \ , \qquad R_2 + H =: R_2' \ ,$$

$$b/V + H = (c/V)/(1+e''^2) + H \approx R_1'/(1+e'^2) \ , \qquad \text{(5-125a)}$$

$$e''^4 \to 0 \qquad \text{(5-125b)}$$

as well as (5-77a,b), (5-79), and

$$e_{ij.} := e_{ij.F} = \begin{bmatrix} -\sin L & -\sin B\cos L & \cos B\cos L \\ \cos L & -\sin B\sin L & \cos B\sin L \\ 0 & \cos B & \sin B \end{bmatrix} \qquad \text{(5-125c)}$$

the transformation equations of (5-124a) can be expressed in more detail as follows:

$$\begin{aligned} \bar{L}-L &= -(e_{i1.}/e_{32.})p_{i.}/R_1' - (1-e''^2(\delta_{(i1).}+\delta_{(i2).}))(e_{i2.}/e_{32.})d_{i.} \ , \\ \bar{B}-B &= \delta B_{\epsilon} - e_{i2.}p_{i.}/R_2' \qquad + (1+e''^2(e^2_{32.}-e^2_{33.}))e_{i1.}\,d_{i.} \ , \\ \bar{H}-H &= \delta H_{\epsilon} - e_{i3.}p_{i.} \qquad + e''^2R_1'e_{32.}e_{33.}e_{i1.}\,d_{i.} \ , \end{aligned} \qquad \text{(5-125d)}$$

δB_{ϵ}, δH_{ϵ} after (5-115b,c) .

In "spherical approximation", i.e.,

$$e''^2 \to 0 \ , \qquad R_1' \approx R_2' \approx R \ , \qquad \text{(5-126a)}$$

these equations become

$$\begin{aligned} \bar{L}-L &= -(e_{i1.}/e_{32.})p_{i.}/R - (e_{i2.}/e_{32.})d_{i.} \ , \\ \bar{B}-B &= \delta B_{\epsilon} - e_{i2.}\,p_{i.}/R \qquad + e_{i1.}\,d_{i.} \ , \\ \bar{H}-H &= \delta H_{\epsilon} - e_{i3.}\,p_{i.} \ . \end{aligned} \qquad \text{(5-126b)}$$

Together with (5-109b)

$$\bar{q}^a_{,b} = \bar{c}^a_{i.}(\delta_{ij.} + \epsilon_{ijk.}d_{k.})c_{j.b} = \bar{c}^a_{i.}c_{i.b} + \epsilon_{ijk.}c^a_{i.}c_{j.b}d_{k.} \qquad (5\text{-}127a)$$

is true for the <u>transformation matrices</u> of (5-13a). On the other hand, they can also be calculated directly based on the transformation equations between q^a and $\bar{q}^a$, which is done here by means of (5-126). Analogous to (5-116b) we use

$$\bar{q}^a_{,b} = \delta^a_b + (\delta\bar{q}^a_{,b})_\epsilon + (\delta\bar{q}^a_{,b})' \qquad (5\text{-}127b)$$

with

$$(\delta\bar{q}^a_{,b})_\epsilon := \delta\bar{q}^a_{,b} \quad \text{after (5-116)} , \qquad (5\text{-}127c)$$

$$(\delta\bar{q}^a_{,b})' := \delta\bar{q}^a_{,b} \quad \text{for} \quad \delta\epsilon^\nu = 0^\nu .$$

Consequently, only $(\delta\bar{q}^a_{,b})'$ is still to be calculated based on (5-126b), for which we yield

$$(\delta\bar{q}^1_{,1})' = -((e_{i2.}e_{33.}-e_{i3.}e_{32.})/e_{32.})p_{i.}/R+(e_{i1.}e_{33.}/e_{32.})d_{i.} ,$$

$$(\delta\bar{q}^1_{,2})' = -(e_{i1.}e_{33.}/e^2_{32.})p_{i.}/R-((e_{i2.}e_{33.}-e_{i3.}e_{32.})/e^2_{32.})d_{i.} , \qquad (5\text{-}127d)$$

$$(\delta\bar{q}^1_{,3})' = 0 ,$$

$$(\delta\bar{q}^2_{,1})' = e_{i1.}e_{33.}p_{i.}/R+(e_{i2.}e_{33.}- e_{i3.}e_{32.})d_{i.} ,$$

$$(\delta\bar{q}^2_{,2})' = e_{i3.}p_{i.}/R , \qquad (\delta\bar{q}^2_{,3})' = 0 , \qquad (5\text{-}127e)$$

$$(\delta\bar{q}^3_{,1})' = -e_{i1.}e_{32.}p_{i.} , \qquad (\delta\bar{q}^3_{,2})' = -e_{i2.}p_{i.} , \qquad (\delta\bar{q}^3_{,3})' = 0 \qquad (5\text{-}127f)$$

observing (5-125c). These results are true in spherical approximation [see (5-126a)].

The <u>local rotation vectors</u> $\overset{F}{d'_{i.}}$ [see (5-109d)] can be calculated analogously to (5-119). Proceeding from (5-13a) or (5-127a)

$$\bar{q}^a_{,b} = \bar{c}^a_{i.}e_{ji.}c_{j.b} , \qquad e_{ij.} := \overset{S}{e}_{ij.\bar{S}} \quad \text{after (5-12a)} \qquad (5\text{-}128a)$$

follows with the assumptions of (5-114a) and (5-127b), i.e.,

$$\bar{q}^a_{,b} = \delta^a_b + (\delta\bar{q}^a_{,b})_\epsilon + (\delta\bar{q}^a_{,b})' = \bar{b}^a_{i.}e_{ji.}b_{j.b} .$$

Observing (5-108e) and (5-109d), the insertion of (5-79a) into the right side of this equation produces

$$\bar{q}^{a}_{,b} = \delta^{a}_{b} + (\delta\bar{q}^{a}_{,b})_{\epsilon} + (\delta\bar{q}^{a}_{,b})' = \delta^{a}_{b} + \delta\bar{q}^{a}_{,b} \tag{5-128b}$$
$$= (\delta_{b(a).} - \epsilon_{b(a)k.}\overset{F}{d}{}'_{k.})((\bar{R}_1\cos B)^{-1},\ \bar{\bar{R}}_2^{-1},\ 1)^{a}(R_1\cos B,\ R_2,\ 1)^{(b)} .$$

For $a \neq b$ the following rotation vector components are obtained:

$$\overset{F}{d}{}'_{i.} = (-\bar{q}^{3}_{,2}/R_2,\ \bar{q}^{3}_{,1}/(R_1\cos B),\ \bar{q}^{1}_{,2}V^2\cos B) . \tag{5-128c}$$

Together with (5-116), (5-115b), and (5-127) this result in spherical approximation [see (5-126a)] reads

$$\overset{F}{d}{}'_{i.} = (-\delta B_{\epsilon} + e_{j2.}p_{j.}/R,\ -e_{j1.}p_{j.}/R\ , \tag{5-128d}$$
$$(-e_{j1.}e_{33.}p_{j.}/R - (e_{j2.}e_{33.} - e_{j3.}e_{32.})d_{j.})/e_{32.}) ,$$
$$e_{ij.} := e_{ij.F} , \qquad \delta B_{\epsilon} \text{ after (5-115b)} .$$

An alternative way of calculating $\overset{F}{d}{}'_{i.}$ proceeds from (5-108e) with (5-109b,d) instead of from (5-128a) as follows:

$$\overset{F}{e}_{ij.\bar{\bar{F}}} = \delta_{ij.} - \epsilon_{ijk.}\overset{F}{d}{}'_{k.} = e_{ki.F}\, e_{kj.\bar{\bar{F}}} - \epsilon_{klm.}e_{ki.F}\, e_{lj.F}d_{m.}$$
$$= e_{ki.F}\, e_{kj.\bar{\bar{F}}} - \epsilon_{ijk.}e_{lk.F}\, d_{l.} . \tag{5-129a}$$

With the same accuracy

$$e_{kj.\bar{\bar{F}}} = e_{kj.F} + e_{kj,\alpha F}\ \delta q^{\alpha} , \qquad \alpha \in \{1,2\} ,$$

whereby

$$e_{ki.F}\, e_{kj.\bar{\bar{F}}} = \delta_{ij.} - \epsilon_{ijk.}(-\delta B,\ e_{32.F}\ \delta L,\ e_{33.F}\ \delta L)_{k.} \tag{5-129b}$$

ensues. Due to (5-129a,b) the <u>local rotation vector $\overset{F}{d}{}'_{i.}$</u> becomes

$$\overset{F}{d}{}'_{i.} = (-\delta B + e_{j1.}d_{j.},\ e_{32.}\delta L + e_{j2.}d_{j.},\ e_{33.}\delta L + e_{j3.}d_{j.}) , \tag{5-129c}$$
$$e_{ij.} := e_{ij.F} .$$

This expression is equivalent to (5-128c). The results of the coordinate transformations of (5-125) are to be substituted for δL and δB, with which the rotation vector components become

$$\overset{F}{d}{}'_{1.} = -\delta B_{\epsilon} + e_{j2.}p_{j.}/R'_2 - e''^2(e^2_{32.} - e^2_{33.})e_{j1.}d_{j.} , \tag{5-129d}$$

$$\overset{F}{d}{}'_{2.} = \qquad - e_{j1.}p_{j.}/R'_1 + e''^2(e_{12.}d_{1.} + e_{22.}d_{2.}) , \tag{5-129e}$$

$$\overset{F}{d}{}'_{3.} = (- e_{33.}e_{j1.}p_{j.}/R'_1 + (e_{21.} + e''^2 e_{33.}e_{12.})d_{1.} \tag{5-129f}$$
$$- (e_{11.} - e''^2 e_{33.}e_{22.})d_{2.})/e_{32.} .$$

(5-129d-f) become (5-128d) in spherical approximation [see (5-126a)].

The <u>transformation equations of the direction vectors between S_F and $S_{\bar{F}}$</u> [see (5-105d)] together with (5-109d,e) assume the form of

$$\delta \overset{F}{r}{}'_{i.} = \overset{\bar{F}}{r}{}'_{i.} - \overset{F}{r}{}'_{i.} = \epsilon_{ijk.} \overset{F}{r}{}'_{j.} \overset{F}{d}{}'_{k.} \quad . \tag{5-130}$$

Inserting $\overset{F}{d}{}'_{k.}$ in accordance with (5-128d) or (5-129d-f) represents $\delta \overset{F}{r}{}'_{i.}$ as a function of the central transformation parameters of (5-107c).

When introducing polar coordinates as <u>local direction parameters in S_F and $S_{\bar{F}}$</u>, i.e.,

$$A',\ z' \ \text{ after (5-102e)} \ , \quad \bar{A}',\ \bar{z}' \ \text{ after (5-103e)} \tag{5-131a}$$

[see also (5-82b) and Fig. 5.4],

$$\begin{aligned} \overset{\bar{F}}{r}{}'_{i.} &= \overset{F}{r}{}'_{i.} + (\partial \overset{F}{r}{}'_{i.}/\partial A')\delta A' + (\partial \overset{F}{r}{}'_{i.}/\partial z')\delta z' \\ \delta A' &= \bar{A}' - A' \ , \qquad \delta z' = \bar{z}' - z' \end{aligned} \tag{5-131b}$$

can be set up. Proceeding from (5-82c) it can be shown that

$$\begin{aligned} &|\overset{F}{r}{}'_{i.}| = 1 \ , \quad |\partial \overset{F}{r}{}'_{i.}/\partial A'| = \sin z' , \quad |\partial \overset{F}{r}{}'_{i.}/\partial z'| = 1 \ , \\ &\overset{F}{r}{}'_{i.} \perp \partial \overset{F}{r}{}'_{i.}/\partial A' \perp \partial \overset{F}{r}{}'_{i.}/\partial z' \perp \overset{F}{r}{}'_{i.} \quad . \end{aligned} \tag{5-131c}$$

(5-130) with (5-131b) becomes

$$(\partial \overset{F}{r}{}'_{i.}/\partial A')\delta A' + (\partial \overset{F}{r}{}'_{i.}/\partial z')\delta z' = \epsilon_{ijk.} \overset{F}{r}{}'_{j.} \overset{F}{d}{}'_{k.} \ ,$$

which yields the transformation equations

$$\begin{aligned} \delta A' = \bar{A}' - A' &= -(\partial \overset{F}{r}{}'_{i.}/\partial z')\overset{F}{d}{}'_{i.}/\sin z' \\ &= -\cot z'(\sin A' \overset{F}{d}{}'_{1.} + \cos A\ \overset{F}{d}{}'_{2.}) + \overset{F}{d}{}'_{3.} \ , \end{aligned} \tag{5-131d}$$

$$\begin{aligned} \delta z' = \bar{z}' - z' &= (\partial \overset{F}{r}{}'_{i.}/\partial A')\overset{F}{d}{}'_{i.}/\sin z' \\ &= \cos A' \overset{F}{d}{}'_{1.} - \sin A' \overset{F}{d}{}'_{2.} \end{aligned} \tag{5-131e}$$

following inner multiplication with $\partial \overset{F}{r}{}'_{i.}/\partial A'$ or $\partial \overset{F}{r}{}'_{i.}/\partial z'$ and observing (5-131c). After substituting (5-128d) or (5-129d-f) for $\overset{F}{d}{}'_{i.}$ we obtain the transformation equations of the local direction parameters as functions of the central transformation parameters of (5-107c). Together with (5-128d), i.e., in spherical approximation [see (5-126a)], these equations read

$$\begin{aligned}\bar{A}' - A' = {} & \sin A' \cot z' \delta B_{\epsilon} \\ & - ((e_{33.}/e_{32.})e_{i1.} - (e_{i1.}\cos A' - e_{i2.}\sin A')\cot z')p_{i.}/R \\ & + (e_{21.}d_{1.} - e_{11.}d_{2.})/e_{32.} \ ,\end{aligned} \tag{5-132a}$$

$$\bar{z}' - z' = -\cos A' \delta B_{\epsilon} + (e_{i1.}\sin A' + e_{i2.}\cos A')\, p_{i.}/R \ , \tag{5-132b}$$

$$\delta B_{\epsilon} \text{ after (5-115b)} \ , \qquad e_{ij.} := e_{ij.F} \text{ after (5-78)} \ .$$

Special local direction vectors are the vertical direction vectors indicating the direction of the zenith, i.e.,

$$l_{i.} \ , \quad |l_{i.}| = 1 \ , \tag{5-133a}$$

for which in S and $\bar{S}$ the vertical direction parameters

$$\left.\begin{matrix} L'', B'' \\ \bar{L}'', \bar{B}'' \end{matrix}\right\} = \text{astronomic longitude and latitude in } \left\{\begin{matrix} S \\ \bar{S} \end{matrix}\right. \tag{5-133b}$$

are used. Corresponding to (4-10a) the vertical direction vectors of (5-133a) together with (5-133b) become

$$\begin{aligned} l_{i.} &= (\cos B''\cos L'', \cos B''\sin L'', \sin B'') \ , \\ \bar{l}_{i.} &= (\cos\bar{B}''\cos\bar{L}'', \cos\bar{B}''\sin\bar{L}'', \sin\bar{B}'') \ . \end{aligned} \tag{5-133c}$$

In S_F and $S_{\bar{F}}$ they have the components

$$\overset{F}{l}_{i.} = e_{ji.F}\, l_{j.} \ , \qquad \overset{\bar{F}}{l}_{i.} = e_{ji.\bar{F}}\, \bar{l}_{j.} \ , \tag{5-133d}$$

for which

$$\begin{aligned} \overset{F}{l}_{i.} &= (\overset{F}{l}_{1.}, \overset{F}{l}_{2.}, 1) \ , & \overset{F}{l}_{\alpha.}\overset{F}{l}_{\beta.} &\to 0 \\ \overset{\bar{F}}{l}_{i.} &= (\overset{\bar{F}}{l}_{1.}, \overset{\bar{F}}{l}_{2.}, 1) \ , & \overset{\bar{F}}{l}_{\alpha.}\overset{\bar{F}}{l}_{\beta.} &\to 0 \ , \quad \alpha \in \{1,2\} \ . \end{aligned} \tag{5-133e}$$

are generally accurate enough presuming a reference ellipsoid close to the geoid. With this accuracy the relationships

$$\begin{aligned} \overset{F}{l}_{1.} &= (L'' - L)\cos B \ , & \overset{F}{l}_{2.} &= B'' - B \ , \\ \overset{\bar{F}}{l}_{1.} &= (\bar{L}'' - \bar{L})\cos B \ , & \overset{\bar{F}}{l}_{2.} &= \bar{B}'' - \bar{B} \ . \end{aligned} \tag{5-133f}$$

for the so-called deflections of the vertical follow from (5-133c,d).

These quantities are used for the vertical direction vector of (5-133e) as local direction parameters instead of (5-131a). (5-130) yields the transformation equations

$$\overset{\bar F}{l}_{\alpha.} - \overset{F}{l}_{\alpha.} = \overset{F}{\epsilon}_{\alpha 3k.}\overset{F}{d}{}'_{k.} = (-\overset{F}{d}{}'_{2.}, \overset{F}{d}{}'_{1.}) , \tag{5-133g}$$

which together with (5-129d-f) become the <u>transformation equations of the deflections of the vertical</u> as functions of the central transformation parameters of (5-107c), i.e.,

$$\begin{aligned} \overset{\bar F}{l}_{1.} - \overset{F}{l}_{1.} &= e_{i1.}p_{i.}/R'_1 - e''^2(e_{12.}d_{1.} + e_{22.}d_{2.}) , \\ \overset{\bar F}{l}_{2.} - \overset{F}{l}_{2.} &= -\delta B_\epsilon + e_{i2.}p_{i.}/R'_2 - e''^2(e^2_{32.} - e^2_{33.})e_{i1.}d_{i.} , \end{aligned} \tag{5-133h}$$

δB_ϵ after (5-115b) , $\quad e_{ij.} := e_{ij.F}$ after (5-78) .

These equations can be written in spherical approximation [see (5-126a)] as follows:

$$\overset{\bar F}{l}_{\alpha.} - \overset{F}{l}_{\alpha.} = e_{i\alpha.}p_{i.}/R - \delta_{\alpha 2.}\delta B_\epsilon , \qquad \alpha \in \{1,2\} . \tag{5-133i}$$

5.5.4 TRANSFORMATIONS BETWEEN ARBITRARY GEODETIC SYSTEMS BASED ON LOCAL TRANSFORMATION PARAMETERS

<u>Local transformation parameters</u> were introduced in (5-107b) as coordinate differences and transformation matrix of a reference point P_o, i.e.,

$$(\delta q^a)_o := (\bar q^a)_o - (q^a)_o , \quad (\bar q^a_{,b})_o := (\partial \bar q^a / \partial q^b)_o . \tag{5-134a}$$

In the following they are termed "curvilinear local transformation parameters". If $(\delta q^a)_o$ is given or determined, it is to be assumed that this is also true for both coordinate triples $(q^a)_o$ and $(\bar q^a)_o$. Due to (5-6b) only three components of $(\bar q^a_{,b})_o$ are independent. The assumptions of (5-109f) are made for all further considerations. Due to $(q^a)_o$, $(\bar q^a)_o$, and $(\bar q^a_{,b})_o$ the "Cartesian local transformation parameters", which are equivalent to (5-134a), are calculated using (5-77a), (5-78), and (5-128c) as follows:

$$\overset{F}{p'_{j.o}} = e_{ij.o}(y_{i.o} - \bar{y}_{i.o}) \ , \ e_{ij.o} = e_{ij.Fo} \ ,$$

$$\overset{F}{d'_{i.o}} = \text{rotation vector of } S_{\bar{F}o} \text{ relative to } S_{Fo} \text{ [see (5-109d)]}. \qquad (5\text{-}134b)$$

It can also be presumed here that both position vectors $y_{i.o}$ and $\bar{y}_{i.o}$ for P_o are given. The parameters of (5-134a) can be derived from (5-134b) by means of (5-77c-f).

The transformation equations for geographically geodetic coordinates and local direction parameters will now be derived. First, the central transformation parameters of (5-107c) are represented as functions of the local parameters of (5-134b). These results are then inserted into the transformation equations of Chapter 5.5.3, leading to the desired transformation equations.

Together with (5-120) $\overset{F}{p'_{i.o}}$ [see (5-134b)] can be represented as follows:

$$\overset{F}{p'_{i.o}} = e_{ji.o}\,(p_{j.} - \epsilon_{jkl.}y_{k.}d_{l.}) \ ,$$

and since $y_{k.} \approx e_{k3.o}R$ can be inserted here with the accuracy as in (5-109), we obtain

$$\overset{F}{p'_{i.o}}/R = e_{ji.o}p_{j.}/R + \epsilon_{ij3.}e_{kj.o}d_{k.} \ . \qquad (5\text{-}135a)$$

According to (5-128d)

$$\overset{F}{d'_{i.o}} = (-\delta B_{\epsilon o} + e_{j2.o}p_{j.}/R,\ -e_{j1.o}p_{j.}/R\ ,\ (-e_{33.o}e_{j1.o}p_{j.}/R + e_{21.o}d_{1.} - e_{11.o}d_{2.})/e_{32.o}) \qquad (5\text{-}135b)$$

is true for the local rotation vectors at P_o [see (5-134b)]. The local Cartesian transformation parameters are represented by (5-135a,b) as linear functions of the central transformation parameters. The inversion of these equations begins with (5-135a) for $i = 3$ and (5-135b) for $i = 1,2$. These equations yield

$$e_{ji.o}p_{j.}/R = (-\overset{F}{d'_{2.o}},\ \overset{F}{d'_{1.o}} + \delta B_{\epsilon o},\ \overset{F}{p'_{3.o}}/R)_{i.} \ , \qquad (5\text{-}136a)$$

and together with (5-135a) for $i = 1,2$ and (5-135b) for $i = 3$ the following condition equations for $d_{i.}$ are obtained:

$$\overset{F}{p'_{1.o}}/R + \overset{F}{d'_{2.o}} = e_{j2.o}d_{j.} \ , \qquad (2\text{-}136b)$$

$$\overset{F}{p}{}'_{2.o}/R - \overset{F}{d}{}'_{1.o} - \delta B_{\epsilon o} = -e_{11.o}d_{1.} - e_{21.o}d_{2.} \;, \tag{5-136b}$$

$$e_{33.o}\overset{F}{d}{}'_{2.o} - e_{32.o}\overset{F}{d}{}'_{3.o} = -e_{21.o}d_{1.} + e_{11.o}d_{2.} \;.$$

Inner multiplication of (5-136a) with $e_{ki.o}$ and inverting (5-136b) produce the inverse equations of (5-135a,b), i.e.,

$$\begin{aligned} p_{i.}/R = {} & e_{ij.o}(-\overset{F}{d}{}'_{2.o} \,,\, \overset{F}{d}{}'_{1.o} + \delta B_{\epsilon o},\, \overset{F}{p}{}'_{3.o}/R)_{j.} \\ = {} & (-e_{11.o}\overset{F}{d}{}'_{2.o} + e_{12.o}(\overset{F}{d}{}'_{1.o} + \delta B_{\epsilon o}) + e_{13.o}\overset{F}{p}{}'_{3.o}/R \,, \\ & -e_{21.o}\overset{F}{d}{}'_{2.o} + e_{22.o}(\overset{F}{d}{}'_{1.o} + \delta B_{\epsilon o}) + e_{23.o}\overset{F}{p}{}'_{3.o}/R \,, \\ & \qquad + e_{32.o}(\overset{F}{d}{}'_{1.o} + \delta B_{\epsilon o}) + e_{33.o}\overset{F}{p}{}'_{3.o}/R\,)_{i.} \,, \end{aligned} \tag{5-137a}$$

$$\begin{aligned} d_{i.} = {} & e_{ij.o}(-\overset{F}{p}{}'_{2.o}/R + \overset{F}{d}{}'_{1.o} - \delta B_{\epsilon o} \,, \\ & \overset{F}{p}{}'_{1.o}/R + \overset{F}{d}{}'_{2.o},\, (e_{33.o}/e_{32.o})\overset{F}{p}{}'_{1.o}/R + \overset{F}{d}{}'_{3.o})_{j.} \\ = {} & (-e_{11.o}(\overset{F}{p}{}'_{2.o}/R - \delta B_{\epsilon o}) + e_{1j.o}\overset{F}{d}{}'_{j.o} \,, \\ & -e_{21.o}(\overset{F}{p}{}'_{2.o}/R - \delta B_{\epsilon o}) + e_{2j.o}\overset{F}{d}{}'_{j.o} \,, \\ & (\overset{F}{p}{}'_{1.o}/R)/e_{32.o} + e_{3j.o}\overset{F}{d}{}'_{j.o})_{i.} \;. \end{aligned} \tag{5-137b}$$

If (5-137) is inserted into (5-126b), (5-132), and (5-133i), we then obtain the linearized <u>transformation equations of the coordinates and local direction parameters</u> as functions of local Cartesian transformation parameters in spherical approximation [see (5-126a)], i.e.,

$$\begin{aligned} (\bar{L}-L)e_{32.} = {} & -(e_{32.}/e_{32.o})\overset{F}{p}{}'_{1.o}/R \\ & + e_{i2.}e_{i1.o}\overset{F}{p}{}'_{2.o}/R - e_{i1.}e_{i3.o}\overset{F}{p}{}'_{3.o}/R \\ & - (e_{i2.}e_{i1.o} + e_{i1.}e_{i2.o})(\overset{F}{d}{}'_{1.o} + \delta B_{\epsilon o}) \\ & + (e_{i1.}e_{i1.o} - e_{i2.}e_{i2.o})\,\overset{F}{d}{}'_{2.o} - e_{i2.}e_{i3.o}\overset{F}{d}{}'_{3.o} \,, \end{aligned} \tag{5-138a}$$

$$\begin{aligned} \bar{B}-B = {} & \delta B_{\epsilon} - e_{i1.}e_{i1.o}\overset{F}{p}{}'_{2.o}/R - e_{i2.}e_{i3.o}\overset{F}{p}{}'_{3.o}/R \\ & - (e_{i2.}e_{i2.o} - e_{i1.}e_{i1.o})(\overset{F}{d}{}'_{1.o} + \delta B_{\epsilon o}) \\ & + (e_{i2.}e_{i1.o} + e_{i1.}e_{i2.o})\,\overset{F}{d}{}'_{2.o} + e_{i1.}e_{i3.o}\overset{F}{d}{}'_{3.o} \,, \end{aligned} \tag{5-138b}$$

$$\bar{H}-H \;=\; \delta H_{\epsilon} - e_{i3.}e_{i3.o}\overset{F}{p}'_{3.o} \tag{5-138c}$$

$$- (e_{i3.}e_{i2.o}(\overset{F}{d}'_{1.o} + \delta B_{\epsilon o}) - e_{i3.}e_{i1.o}\overset{F}{d}'_{2.o})R \;,$$

$$\bar{A}'-A' = \sin A \cot z\; \delta B_{\epsilon o} \tag{5-138d}$$

$$- ((e_{21.}e_{11.o} - e_{11.}e_{21.o})/e_{32.})\overset{F}{p}'_{2.o}/R$$

$$- ((e_{33.}/e_{32.})e_{i1.} - (e_{i1.}\cos A - e_{i2.}\sin A)\cot z)e_{i3.o}\overset{F}{p}'_{3.o}/R$$

$$+ ((e_{21.}e_{11.o} - e_{22.}e_{12.o} - e_{11.}e_{21.o} + e_{12.}e_{22.o})/e_{32.}$$

$$-(e_{i1.}\cos A - e_{i2.}\sin A)e_{i2.o}\cot z)(\overset{F}{d}'_{1.o} + \delta B_{\epsilon o})$$

$$+ ((e_{22.}e_{11.o} + e_{21.}e_{12.o} - e_{12.}e_{21.o} - e_{11.}e_{22.o})/e_{32.}$$

$$-(e_{i1.}\cos A - e_{i2.}\sin A)e_{i1.o}\cot z)\overset{F}{d}'_{2.o}$$

$$+ ((e_{21.}e_{13.o} - e_{11.}e_{23.o})/e_{32.})\overset{F}{d}'_{3.o} \;,$$

$$\bar{z}'-z' = -\cos A\; \delta B_{\epsilon} + (e_{i1.}\sin A + e_{i2.}\cos A) \tag{5-138e}$$

$$\cdot\; (e_{i3.o}\overset{F}{p}'_{3.o}/R + e_{i2.o}(\overset{F}{d}'_{1.o} + \delta B_{\epsilon o}) - e_{i1.o}\overset{F}{d}'_{2.o}) \;,$$

$$\bar{\overset{F}{l}}_{\alpha.} - \overset{F}{l}_{\alpha.} = -\delta_{\alpha 2.}\,\delta B_{\epsilon} + e_{i\alpha.}e_{i3.}\overset{F}{p}'_{3.a}/R \tag{5-138f}$$

$$+ e_{i\alpha.}(e_{i2.o}(\overset{F}{d}'_{1.o} + \delta B_{\epsilon o}) - e_{i1.o}\overset{F}{d}'_{2.o}) \;, \qquad \alpha \in \{1,2\} \;,$$

δB_{ϵ}, δH_{ϵ} after (5-115b,c) , $\qquad e_{ij.o} = e_{ij.Fo}$ after (5-78) .

When estimating the <u>order of magnitude of the coefficients in (5-138)</u>, it must be noted that

$$e_{ij.}e_{ik.o} \rightarrow \delta_{jk.} \quad \text{for} \quad P \rightarrow P_a \;. \tag{5-139a}$$

If

$$s_m = \text{maximum distance } \; P - P_a \text{ on the ellipsoid} \tag{5-139b}$$

for a set of points on the earth's surface $\{P\}$ to be transformed, we obtain

$$1 \geq |e_{ij.}e_{ik.o}| > \cos(s_m/R) \quad \text{for} \quad j = k \;,$$
$$0 \leq |e_{ij.}e_{ik.o}| < \sin(s_m/R) \quad \text{for} \quad j \neq k \;. \tag{5-139c}$$

For $s_m \approx 50$ km

$$\cos(s_m/R) = 1 - 3\cdot 10^{-5}, \qquad \sin(s_m/R) = 8\cdot 10^{-3} \;. \tag{5-139d}$$

In (5-138)

$$\cot z = 0 \tag{5-139e}$$

can be used just as well in many applications.

The transformation equations of (5-138) were derived on the basis of the local Cartesian transformation parameters of (5-134b). An alternative to this method will now be given in the form presented for (5-7) to (5-9), which is based on the local curvilinear transformation parameters of (5-8) and (5-134a). The presuppositions of (5-109) are also made for these parameters so that with

$$\bar{q}^a_{,b} = \delta^a_b + \delta\bar{q}^a_{,b} \tag{5-140a}$$

and

$$\delta q^a = \Delta\bar{q}^a - \Delta q^a + (\delta q^a)_o \tag{5-140b}$$

series (5-7) assumes the following form for $n' = 2$:

$$\begin{aligned} \delta q^a &= (\bar{L}-L,\ \bar{B}-B,\ \bar{H}-H)^a \\ &= (\delta q^a)_o + (\delta\bar{q}^a_{,b})_o \Delta q^b + (1/2)(\delta\bar{q}^a_{,bc})_o \Delta q^b \Delta q^c + \ldots \end{aligned} \tag{5-140c}$$

$$\Delta q^a = (L-L_o,\ B-B_o,\ H-H_o)^a .$$

Due to (5-9e) and observing (5-109f) we obtain

$$(\delta\bar{q}^a_{,bc})_o = (\Gamma^a_{bc} - \bar{\Gamma}^a_{bc})_o + (\delta^a_d\, \Gamma^e_{bc} - \delta^e_b\, \Gamma^a_{cd} - \delta^e_c\, \Gamma^a_{bd})_o (\delta\bar{q}^d_{,e})_o , \tag{5-141a}$$

in which

$$(\bar{\Gamma}^a_{bc})_o = (\Gamma^a_{bc})_o + (\Gamma^a_{bc,d})_o (\delta q^d)_o + (\partial\Gamma^a_{bc}/\partial\varepsilon^\nu)_o \delta\varepsilon^\nu , \tag{5-141b}$$

$\delta\varepsilon^\nu$ after (5-104c,d)

is accurate enough. Thus, (5-141a) can be written as follows:

$$(\delta\bar{q}^a_{,bc})_o = (\delta\bar{q}^a_{,bc})_\varepsilon - (\Gamma^a_{bc,d})_o (\delta q^d)_o + (\delta^a_d \Gamma^e_{bc} - \delta^e_b \Gamma^a_{cd} - \delta^e_c \Gamma^a_{bc})_o (\delta\bar{q}^d_{,e})_o \tag{5-141c}$$

with

$$(\delta\bar{q}^a_{,bc})_\varepsilon := -(\partial\Gamma^a_{bc}/\partial\varepsilon^\nu)_o \delta\varepsilon^\nu = (\bar{q}^a_{,bc})_o \text{ after (5-117b-d)} . \tag{5-141d}$$

When (5-141c) is inserted into (5-140c), the general form of the <u>transformation equations between the geodetic coordinate differences Δq^a and $\bar{q}^a$</u> becomes

$$\begin{aligned} \delta q^a &= (\delta\bar{q}^a_{,bc})_\varepsilon \Delta q^b \Delta q^c + \ldots \\ &+ (\delta^a_d - (1/2)(\Gamma^a_{bc,d})_o \Delta q^b \Delta q^c + \ldots)(\delta q^d)_o \\ &+ (\delta^a_d \Delta q^e + (1/2)(\delta^a_d \Gamma^e_{bc} - \delta^e_b \Gamma^a_{cd} - \delta^e_c \Gamma^a_{bd})_o \Delta q^b \Delta q^c + \ldots)(\delta\bar{q}^d_{,e})_o \end{aligned} \tag{5-142a}$$

based on the local curvilinear transformation parameters

$$(\delta q^a)_o \, , \quad (\delta \bar{q}^a_{\;,b})_o \, . \tag{5-142b}$$

Due to (5-6b) only three components of the transformation matrices at P_o are independent, i.e.,

$$(\bar{q}^a_{\;,b})_o = \delta^a_b + (\delta \bar{q}^a_{\;,b})_o \, . \tag{5-143a}$$

Observing (5-109b) and (5-80b), (5-6b) yields the condition equations

$$(\bar{g}_{a(a)})_o (\delta \bar{q}^{(a)}_{\;,b})_o + (\bar{g}_{b(b)})_o (\delta \bar{q}^{(b)}_{\;,a})_o = (g_{ab} - \bar{g}_{ab})_o \, . \tag{5-143b}$$

For $a = b$ we then obtain

$$(\delta \bar{q}^{(a)}_{\;,(a)})_o = (1/2)(g_{(aa)} - \bar{g}_{(aa)})_o / (\bar{g}_{(aa)})_o \tag{5-143c}$$

so that these components are thus completely determined by the coordinates of P_o or the transformation parameters $(\delta q^a)_o$. For $a \neq b$ the right side of (5-143b) vanishes, and

$$(\delta \bar{q}^{(a)}_{\;,(b)})_o = -(\bar{g}_{(bb)}/\bar{g}_{(aa)})_o (\delta \bar{q}^{(b)}_{\;,(a)})_o \, , \quad a \neq b \tag{5-143d}$$

remains true. It follows from (5-143c,d) that only three components of $(\delta q^a_{\;,b})_o$ for $a \neq b$ can be chosen freely, e.g.,

$$(\delta \bar{q}^1_{\;,2})_o \, , \quad (\delta \bar{q}^3_{\;,1})_o \, , \quad (\delta q^3_{\;,2})_o \, . \tag{5-143e}$$

The transformation equations of (5-142) are equivalent to the equations of (5-138a-c). The latter equations can be converted into those of (5-142a) by means of (5-128c), (5-134b), and power series expansions for the coefficients of $\overset{F}{p}{}'_{i.o}$ and $\overset{F}{d}{}'_{i.o}$. If the local Cartesian transformation parameters of (5-134b) are given, then the curvilinear parameters of (5-142b) and (5-143a) can be calculated as follows. Based on (5-138a-c) we obtain due to $e_{ij.} = e_{ij.o}$

$$\begin{aligned}(\delta q^a)_o &= (\bar{L}_o - L_o, \; \bar{B}_o - B_o, \; \bar{H}_o - H_o) \\ &= (-(1/e_{32.o})\overset{F}{p}{}'_{1.o}/R, \; \delta B_\epsilon - \overset{F}{p}{}'_{2.o}/R, \; \delta H_\epsilon - \overset{F}{p}{}'_{3.o}) \, ,\end{aligned} \tag{5-144a}$$

δB_ϵ, δH_ϵ after (5-115b,c) ,

and

$$(\bar{q}^a_{\;,b})_o = \bar{q}^a_{\;,b} \text{ after (5-128b) for } P_o \, . \tag{5-144b}$$

The Christoffel's symbols of the second kind and their partial derivatives in (5-142a) are to be calculated based on (5-9f) with the metric tensor g'_{ab} [see (5-80a,b)]. In many applications this can be done in spherical approximation [see (5-126a)], by which we obtain

$$(\Gamma^1_{12})_o = (\Gamma^1_{21})_o = -\tan B_o\ , \qquad (\Gamma^1_{13})_o = (\Gamma^1_{31})_o = 1/R\ ,$$
$$(\Gamma^2_{11})_o = \sin B_o \cos B_o\ , \qquad (\Gamma^2_{23})_o = (\Gamma^2_{32})_o = 1/R\ , \qquad (5\text{-}145a)$$
$$(\Gamma^3_{11})_o = -R\cos^2 B_o\ , \qquad (\Gamma^3_{22})_o = -R\ ;$$

$$(\Gamma^1_{12,2})_o = (\Gamma^1_{21,2})_o = -1/\cos^2 B_o,\ (\Gamma^2_{11,2})_o = \cos^2 B_o - \sin^2 B_o\ ,$$
$$(\Gamma^3_{11,2})_o = 2R\sin B_o \cos B_o\ , \qquad (5\text{-}145b)$$
$$(\Gamma^1_{13,3})_o = (\Gamma^1_{31,3})_o = -1/R^2\ , \qquad (\Gamma^2_{23,3})_o = (\Gamma^2_{32,3})_o = -1/R^2\ ,$$
$$(\Gamma^3_{11,3})_o = -\cos^2 B_o\ , \qquad (\Gamma^3_{22,3})_o = -1\ ;$$

the components not listed here vanish.

5.5.5 DETERMINING TRANSFORMATION PARAMETERS

The six central transformation parameters of (5-107c) and (5-109b), i.e.,

$$p_{i.}\ ,\ d_{i.}\ , \qquad (5\text{-}146a)$$

as well as the six local transformation parameters of (5-134b), i.e.,

$$\overset{F}{p}{}'_{i.o}\ ,\ \overset{F}{d}{}'_{i.o}\ , \qquad (5\text{-}146b)$$

or of (5-134a) with (5-143), i.e.,

$$(\delta q^a)_o,\ (\delta \bar{q}^1_{,2})_o,\ (\delta \bar{q}^3_{,1})_o,\ (\delta \bar{q}^3_{,2})_o\ , \qquad (5\text{-}146c)$$

with which the relative movements of the geodetic GC and $\overline{\text{GC}}$ systems [see (5-102) and (5-103)] are determined, can generally be determined based on six independent transformation equations for coordinates and local direction parameters when these quantities are given in both systems. If there is a redundant number of condition equations of this type, the parameters are usually estimated by least-squares adjustment in the Gauss-Markoff model [see Koch 1988]. For this purpose, the coordinates and direction parameters are usually regarded independently, whereby the transformation equations of (5-125), (5-132), (5-133), (5-138), or

(5-142)ff are the necessary linearized observation equations. In the following several relevant examples for uniquely determining the transformation parameters of (5-146) are dealt with.

A method for uniquely determining transformation parameters based on coordinates alone requires the knowledge of six independent coordinates from three noncollinear points, i.e.,

$$P_B \,, \quad B \in \{a,b,c\} \,, \tag{5-147}$$

in the $\overline{GC}$ system whose coordinates in the GC system are all known. The case of the Cartesian coordinates of (5-102c) and (5-103c) in both systems will be treated here first, i.e.,

$$y_{i.B} \text{ in } \underset{\circ}{S} \,, \qquad \bar{y}_{i.B} \text{ in } \bar{S} \,. \tag{5-148a}$$

Let the coordinates

$$\begin{aligned} &\bar{y}_{i.a} \,;\; \bar{y}_{2.b}, \bar{y}_{3.b} \,;\; \bar{y}_{3.c} \quad \text{in } \bar{S} \,, \\ &y_{i.a} \,;\; y_{i.b} \qquad ;\; y_{i.c} \quad \text{in } S \end{aligned} \tag{5-148b}$$

be given, for which we obtain the following six linear equations for the six central transformation parameters of (5-146a) due to (5-120):

$$\begin{aligned} &p_{i.} - \epsilon_{ijk.} y_{j.a} d_{k.} && = y_{i.a} - \bar{y}_{i.a} \,, \qquad i = 1,2,3 \\ &p_{2.} - y_{3.b} d_{1.} \qquad + y_{1.b} d_{3.} && = y_{2.b} - \bar{y}_{2.b} \,, \\ &p_{3.} + y_{2.b} d_{1.} - y_{1.b} d_{2.} && = y_{3.b} - \bar{y}_{3.b} \,, \\ &p_{3.} + y_{2.c} d_{1.} - y_{1.c} d_{2.} && = y_{3.c} - \bar{y}_{3.c} \,. \end{aligned} \tag{5-148c}$$

$p_{2.}$ and $p_{3.}$ can be eliminated from the last three equations by means of the second and third equations for P_a, i = 2,3. Together with

$$\Delta y_{i.B} = y_{i.B} - y_{i.a}, \quad \Delta\bar{y}_{i.B} = \bar{y}_{i.B} - \bar{y}_{i.a}, \quad B \in \{b,c\} \tag{5-148d}$$

these equations then become

$$\begin{aligned} &-\Delta y_{3.b} d_{1.} \qquad + \Delta y_{1.b} d_{3.} && = \Delta y_{2.b} - \Delta\bar{y}_{2.b} \\ &\Delta y_{2.b} d_{1.} - \Delta y_{1.b} d_{2.} && = \Delta y_{3.b} - \Delta\bar{y}_{3.b} \\ &\Delta y_{2.c} d_{1.} - \Delta y_{1.c} d_{2.} && = \Delta y_{3.c} - \Delta\bar{y}_{3.c} \,. \end{aligned} \tag{5-148e}$$

The inversion of these equations yields $d_{i.}$, with which the parallel displacement can then be calculated from the first vector equation of (5-148c), i.e.,

$$p_{i.} = y_{i.a} - \bar{y}_{i.a} + \epsilon_{ijk.} y_{j.a} d_{k.} \ . \tag{5-148f}$$

In the <u>case of the geographically geodetic coordinates</u> of (5-102b) and (5-103b) in both systems of

$$(q^a)_B = (L_B, B_B, H_B) \text{ in } S \ , \qquad (\bar{q}^a)_B = (\bar{L}_B, \bar{B}_B, \bar{H}_B) \text{ in } \bar{S} \tag{5-149a}$$

we will assume that the coordinates

$$\begin{array}{llll} (\bar{q}^a)_a \ ; & \bar{L}_b \text{ or } \bar{B}_b, \bar{H}_b \ ; & \bar{H}_c & \text{in } \bar{S} \ , \\ (q^a)_a \ ; & (q^a)_b & ; (q^a)_c & \text{in } S \end{array} \tag{5-149b}$$

are given analogous to (5-148b). It is meaningful in most applications when $\bar{H}_B$ is given in all three points of (5-147). If, for instance, we select

$$\bar{H}_B = H_{oB} = \text{orthometric height of } P_B \ , \tag{5-149c}$$

the reference ellipsoid then lies close to the geoid.

The equations of (5-125d) or, in spherical approximation, (5-126b) are used for <u>determining the central transformation parameters</u> of (5-146a). Due to the ellipsoidal heights the equations of (5-126b) directly yield three linear condition equations for $p_{i.}$, i.e.,

$$e_{i3.B} p_{i.} = \delta H_{\epsilon B} - \bar{H}_B + H_B \ , \quad B \in \{a,b,c\} \ . \tag{5-150}$$

Then we also obtain three linear condition equations for the rotation vector $d_{i.}$ via the first two equations of (5-126b) based on the given ellipsoidal longitudes and latitudes.

If, on the other hand, the <u>determination of the local Cartesian transformation parameters</u> of (5-146b) and (5-134b) is required, then the equations of (5-138a-c) are to be used instead of (5-126a). The three height equations are to be solved first in this case as well, i.e.,

$$\begin{aligned} & e_{i3.B}(e_{i3.o} \overset{F}{p'_{3.o}} \\ & \quad + e_{i2.o}(\overset{F}{d'_{1.o}} + \delta B_{\epsilon o})R - e_{i1.o} \overset{F}{d'_{2.o}} R) = \delta H_{\epsilon B} - \bar{H}_B + H_B \ , \end{aligned} \tag{5-151a}$$

leading to the results for $\overset{F}{p'_{3.o}}$, $\overset{F}{d'_{1.o}}$ and $\overset{F}{d'_{2.o}}$. Based on the given longitudes and latitudes we then obtain $\overset{F}{p'_{1.o}}$, $\overset{F}{p'_{2.o}}$ and $\overset{F}{d'_{3.o}}$ via (5-138a,b).

If we choose

$$P_a \equiv P_o = \text{reference point of the local transformation parameters of (5-146b)}\ , \tag{5-151b}$$

(5-138a-c) directly yields the following condition equations for $\overset{F}{p}{}'_{i.o}$ due to (5-139a):

$$\overset{F}{p}{}'_{i.o} = -(R(\bar{L}_a-L_a)e_{32.a}\ ,\ R(\bar{B}_a-B_a-\delta B_{\epsilon a})\ ,\ \bar{H}_a-H_a-\delta H_{\epsilon a})\ . \tag{5-151c}$$

$\overset{F}{d}{}'_{1.o}$ and $\overset{F}{d}{}'_{2.o}$ are then determined via (5-151a) for $B \in \{b,c\}$, and $\overset{F}{d}{}'_{3.o}$ via (5-138a) and/or via (5-138b) based on $\bar{L}_b$ or $\bar{B}_b$.

Analogous to (5-151) the <u>determination of the local curvilinear transformation parameters</u> of (5-146c) can be done based on the transformation equations of (5-142)ff. With (5-151b) we directly obtain

$$(\delta q^a)_o = (\bar{L}_a-L_a,\ \bar{B}_a-B_a,\ \bar{H}_a-H_a)\ ,\quad P_a \equiv P_o\ . \tag{5-152a}$$

Since the coordinates of P_o in both systems are known,

$$(\delta\bar{q}^{(a)}_{,(a)})_o \ \text{after (5-143c)} \tag{5-152b}$$

can be calculated. Observing (5-143d,e)

$$(\delta\bar{q}^{(a)}_{,(b)})_o \quad \text{for} \quad a \neq b \tag{5-152c}$$

is obtained by applying the equations of (5-142a) to $\bar{q}^a$ known at P_b and P_c [see (5-149b)].

The previous observations were based on six coordinates at three points [see (147)]. We will now go into uniquely <u>determining the transformation parameters based on coordinates and local direction parameters</u> at one point, i.e.,

$$P_a \equiv P_o \quad \text{after (5-151b)}\ . \tag{5-153a}$$

Let

$$\begin{aligned} &\bar{L}_a,\ \bar{B}_a,\ \bar{H}_a;\ \ \overset{\bar{F}}{l}_{1.a},\ \overset{\bar{F}}{l}_{2.a},\ \bar{A}'_a \quad \text{in } \bar{S}\ , \\ &L_a,\ B_a,\ H_a;\ \ \overset{F}{l}_{1.a},\ \overset{F}{l}_{2.a},\ A'_a \quad \text{in } S \end{aligned} \tag{5-153b}$$

be given at P_a. Thus, besides the geodetic coordinates of P_a in $\bar{S}$ and S the deflections of the vertical at P_a and the ellipsoidal azimuth of an arbitrary direction in $S_{\bar{F}a}$ and S_{Fa} are known.

For determining the local Cartesian transformation parameters of (5-146b) we first have

$$\overset{F}{p}{}'_{i.o} \text{ after (5-151c),} \tag{5-154a}$$

and together with

$$\delta \overset{F}{l}_{\alpha.a} := \overset{\bar{F}}{l}_{\alpha.a} - \overset{F}{l}_{\alpha.a}, \qquad \alpha \in \{1,2\} \tag{5-154b}$$

(5-131d) and (5-133g) yield

$$\overset{F}{d}{}'_{i.o} = (\delta \overset{F}{l}_{2.a}, -\delta \overset{F}{l}_{1.a}, \bar{A}'_a - A'_a - (\cos A'_a \delta \overset{F}{l}_{1.a} - \sin A'_a \delta \overset{F}{l}{}'_{2.a}) \cot z_a)_i. \tag{5-154c}$$

for the local rotation vector. An initial result of the <u>determination of the local transformation parameters</u> of (5-146c) is given here as well by (5-152a), i.e.,

$$(\delta q^a)_o = (\bar{L}_a - L_a, \bar{B}_a - B_a, \bar{H}_a - H_a), \quad P_a \equiv P_o . \tag{5-155a}$$

The independent components of $(\delta q^{(a)}_{,(b)})_o$ for $a \neq b$ are produced by inserting (5-154c) into (5-128b). In spherical approximation

$$(\delta \bar{q}^1_{,2})_o = (\bar{A}_a - A'_a - (\cos A'_a \delta \overset{F}{l}_{1.a} - \sin A'_a \delta \overset{F}{l}_{2.a}) \cot z_a)/\cos B_a,$$

$$(\delta \bar{q}^3_{,1})_o = -R \cos B\, \delta \overset{F}{l}_{1.a}, \quad (\delta \bar{q}^3_{,2})_o = -R\, \delta \overset{F}{l}_{2.a} . \tag{5-155b}$$

All remaining components of $(\delta \bar{q}^a_{,b})_o$ are to be calculated based on (5-143c,d).

For <u>determining the central transformation parameters</u> of (5-146a) we obtain the following equations by inserting (5-154a,c) into (5-137a,b):

$$p_{i.} = e_{ij.a}(R\, \delta \overset{F}{l}_{1.a}, R\, \delta \overset{F}{l}_{2.a} + \delta B_{\epsilon a}, -\bar{H}_a + H_a - \delta H_{\epsilon a})_j. , \tag{5-156a}$$

$$\begin{aligned} d_{i.} = e_{ij.a}(&\bar{B}_a - B_a + \delta \overset{F}{l}_{2.a}, -(\bar{L}_a - L_a) e_{32.a} - \delta \overset{F}{l}_{1.a}, \\ &- (\bar{L}_a - L_a) e_{33.a} - (\bar{A}'_a - A'_a) + (\cos A_a \delta \overset{F}{l}_{1.a} - \sin A_a \delta \overset{F}{l}_{2.a}) \cos z'_a)_j. \end{aligned} \tag{5-156b}$$

These equations can also be obtained by applying (5-126b), (5-132a), and (5-133i) to P_a.

5.5.6 DETERMINING MEAN REFERENCE ELLIPSOIDS

The method given in Chapter 5.5.5 for determining the transformation parameters of (5-146) between the two geodetic systems of GC and $\overline{GC}$ [see (5-102) and (5-103)] presumes that the ellipsoid parameters of both systems are given. In the following, however, only the ellipsoid parameters of (5-102a) in the form of (5-104a) will be given, i.e.,

$$\epsilon^{\upsilon} = (c,\ e''^2)\ , \tag{5-157a}$$

and the corresponding ellipsoid parameters of (5-103a), i.e.,

$$\bar{\epsilon}^{\upsilon} = (\bar{c},\ \bar{e}''^2) \quad \text{or} \quad \delta\epsilon^{\upsilon} = \bar{\epsilon}^{\upsilon} - \epsilon^{\upsilon} = (\delta c,\ \delta e''^2)\ , \tag{5-157b}$$

are the parameters to be estimated, in addition to the transformation parameters of (5-146). Thus, the number of unknowns increases from six to eight, which can be written together with (5-146a) as the vector

$$z^k = (\delta c,\ \delta e''^2,\ p_{i.},\ d_{i.}),\ k \in \{1,2...8\}\ . \tag{5-157c}$$

Analogous to the comments for (5-146) it is also true here that the parameters of (5-157c) can generally be determined based on eight independent transformation equations for coordinates and local direction parameters when these quantities are given in both systems. When there is a redundant number of condition equations, the parameters are best estimated by least-squares adjustment in the Gauss-Markoff model assuming independent coordinates and direction parameters [see Koch 1988]. The accompanying linearized observation equations are represented by the transformation equations of (5-125), (5-132), and (5-133), whereby the dependency of δc and $\delta e''^2$ is given through δB_{ϵ} and δH_{ϵ} in accordance with (5-115b,c).

Variations in the ellipsoid parameters of (5-157b) are generally used for <u>determining a mean reference ellipsoid or model geoid $\overline{MG}$</u> corresponding to the explanations for (1-1) and (1-2). These ellipsoids are usually defined in a way that they give the best possible, regional or global, mean approximation of the true geoid, which is why they are referred to as model geoids. Hereby, for the area of approximation, i.e.,

$$\Delta F \subseteq \overline{MG} \tag{5-158a}$$

we directly obtain the requirement

$$\bar{H}_B \approx H_{oB} = \text{orthometric heights of the } P_B \in \Delta F \tag{5-158b}$$

corresponding to (5-149c). Due to the very slight curvature of the plumb lines it follows from (5-158) that the deflections of the vertical in the $\overline{GC}$ system [see (5-133f)] must have minimum values, i.e.,

$$\overset{\bar{F}}{l}_{\alpha.B} = ((\bar{L}''-\bar{L})\cos\bar{B},\ \bar{B}''-\bar{B})_B \ \rightarrow\ \text{minimum},\ P_B \in \Delta F\ . \tag{5-158c}$$

The results of (5-158b,c) are the frequently used <u>minimum requirements</u>

$$M_H^2 := \sum_{\Delta F} (\bar{H}-H_o)^2 = \text{minimum}\ , \tag{5-159a}$$

$$M_l^2 := \sum_{\Delta F} (\overset{\bar{F}}{l}_{\alpha.})^2 = \text{minimum}\ , \tag{5-159b}$$

or

$$M_{Hl}^2 := \sum_{\Delta F} (p_H(\bar{H}-H_o)^2 + p_l(\overset{\bar{F}}{l}_{\alpha.})^2) = \text{minimum}\ , \tag{5-159c}$$

p_H, p_l = suitable selected weights

<u>for adjustments</u> corresponding to the explanations following (5-157c) based on a redundant number of H_o values and/or L" and B" values. The eight "normal equations" for (5-159c) read

$$\partial(M_{Hl}^2)/\partial z^k = 0\ , \qquad z^k \text{ after (5-157c)}$$

or

$$\sum_{\Delta F} (p_H(\bar{H}-H_o)\partial(\bar{H}-H_o)/\partial z^k + p_l \overset{\bar{F}}{l}_{\alpha.}\, \partial \overset{\bar{F}}{l}_{\alpha.}/\partial z^k) = 0\ ,\quad k \in \{1,2\ldots 8\}\ , \tag{5-159d}$$

from which the normal equations for (5-159a,b) with $p_H = 1$ and $p_l = 0$ or $p_H = 0$ and $p_l = 1$ ensue. The accompanying linearized "observation equations" are the third transformation equation of (5-125d) and the equations of (5-133h), which are written here as follows:

$$\bar{H}-H_o = H-H_o + \delta H_{\epsilon,1}\delta c + \delta H_{\epsilon,2}\delta e''^2 \tag{5-159e}$$
$$- e_{i3.}p_{i.} + e''^2\, c\, e_{32.}e_{33.}e_{i1.}d_{i.}\ ,$$

$$\overset{\bar{F}}{l}_{1.} = \overset{F}{l}_{1.} + (e_{i1.}/R_1')p_{i.} - e''^2\,(e_{12.}d_{1.} + e_{22.}d_{2.})\ , \tag{5-159f}$$

$$\overset{\bar{F}}{l}_{2.} = \overset{F}{l}_{2.} - \delta B_{\epsilon,1}\delta c - \delta B_{\epsilon,2}\delta e''^2 \tag{5-159g}$$
$$+ (e_{i2.}/R_2')p_{i.} - e''^2(e_{32.}^2 - e_{33.}^2)e_{i1.}d_{i.}$$

with $e_{ij.}$, R_1', R_2' according to (5-125) and

$$\delta H_{\epsilon,\alpha} = (-1+e''^2(1-e_{32.}^2/2),\ c(1-e_{32.}^2/2-e''^2(2-e_{32.}^2-e_{32.}^4/4))) \quad , \qquad (5\text{-}159h)$$

$$\delta B_{\epsilon,\alpha} = e_{32.}e_{33.}(e''^2/c,\ 1-e''^2(2-e_{32.}^2/2)) \qquad (5\text{-}159i)$$

based on (5-115). The coordinates in the GC system

$$L,\ B,\ H \qquad (5\text{-}159j)$$

and the "observations"

$$H_o,\ \overset{F}{l}_{\alpha.} \quad \text{or} \quad L'',\ B'' \qquad (5\text{-}159k)$$

are presumed to be known for all stations. Regarding the general adjustment procedures to be used, see, for example, Koch (1988).

The observation equations of (5-159e-g) are only slightly dependent on the rotation vector $d_{i.}$. This vector can be determined with enough accuracy using these equations, i.e., purely geometrically, only based on globally distributed observation values of (5-159k). This will be demonstrated in the following using an example for determining a mean reference ellipsoid based on H_o values. It is presumed that both the orthometric heights H_o [see (5-158b)] and the ellipsoidal heights H [see (5-159j)] are known for a sufficiently dense point field on the earth's surface so that the differences $H - H_o$ can be expanded into a series of spherical harmonics, i.e.,

$$H(L,B) - H_o(L,B) = \sum_{n=0}^{n'} Y_n(L,B) \ ,$$
$$Y_n(L,B) = \sum_{m=0}^{n} (A_{nm}\cos mL + B_{nm}\sin mL)P_{nm}(\sin B) \qquad (5\text{-}160a)$$

[see, e.g., Heiskanen & Moritz 1967, Chapter 1-13; Heitz 1980-1983, Chapter 12.3]. The presently unknown differences $\bar{H} - H_o$ on the left side of (5-159c) can be represented in the same way, i.e.,

$$\bar{H}(L,B) - H_o(L,B) = \sum_{n=0}^{n'} \bar{Y}_n(L,B) \ ,$$
$$\bar{Y}_n(L,B) = \sum_{m=0}^{n} (\bar{A}_{nm}\cos mL + \bar{B}_{nm}\sin mL)P_{nm}(\sin B) \ . \qquad (5\text{-}160b)$$

The coefficients of z^k [see (5-157c)] on the right side of (5-159e) are polynomials in $e_{ij.}$ [see (5-125c)] so that they can be completely approximated by series of spherical harmonics. With them and (5-160a,b), (5-159e) changes into the following expansion into spherical harmonics:

$$\bar{H}-H_o = \sum_{n=0}^{n'} \bar{Y}_n = \sum_{n=0}^{n'} Y_n$$
$$- ((1-(2/3)e''^2)\delta c - (c/3)(2-(18/5)e''^2)\delta e''^2)P_{00} - p_{3.}P_{10} \qquad (5\text{-}160c)$$
$$- (p_{1.}\cos L + p_{2.}\sin L)P_{11} + (1/3)(e''^2\delta c + c(1-(18/7)e''^2)\delta e''^2 P_{20}$$
$$+ (e''^2 c/3)(d_{2.}\cos L - d_{1.}\sin L)P_{21} + 2(e''^2 c/35)\delta e''^2 P_{40} .$$

In this equation only the first seven parameters of (5-157c) remain because $\bar{H}$ is independent of $d_{3.}$ due to the symmetry of the reference ellipsoid with respect to the $\bar{3}.$ axis. The last four lines of (5-160c) form parts of $\bar{Y}_0$, $\bar{Y}_1$, $\bar{Y}_2$, and $\bar{Y}_4$ so that a comparison of the coefficients of the spherical harmonics yields

$$\begin{aligned}
&\bar{A}_{00} = A_{00} - (1-(2/3)e''^2)\delta c + (c/3)(2-(18/5)e''^2)\delta e''^2 ,\\
&\bar{A}_{10} = A_{10} - p_{3.} , \quad \bar{A}_{11} = A_{11} - p_{1.} , \quad \bar{B}_{11} = B_{11} - p_{2.} ,\\
&\bar{A}_{20} = A_{20} + (e''^2/3)\delta c + (c/3)(1-(18/7)e''^2)\delta e''^2 , \qquad (5\text{-}160d)\\
&\bar{A}_{21} = A_{21} + (e''^2 c/3)d_{2.} , \quad \bar{B}_{21.} = B_{21.} - (e''^2 c/3)d_{1.} ,\\
&\bar{A}_{40} = A_{40} + 2(e''^2 c/35)\delta e''^2 ,\\
&\bar{A}_{nm} = A_{nm} , \quad \bar{B}_{nm} = B_{nm} \quad \text{for all other values of } nm.
\end{aligned}$$

In this case the normal equations of (5-159d) have the form of

$$\int_{MG} (\bar{H}-H_o)(\partial(\bar{H}-H_o)/\partial z^k)do = 0 , \qquad (5\text{-}160e)$$

in which

$$do = \cos B \, dL \, dB = dL \, d\sin B$$

can be used in spherical approximation. The partial derivatives of the differences $\bar{H} - H_o$ with respect to z^k become

$$\partial(\bar{H}-H_o)/\partial z^k = (-(1-(2/3)e''^2)P_{00}+(1/3)e''^2 P_{20}, \ (c/3)(2-(18/5)e''^2)P_{00}$$
$$+(c/3)(1-(18/7)e''^2)P_{20}+2(e''^2 c/35)P_{40}, \ -\cos L \ P_{11}, \ -\sin L \ P_{11}, \qquad (5\text{-}160f)$$
$$-P_{10}, \ -(e''^2 c/3)\sin L \ P_{21}, \ (e''^2 c/3)\cos L \ P_{21}, \ 0)$$

based on (5-160c). Observing the orthogonality of the spherical harmonics, inserting (5-160b,f) into (5-160e) yields seven condition equations for the eight coefficients $\bar{A}_{00}$, $\bar{A}_{20}$, $\bar{A}_{40}$, $\bar{A}_{10}$, $\bar{A}_{11}$, $\bar{B}_{11}$, $\bar{A}_{21}$, and $\bar{B}_{21}$:

$$(1-(2/3)e''^2)\bar{A}_{00} - (1/15)e''^2\bar{A}_{20} = 0 ,$$
$$(2-(18/5)e''^2)\bar{A}_{00} + (1/5)(1-(18/7)e''^2)\bar{A}_{20} + (2/105)e''^2\bar{A}_{40} = 0 , \quad (5\text{-}160g)$$
$$\bar{A}_{10} = \bar{A}_{11} = \bar{B}_{11} = \bar{A}_{21} = \bar{B}_{21} = 0 .$$

In spherical approximation [see (5-126a)]

$$\bar{A}_{00} = \bar{A}_{20} = 0 \quad \text{for} \quad e''^2 \to 0 \quad (5\text{-}161a)$$

results from the first two equations, whereby series (5-160b) assumes the form of

$$\bar{H}-H_o = (\bar{A}_{22}\cos 2L + \bar{B}_{22}\sin 2L)P_{22}(\sin B) + \sum_{n=3}^{n'} \bar{Y}_n(L,B) . \quad (5\text{-}161b)$$

Based on the expansion into spherical harmonics of (5-160a) with the presumed known coefficients A_{nm} and B_{nm} the ellipsoid and transformation parameters of z^k [see (5-157)] up to $k = 7$ can be calculated by inserting (5-160d) into (5-160g) and solving with respect to z^k. With the accuracy of (5-114a) the results are

$$\delta c = A_{00} - (2+(54/35)e''^2)A_{20} - (4/21)e''^2A_{40} , \quad (5\text{-}162a)$$
$$\delta e''^2 = -e''^2A_{00}/c - (3+(40/7)e''^2)A_{20}/c - (2/7)e''^2A_{40}/c ,$$
$$p_{i.} = (A_{11}, B_{11}, A_{10}) , \quad (5\text{-}162b)$$
$$d_{1.} = 3\, B_{21}/(e''^2c) , \qquad d_{2.} = -3\, A_{21}/(e''^2c) . \quad (5\text{-}162c)$$

The purely geometric method described for (5-160) to (5-162) for determining a mean reference ellipsoid is presently possible. Satellite methods namely allow the positioning of continental and oceanic stations, the latter via the so-called satellite altimetry. In this way we obtain the necessary $H - H_o$ values if the orthometric heights H_o in continental areas are known with sufficient accuracy. As an alternative, satellite methods also enable the determination of a dynamically defined mean reference ellipsoid or model geoid, which suffices at least in spherical approximation the minimum condition of (5-159a) and thus (5-160e) [see also Heiskanen & Moritz 1967, Chapters 2-21, 5-11)].

BIBLIOGRAPHY

MATHEMATICAL FUNDAMENTALS

Ahlfors L W (1966) Complex analysis, McGraw-Hill, New York

Bäschlin C F, Höhn W (1947) Einführung in die Kurven- und Flächentheorie auf vektorieller Grundlage, Orell Füssli Verlag, Zürich

Blaschke W, Leichtweiss K (1973) Elementare Differentialgeometrie, Springer, Berlin Heidelberg New York

Derrick W R (1972) Introductory complex analysis and applications, Academic Press, New York London

Goetz A (1970) Introduction to differential geometry, Addison Wesley, Reading, Massachusetts

Heitz S (1980-1983) Mechanik fester Körper, mit Anwendungen in Geodäsie, Geophysik und Astronomie, 2 Vols, Ferdinand Dümmlers Verlag, Bonn

Hotine M (1969) Mathematical geodesy, ESSA Monographs No. 2, US Department of Commerce, Washington, D.C.

Jeffreys H (1974) Cartesian tensors, Cambridge Univ Press

Knopp K (1947) Theory of functions, Parts I and II, Dover Publications, New York

Koch K R (1988) Parameter estimation and hypothesis testing in linear models, Springer, Berlin Heidelberg New York

Kreyszig E (1975) Introduction to differential geometry and Riemannian geometry, Univ Toronto Press, Toronto Buffalo

Laugwitz D (1965) Differential and Riemannian geometry, Academic Press, Orlando London

Levi-Civita T (1977) The absolute differential calculus, 1926; Reprint by Dover Publications, New York

McConnell A J (1957) Applications of tensor analysis; Reprint by Dover Publications, New York

Sokolnikoff I S (1964) Tensor analysis, theory and applications to geometry and mechanics of continua, John Wiley & Sons Inc, New York Chichester Brisbane Toronto

Taschner R J (1977) Differentialgeometie für Geodäten, Manz Verlag, Wien

GEODETIC APPLICATIONS

Bäschlin C F (1948) Lehrbuch der Geodäsie, Orell Füssli Verlag, Zürich

Bomford G (1975) Geodesy, Univ Press, Oxford

Dragomir V, Ghitau D, Mihailescu M, Rotaru M (1982) Theory of the earth's shape, Elsevier Scientific Publishing Co., Amsterdam Oxford New York

Glasmacher H, Krack K (1984) Umkehrung von vollständigen Potenzreihen mit zwei Veränderlichen, Geodetic Reports of the University of the Bundeswehr, Vol 10, München, p 49-69

Glasmacher H (1987) Die Gausssche Ellipsoid-Abbildung mit komplexer Arithmetik und numerischen Näherungsverfahren, Dissertation, University of the Bundeswehr München

Grossmann W (1976) Geodätische Rechnungen und Abbildungen in der Landesvermessung, Konrad Wittwer Verlag, Stuttgart

Heck B (1987) Rechenverfahren und Auswertemodelle der Landesvermessung, Herbert Wichmann Verlag, Karlsruhe

Heiskanen W A, Moritz H (1967) Physical geodesy, Freeman & Co., San Francisco London

Heitz S (1983) Geometrische Modelle der Geodäsie, Reports of the Geodetic Institutes of the University Bonn, No 64, Bonn

Helmert F R (1962) Die mathematischen und physikalischen Theorien der höheren Geodäsie, Vol 1: Die mathematischen Theorien, B G Teubner Verlag, Leipzig 1880. Reprint by Minerva GmbH, Frankfurt am Main

Hopfner F (1949) Grundlagen der höheren Geodäsie, Springer, Wien

Hristow W K (1955) Die Gaussschen und geographischen Koordinaten auf dem Ellipsoid von Krassowsky, VEB Verlag Technik, Berlin

Hubeny K (1953) Isotherme Koordinatensysteme und konforme Abbildungen des Rotationsellipsoids, Special Publication No 13 of Österr. Zeitschrift für Vermessungswesen, Wien
2nd Edition (1977): Reports of the Geodetic Institutes of the Technical University Graz, No 27, Graz

Hubeny K (1954) Zur Entwicklung der Gaussschen Mittelbreitenformeln, Österr. Zeitschrift für Vermessungswesen, p 8-17

Hubeny K (1959) Weiterentwicklung der Gaussschen Mittelbreitenformeln, Zeitschrift für Vermessungswesen, p 159-163

Kneissl M (1958/59) Mathematische Geodäsie, Vol IV (2 Parts) of Handbuch der Vermessungskunde of Jordan/Eggert/Kneissl, Metzlersche Verlagsbuchhandlung, Stuttgart

König R, Weise K H (1951) Mathematische Grundlagen der höheren Geodäsie und Kartographie, Springer, Berlin Göttingen Heidelberg

Krack K (1982a) Rechnerunterstützte Ableitung der Legendreschen Reihen und Abschätzung ihrer ellipsoidischen Anteile zur Lösung der ersten geodätischen Hauptaufgabe auf Bezugsellipsoiden, Zeitschrift für Vermessungswesen, p 118-125

Krack K (1982b) Rechnerunterstützte Entwicklung der Mittelbreitenformeln und Abschätzung ihrer ellipsoidischen Anteile zur Lösung der zweiten geodätischen Hauptaufgabe auf dem Rotationsellipsoid, Zeitschrift für Vermessungswesen, p 502-513

Kuntz E (1983) Kartennetzentwurfslehre, Grundlagen und Anwendungen, Herbert Wichmann Verlag, Karlsruhe

Marussi A (1985) Intrinsic geodesy, Springer, Berlin Heidelberg New York Tokyo

Mittermayer E (1963) Berechnung der Meridianbogenlänge durch ein Polynom, Zeitschrift für Vermessungswesen, p 436-439

Mittermayer E (1965) Formeln zur Berechnung der ellipsoidiscchen geographischen Endbreite für Meridianbögen beliebiger Länge, Zeitschrift für Vermessungswesen, p 403-408

Möhle A (1943) Die Verwendung von geographischen Koordinaten in der Theorie allgemeiner Flächen, Dissertation, University Bonn

Rapp R H (1984-1987) Geometric geodesy, 2 parts, The Ohio State University, Department of Geodetic Science and Surveying, Columbus, Ohio

Sakatow P S (1957) Lehrbuch der höheren Geodäsie, VEB Verlag Technik Berlin

Schödlbauer A (1963) Über eine neue numerische Lösung der 1. geodätischen Hauptaufgabe auf einem Referenz-Rotationsellipsoid der Erde für Seitenlängen bis 120 km, Publication of Deutsche Geodätische Kommission, Series C, No 58, München

Schödlbauer A (1981-1982-1984) Rechenformeln und Rechenbeispiele zur Landesvermessung, 3 Vols, Herbert Wichmann Verlag, Karlsruhe

Sigl R (1977) Ebene und sphärische Trigonometrie, mit Anwendungen auf Kartographie, Geodäsie und Astronomie, Herbert Wichmann Verlag, Karlsruhe

Urmajew N A (1958) Sphäroidische Geodäsie, VEB Verlag Technik, Berlin

Wolfrum O (1983) Zur Theorie der ellipsoidischen Normalschnitte und ihren Anwendungen, Reports of the Geodetic Institute of the Technical University Darmstadt No 2, Darmstadt

Wolfrum O (1984) Die Reduktion einer Raumstrecke auf den ellipsoidischen Normalschnittbogen, Allgemeine Vermessungsnachrichten, p 23-33

I N D E X